T0183170

Communications in Computer and Information Science 672

Commenced Publication in 2007
Founding and Former Series Editors:
Alfredo Cuzzocrea, Dominik Ślęzak, and Xiaokang Yang

More information about this series at http://www.springer.com/series/7899

Emmanouel Garoufallou · Imma Subirats Coll
Armando Stellato · Jane Greenberg (Eds.)

Metadata and Semantics Research

10th International Conference, MTSR 2016
Göttingen, Germany, November 22–25, 2016
Proceedings

 Springer

Editors
Emmanouel Garoufallou
Alexander Technological Educational
  Institute of Thessaloniki
Thessaloniki
Greece

Imma Subirats Coll
Food and Agriculture Organization of the
  United Nations (FAO)
Rome
Italy

Armando Stellato
University of Rome Tor Vergata
Rome
Italy

Jane Greenberg
Drexel University
Philadelphia, PA
USA

ISSN 1865-0929   ISSN 1865-0937   (electronic)
Communications in Computer and Information Science
ISBN 978-3-319-49156-1   ISBN 978-3-319-49157-8 (eBook)
DOI 10.1007/978-3-319-49157-8

Library of Congress Control Number: 2016956613

Printed on acid-free paper

This Springer imprint is published by Springer Nature
The registered company is Springer International Publishing AG
The registered company address is: Gewerbestrasse 11, 6330 Cham, Switzerland

Preface

Metadata and semantics are integral to any information system and important to the sphere of Web data. Research and development addressing metadata and semantics are crucial to advancing how we effectively discover, use, archive, and repurpose information. In response to this need, researchers are actively examining methods for generating, reusing, and interchanging metadata. Integrated with these developments is research on the application of computational methods, linked data, and data analytics. A growing body of literature also targets conceptual and theoretical designs providing foundational frameworks for metadata and semantic applications. There is no doubt that metadata weaves its way through nearly every aspect of our information ecosystem, and there is great motivation for advancing the current state of understanding in the fields of metadata and semantics. To this end, it is vital that scholars and practitioners convene and share their work.

Since 2005, the International Metadata and Semantics Research Conference (MTSR) has served as a significant venue for dissemination and sharing of metadata and semantic-driven research and practices. This year, 2016, marked the tenth anniversary of the MTSR, drawing scholars, researchers, and practitioners who are investigating and advancing our knowledge on a wide range of metadata and semantic-driven topics. The tenth anniversary of the International Conference on Metadata and Semantics Research (MTSR 2016) was held at the Universitäts Bibliothek of Göttingen (Germany) during November 22–25, 2016. MTSR 2016 celebrated its tenth anniversary with the theme of "Bridging the Past, Present, and Future of Metadata, Data, and Semantic Technologies" and reflected on the following questions: (1) How can the documented evidence produced over the past years be used as a driver for innovating management and processing of data and information? (2) How close are we from the vision of building powerful learning systems that will meet the needs of modern societies through high-quality data infrastructures and data-driven interfaces? (3) What are the main challenges yet to be addressed by modern metadata and semantics research?

MTSR conferences have grown in number of participants and paper submission rates over the past decade, marking it as a leading, international research conference. Continuing in the successful legacy of previous MTSR conferences (MTSR 2005, MTSR 2007, MTSR 2009, MTSR 2010, MTSR 2011, MTSR 2012, MTSR 2013, MTSR 2014, and MTSR 2015), MTSR 2016 brought together scholars and practitioners who share a common interest in the interdisciplinary field of metadata, linked data, and ontologies.

The MTSR 2016 program and the following proceedings show a rich diversity of research and practices from metadata and semantically focused tools and technologies, linked data, cross-language semantics, ontologies, metadata models, semantic systems, and metadata standards. The general session of the conference included nine papers covering a broad spectrum of topics, proving the interdisciplinary field of metadata, and was divided into three main themes: Semantic Data Management, Big Data, Scalability; Synthesis of Semantic Models; and Information Extraction and Retrieval. Metadata as a research topic is maturing, and the conference also supported the following five tracks: Digital Libraries, Information Retrieval, Big, Linked and Social Data; Metadata and Semantics for Open Repositories, Research Information Systems and Data Infrastructures; Metadata and Semantics for Agriculture, Food, and Environment; Metadata and Semantics for Cultural Collections and Applications; and European and National Projects. Each of these tracks had a rich selection of short and full research papers, in total 23, giving broader diversity to MTSR, and enabling deeper exploration of significant topics.

All the papers underwent a thorough and rigorous peer-review process. The review and selection for this year were highly competitive and only papers containing significant research results, innovative methods, or novel and best practices were accepted for publication. From the general session, only nine submissions were accepted as full research papers. An additional 17 contributions from tracks covering noteworthy and important results were accepted as full research papers, and six as short papers, totaling 32 accepted contributions for this year's MTSR. The full papers represent 38.8 % of the total number of submissions.

Göttingen State and University Library (or the "SUB") is not only a historic site with precious special collections and one of Germany's largest academic libraries but also a hub for innovative library development, specifically with respect to metadata. Ranging from early involvement in generic metadata standards such as Dublin Core to substantial contributions to the highly sophisticated vocabularies, ontologies, and provenance models of the Consortium of European Research Libraries (CERL), it takes part in significant developments in the library world. Beyond the library world, it has established a metadata system for historic object collections and has been a forerunner in open access and research data management, contributing to metadata developments in the area of current research information systems (CRIS), data curation, usage data, and alternative metrics for scholarly communication.

This year the MTSR conference was pleased to host two remarkable keynote presentations by internationally known leaders in academia and multinational organizations. Prof. Dr. Philipp Cimiano, a professor of computer science at Bielefeld University, introduced: "Lemon: A Lexicon Model for Ontologies," a model recently synthesized into a suite of ontologies. This work is the output of an initiative of over three years carried out within the W3C community group focusing on the lexicon–ontology interface (ontolex). The initiative further incorporates earlier research in the field. Natural language is increasingly being recognized as an important facet in the field of ontologies, and it is of vital importance in the fields of digital libraries. The Lemon/Ontolex model presents innovative work in this area. Dr. Johannes Keizer has been working for the Food and Agriculture Organization of the United Nations (FAO of the UN) for the past 16 years, guiding the development of important digital resources such as AGROVOC and AGRIS, and heading a restructuring operation of the "International Information System for Agricultural Science and Technology." In his presentation, "Investing into Metadata and Semantics," Dr. Keizer shared his extensive experiences and insights reflecting the whys and wherefores of adopting semantic solutions in the full array of knowledge management, dissemination, and sharing. His words emphasize the pivotal role that metadata and semantics can play in modern information systems.

We conclude this preface by thanking the many people who contributed their time and efforts to MTSR 2016 and made this year's conference possible. We also thank all the organizations that supported this conference. We extend a sincere gratitude to members of the Program Committees of both main and special tracks, the Steering Committee and the Organizing Committees (both general and local), and the conference reviewers. A special thank you to Dr. Wolfram Horstmann, Director of the SUB Göttingen of Göttingen, for hosting and supporting MTSR 2016, to Karna Wegner from FAO, UN, for supporting us throughout this event, to D. Koutsomiha, who assisted us with the proceedings, and to Stavroula, Vasiliki, and Nikoleta for their endless support and patience.

September 2016

Emmanouel Garoufallou
Imma Subirats Coll
Armando Stellato
Jane Greenberg

Organization

General Chair

Emmanouel Garoufallou   Alexander Technological Educational Institute (ATEI) of Thessaloniki, Greece

Program Co-chairs

Imma Subirats   Food and Agriculture Organization (FAO) of the United Nations, Italy
Armando Stellato   University of Rome Tor Vergata, Italy
Jane Greenberg   Drexel University, USA

Special Track Chairs

Nikos Houssos   RedLink, Greece
Imma Subirats   Food and Agriculture Organization (FAO) of the United Nations, Italy
Juliette Dibie   AgroParisTech and INRA, France
Liliana Ibanescu   AgroParisTech and INRA, France
Michalis Sfakakis   Ionian University, Greece
Lina Bountouri   Ionian University and EU Publications Office, Luxembourg
Emmanouel Garoufallou   Alexander Technological Educational Institute (ATEI) of Thessaloniki, Greece
Sirje Virkus   Tallinn University, Estonia
R.J. Hartley   Manchester Metropolitan University, UK
Rania Siatri   Alexander Technological Educational Institute (ATEI) of Thessaloniki, Greece
Stavroula Antonopoulou   American Farm School, Greece
Panorea Gaitanou   Ionian University, Greece

Steering Committee

Juan Manuel Dodero   University of Cádiz, Spain
Emmanouel Garoufallou   Alexander Technological Educational Institute (ATEI) of Thessaloniki, Greece
Nikos Manouselis   AgroKnow, Greece
Fabio Sartori   Università degli Studi di Milano-Bicocca, Italy
Miguel-Angel Sicilia   University of Alcalá, Spain

Organizing Committee

Karna, Wegner (Chair)	Food and Agriculture Organization (FAO) of the United Nations, Italy
Damiana Koutsomiha	American Farm School, Greece
Anxhela Dani	Alexander Technological Educational Institute (ATEI) of Thessaloniki, Greece
Chrysanthi Chatzopoulou	Alexander Technological Educational Institute (ATEI) of Thessaloniki, Greece
Eleni Tzoura	Alexander Technological Educational Institute (ATEI) of Thessaloniki, Greece

Technical Support Staff

Ilias Nitsos	Alexander Technological Educational Institute (ATEI) of Thessaloniki, Greece

Program Committee

Rajendra Akerkar	Western Norway Research Institute, Norway
Arif Altun	Hacetepe University, Turkey
Ioannis N. Athanasiadis	Democritus University of Thrace, Greece
Panos Balatsoukas	University of Manchester, UK
Tomaz Bartol	University of Ljubljana, Slovenia
Ina Bluemel	German National Library of Science and Technology TIBm, Germany
Derek Bousfield	Manchester Metropolitan University, UK
Gerhard Budin	University of Vienna, Austria
Caterina Caracciolo	Food and Agriculture Organization (FAO) of the United Nations, Italy
Ozgu Can	Ege University, Turkey
Christian Cechinel	Federal University of Pelotas, Brazil
Fabrizio Celli	Food and Agriculture Organization (FAO) of the United Nations, Italy
Artem Chebotko	University of Texas – Pan American, USA
Sissi Closs	Karlsruhe University of Applied Sciences, Germany
Ricardo Colomo-Palacios	Universidad Carlos III, Spain
Constantina Costopoulou	Agricultural University of Athens, Greece
Sally Jo Cunningham	Waikato University, New Zealand
Sándor Darányi	University of Borås, Sweden
Ernesto William De Luca	Georg Eckert Institute, Germany
Milena Dobreva	University of Malta, Malta
Juan Manuel Dodero	University of Cádiz, Spain
Erdogan Dogdu	TOBB Teknoloji ve Ekonomi University, Turkey
Juan José Escribano Otero	Universidad Europea de Madrid, Spain
Muriel Foulonneau	Tudor Public Research Centre, Luxembourg

Panorea Gaitanou	Ionian University, Greece
Emmanouel Garoufallou	Alexander Technological Educational Institute (ATEI) of Thessaloniki, Greece
Manolis Gergatsoulis	Ionian University, Greece
Jorge Gracia Del Río	Universidad Politécnica de Madrid, Spain
Jane Greenberg	Drexel University, USA
Jill Griffiths	Manchester Metropolitan University, UK
R.J. Hartley	Manchester Metropolitan University, UK
Nikos Houssos	RedLink, Greece
Carlos A. Iglesias	Universidad Politecnica de Madrid, Spain
Frances Johnson	Manchester Metropolitan University, UK
Brigitte Jörg	Thomson Reuters, UK
Dimitris Kanellopoulos	University of Patras, Greece
Sarantos Kapidakis	Ionian University, Greece
Pinar Karagoz	Middle East Technical University (METU), Turkey
Pythagoras Karampiperis	AgroKnow, Greece
Brian Kelly	CETIS, University of Bolton, UK
Christian Kop	University of Klagenfurt, Austria
Rebecca Koskela	University of New Mexico, USA
Daniela Luzi	National Research Council, Italy
Paolo Manghi	Institute of Information Science and Technologies (ISTI), National Research Council (CNR), Italy
John McCrae	National University of Ireland Galway, Ireland
Xavier Ochoa	Centro de Tecnologias de Informacion Guayaquil, Ecuador
Mehmet C. Okur	Yaşar University, Turkey
Matteo Palmonari	University of Milano-Bicocca, Italy
Manuel Palomo Duarte	University of Cádiz, Spain
Laura Papaleo	University of Genoa, Italy
Christos Papatheodorou	Ionian University, Greece
Marios Poulos	Ionian University, Greece
T.V. Prabhakar	Indian Institute of Technology Kanpur, India
Athena Salaba	Kent State University, USA
Salvador Sanchez-Alonso	University of Alcalá, Spain
Fabio Sartori	Università degli Studi di Milano-Bicocca, Italy
Cleo Sgouropoulou	Technological Educational Institute of Athens, Greece
Rania Siatri	Alexander Technological Educational Institute (ATEI) of Thessaloniki, Greece
Miguel-Ángel Sicilia	University of Alcalá, Spain
Armando Stellato	University of Rome Tor Vergata, Italy
Imma Subirats	Food and Agriculture Organization (FAO) of the United Nations, Italy
Shigeo Sugimoto	University of Tsukuba, Japan
Stefaan Ternier	Open University of the Netherlands, The Netherlands
Emma Tonkin	King's College London, UK
Effie Tsiflidou	AgroKnow, Greece

Andrea Turbati   University of Rome Tor Vergata, Italy
Fabio Massimo Zanzotto   University of Rome Tor Vergata, Italy
Thomas Zschocke   United Nations University, Germany

Additional Reviewers

Stavroula Antonopoulou   American Farm School, Greece
Alessia Bardi   Istituto di Scienza e Tecnologie dell' Informazione A.
                                   Faedo — CNR, Italy
Damiana Koutsomiha   American Farm School, Greece
Christopher Munro   University of Manchester, UK
Dimitris Rousidis   University of Alcalá, Spain
Evgenia Vasilakaki   National Library of Greece
Burley Zhong Wang   Sun Yat-Sen University, China
Georgia Zafeiriou   University of Macedonia, Greece
Sofia Zapounidou   Aristotle University of Thessaloniki

Track on Digital Libraries, Information Retrieval, Linked and Social Data

Special Track Chairs

Emmanouel Garoufallou   Alexander Technological Educational Institute (ATEI)
                                   of Thessaloniki, Greece
Sirje Virkus   Tallinn University, Estonia
R.J. Hartley   Manchester Metropolitan University, UK
Rania Siatri   Alexander Technological Educational Institute (ATEI)
                                   of Thessaloniki, Greece

Program Committee

Panos Balatsoukas   University of Manchester, UK
Ozgu Can   Ege University, Turkey
Sissi Closs   Karlsruhe University of Applied Sciences, Germany
Mike Conway   University of North Carolina at Chapel Hill, USA
Phil Couch   University of Manchester, UK
Milena Dobreva   University of Malta, Malta
Ali Emrouznejad   Aston University, UK
Panorea Gaitanou   Ionian University and Benaki Museum, Greece
Jane Greenberg   Drexel University, USA
Jill Griffiths   Manchester Metropolitan University, UK
R.J. Hartley   Manchester Metropolitan University, UK
Frances Johnson   Manchester Metropolitan University, UK
Nikos Korfiatis   University of East Anglia, UK
Rebecca Koskela   University of New Mexico, USA

Valentini Moniarou-Papaconstantinou	Technological Educational Institute of Athens, Greece
Ioanna Pervolaraki	University of Crete, Greece
Dimitris Rousidis	University of Alcalá, Spain
Athena Salaba	Kent State University, USA
Miguel-Angel Sicilia	University of Alcalá, Spain
Christine Urquhart	Aberystwyth University, UK
Evgenia Vassilakaki	Technological Educational Institute of Athens, Greece
Sirje Virkus	Tallinn University, Estonia
Georgia Zafeiriou	University of Macedonia, Greece

Track on Metadata and Semantics for Open Repositories, Research Information Systems, and Data Infrastructures

Special Track Chairs

Imma Subirats	Food and Agriculture Organization (FAO) of the United Nations, Italy
Nikos Houssos	RedLink, Greece

Program Committee

Sophie Aubin	Institut National de la Recherche Agronomique, France
Thomas Baker	Sungkyunkwan University, Korea
Hugo Besemer	Wageningen UR Library, The Netherlands
Gordon Dunshire	University of Strathclyde, UK
Jan Dvorak	Charles University of Prague, Czech Republic
Jane Greenberg	Drexel University, USA
Siddeswara Guru	University of Queensland, Australia
Keith Jeffery	Keith G. Jeffery Consultants, UK
Nikolaos Konstantinou	University of Manchester, UK
Rebecca Koskela	University of New Mexico, USA
Jessica Lindholm	Malmö University, Sweden
Paolo Manghi	Institute of Information Science and Technologies — Italian National Research Council (ISTI-CNR), Italy
Brian Matthews	Science and Technology Facilities Council, UK
Jochen Schirrwagen	University of Bielefeld, Germany
Birgit Schmidt	University of Göttingen, Germany
Joachim Schöpfel	University of Lille, France
Kathleen Shearer	Confederation of Open Access Repositories (COAR), Germany
Chrisa Tsinaraki	European Commission, Joint Research Centre, Italy
Yannis Tzitzikas	University of Crete and ICS-FORTH, Greece
Zhong Wang	Sun-Yat-Sen University, China
Marcia Zeng	Kent State University, USA

Track on Metadata and Semantics for Agriculture, Food, and Environment

Special Track Chairs

Juliette Dibie	AgroParisTech and INRA, France
Liliana Ibanescu	AgroParisTech and INRA, France

Program Committee

Ioannis Athanasiadis	Wageningen University, The Netherlands
Rainer Baritz	Food and Agriculture Organization (FAO) of the United Nations, Italy
Lilia Berrahou	LIRMM, France
Christopher Brewster	TNO, The Netherlands
Patrice Buche	INRA (Institut National de Recherche Agronomique), France
Caterina Caracciolo	Food and Agriculture Organization (FAO) of the United Nations, Italy
Johannes Keizer	Food and Agriculture Organization (FAO) of the United Nations, Italy
Stasinos Konstantopoulos	NCSR Demokritos, Greece
Daniel Martini	Kuratorium für Technik und Bauwesen in der Landwirtschaft e. V. (KTBL), Germany
Ajit Maru	Global Forum on Agricultural Research, Italy
Claire Nedellec	INRA (Institut National de Recherche Agronomique), France
Ivo Jr. Pierozzi	Embrapa Agricultural Informatics, Brazil
Vassilis Protonotarios	AgroKnow, Greece
Mathieu Roche	CIRAD, France
Catherine Roussey	IRSTEA, France
Ben Schaap	Global Open Data in Agriculture and Nutrition (GODAN), The Netherlands
Miguel-Ángel Sicilia	University of Alcalá, Spain
Marc Taconet	Food and Agriculture Organization (FAO) of the United Nations, Italy
Jan Top	Vrije Universiteit, The Netherlands

Track on Metadata and Semantics for Cultural Collections and Applications

Special Track Chairs

Michalis Sfakakis	Ionian University, Greece
Lina Bountouri	Ionian University, Greece and EU Publications Office, Luxembourg

Program Committee

Trond Aalberg	Norwegian University of Science and Technology (NTNU), Norway
Karin Bredenberg	The National Archives of Sweden, Sweden
Costis Dallas	University of Toronto, Canada
Enrico Francesconi	EU Publications Office, Luxembourg, and Consiglio Nazionale delle Ricerche, Firenze, Italy
Patrick Gratz	Infeurope S.A., Luxembourg
Antoine Isaac	Vrije Universiteit Amsterdam, The Netherlands
Sarantos Kapidakis	Ionian University, Greece
Irene Lourdi	National and Kapodistrian University of Athens, Greece
Christos Papatheodorou	Ionian University and Digital Curation Unit, IMIS, Athena RC, Greece
Chrisa Tsinaraki	Joint Research Centre, European Commission, Italy
Andreas Vlachidis	University of South Wales, UK
Maja Žumer	University of Ljubljana, Slovenia

Track on European and National Projects

Special Track Chairs

Emmanouel Garoufallou	Alexander Technological Educational Institute (ATEI) of Thessaloniki, Greece
Panorea Gaitanou	Ionian University, Greece
Stavroula Antonopoulou	American Farm School, Greece

Program Committee

Panos Balatsoukas	University of Manchester, UK
Mike Conway	University of North Carolina at Chapel Hill, USA
Panorea Gaitanou	Ionian University and Benaki Museum, Greece
Jane Greenberg	Drexel University, USA
R.J. Hartley	Manchester Metropolitan University, UK
Nikos Houssos	RedLink, Greece

Damiana Koutsomiha   American Farm School, Greece
Paolo Manghi   Institute of Information Science and Technologies
                              (ISTI), National Research Council, Italy
Dimitris Rousidis   University of Alcalá, Spain
Miguel-Angel Sicilia   University of Alcalá, Spain
Sirje Virkus   Tallinn University, Estonia

Contents

Track on Cultural Collections and Applications

Track on Digital Libraries, Information Retrieval, Big, Linked and Social Data

Track on Open Repositories, Research Information Systems and Data Infrastructures

General Session: Semantic Data Management, Big Data, Scalability

Sharing Linked Open Data over Peer-to-Peer Distributed File Systems: The Case of IPFS

Miguel-Angel Sicilia$^{(\boxtimes)}$, Salvador Sánchez-Alonso,
and Elena García-Barriocanal

Computer Science Department, University of Alcalá, Polytechnic Building, Ctra.
Barcelona Km. 33.6, 28871 Alcalá de Henares, Madrid, Spain
{msicilia,salvador.sanchez,elena.garciab}@uah.es

Abstract. Linked Open Data (LOD) is a method of publishing machine-readable open data so that it can be interlinked and become more useful through semantic querying. The decentralized nature of the current LOD cloud relies on location-specific services, which is known to result in problems of availability and broken links. Current approaches to peer-to-peer (P2P) decentralized file systems could be used to support better availability and performance and provide permanent data, while preserving LOD principles. Applications would also benefit from mechanisms that ensure that LOD entities are permanent and immutable, independently of their original publishers. This paper outlines a first prototype design of LOD over the *Interplanetary File System* (IPFS), a P2P system based on Merkel DAGs and a content-addressed block storage model. The fundamental ideas on that implementation are discussed and an example implementation on the early version of IPFS is described, showing the feasibility of such approach and its main differentiating features.

Keywords: IPFS · Linked Open Data · P2P file systems

1 Introduction

Linked Open Data (LOD) is a method of publishing machine-readable structured open data so that it can be interlinked and become more useful and actionable through semantic querying. The current implementation of LOD has resulted in a growing cloud of interlinked *datasets* or *"Web of Data"*, that diverse kind of providers (from governments to individuals) expose and give support to, typically using RDF [3]. These providers organize open data exposure around a number of good practices that become progressively adopted at least to a certain extent [18].

LOD is considered by many as an approach to implement open data. In 2010, the Sunlight Foundation collected in ten principles the desirable properties of open (government) data[1], which included accessibility, non-discrimination and permanence. These and other requirements for the public require a robust infrastructure that guarantees sustainability, and it has proven difficult to

---

[1] http://sunlightfoundation.com/policy/documents/ten-open-data-principles/.

© Springer International Publishing AG 2016
E. Garoufallou et al. (Eds.): MTSR 2016, CCIS 672, pp. 3–14, 2016.
DOI: 10.1007/978-3-319-49157-8_1

achieve it in the current Web of Data, since the LOD cloud is a fully decentralized system that does not feature any built-in redundancy. While open data is in some cases sustained by government policies and programs [21], there is a diversity of providers of diverse size and nature, and availability is not guaranteed for every case.

The current implementation of LOD on top of common Web technology is known to be subject to inherent problems of lack of reliable availability [17,20], which obviously hamper accessibility. This is a natural consequence of the decentralized but location-based approach to publishing in the Web. Many datasets become abandoned or their support discontinued due to a variety of reasons. This is also problematic as machine-readable data is often access via autonomous software programs and not humans, and applications on top of LOD require high availability to avoid those programs to cease function or to be forced to maintain their own proprietary and expensive systems of data caching. Further, many LOD datasets are maintained by organizations that do not have the capacity to sustain the effort of providing the service beyond some project funding. Even worse, many of the datasets are easy targets for different attacks as denial of service [19], as in many cases they do not provide the mechanisms to protect their data against them. Further, the reliance on servers at particular locations may in the future compromise non-discrimination and permanence [10] if some organization decides to revert their open police and restrict access in some way, for example, via throttling. This is also controversial since open data policies [11] are subject to some issues, some of them revolving around the property of the service and its deployment on a particular hosting service.

The above described known issues represent a threat to the success of LOD as an approach for sharing machine-readable data. This situation calls for more robust approaches to data sharing that do not trade the decentralized nature of the LOD cloud. According to its proponents, the *InterPlanetary File System* (IPFS) is a peer-to-peer distributed file system that seeks to connect all computing devices with the same system of files [2][2]. IPFS and other similar frameworks bring a disruptive approach to the archival of digital resources that is based essentially on independence of location and decentralized storage by networks of untrusted peers (*swarms*). These technologies feature important implications from the technical perspective (as built-in de-duplication), but also from the view of the governance of open data and the current reliance on trusted data providers.

In this paper, we report a first design rationale of deployment mechanisms of LOD graphs on IPFS. We consider the problems of interlinking using content-based storage, versioning and the techniques for bootstraping that kind of alternative LOD version. We also sketch the main envisaged implications of a IPFS based LOD backbone.

The rest of this paper is structured as follows. Section 2 provides some brief background on the technologies involved and the practice of LOD and P2P file systems. Then, Sect. 3 discusses the approach devised to deploy a LOD cloud on

---

[2] Note this is a reference for the first draft, now superseded by more complete versions.

top of IPFS. Section 4 gives details on how that could be realized with the current state of the tools. Finally, conclusions and outlook are provided in Sect. 5.

2 Background

2.1 Linked Open Data

Linked Open Data (LOD) is a set of conventions to expose open data on the Web, based on adapting the idea of Web links to structured, machine-readable formats as RDF(S) or JSON-LD [14]. A number of tools have been developed to aid in the conversion and exposure of LOD [12] and this has resulted in a diversity of technologies supporting Linked Data.

A number of perceived technical barriers have been identified for the adoption of open data, ranging from the unavailability of a supporting infrastructure to the lack of standards, fragmentation and legacy [9]. These are not different from the problems on reliance on central actors in the Web in general, and they are exacerbated in the case of centralization of providers [4].

Particularly, the fragmentation and heterogeneity of providers results in unavailability of entire datasets [17], non-announced changes in formats or interfaces or simply services being abandoned. This together with lack of performance, scalability or robustness represents a serious risk to application developers and more in general to the accessibility of data. These problems are not inherent to the idea of LOD, but to its current technical and organizational deployment, so that alternatives in the infrastructure layer may bring the required level of robustness both from the perspective of service deployment and of data curation and custody. The read-only and self-mangement nature of P2P networks was identified as a future potential for managing LOD data by Hausenblas and Karnstedt [8], with an understanding that the use case would be different from that of centralized repositories or live look-up systems.

2.2 Peer to Peer Decentralized File Systems

The main concept of P2P technologies is that users contribute part of their computing resources and receive content-centric services in return. Most of the time, P2P services are free of charge, distributed, and there is no concept of operator or central manager. For example, in BitTorrent, each peer contributes parts of the files it locally stores, and is granted download bandwidth by other peers based on how much upload bandwidth it is contributing. Optimizations based on content can also be devised on top of P2P file systems, e.g. by grouping or clustering nodes [15].

The fact that P2P sharing systems rely on a decentralized network of machines with no barriers to entry or exit have resulted in different approaches to incentivizing users while preserving a degree of accessibility.

The InterPlanetary File System (IPFS) is a content-addressable, globally distributed protocol for sharing content that aims to provide a permanent Web.

It combines elements of file-sharing applications such as BitTorrent and version control systems like Git, IPFS can be described as a P2P version controlled file system. It allows for mounting on POSIX file systems also, supporting a seamless interface to applications. The use of P2P technologies for sharing data has already been proposed, as in the case of Biotorrent [13], but more generic frameworks as IPFS would bring better universal management capabilities for it.

3   Proposed Approach

3.1   Publishing Datasets

In the LOD cloud, the common approach to organize data is that of publishing datasets. What one considers as a *dataset* is a matter of convention, but the VoID vocabulary[3] provides a useful account if that concept as "a set of RDF triples that are published, maintained or aggregated by a single provider.". The use of RDF is actually an implementation detail, as other formats as JSON-LD are nowadays also commonly used and accepted.

There are two obvious approaches to publish LOD datasets from an archival perspective:

- Publishing entire graphs (whole datasets with many records) as a single object. For example, a georeferenced set of bus stops could be published as a single unit.
- Publishing documents of each independent dereferenceable entity as a IPFS object. For example, in a drug database, the information on each different drug is usually an independently addressable entity in LOD.

The *graph as an object* has the benefit of easing the publishing process for datasets that are immutable (e.g. the data from a project that is completely frozen and will never change) or for the cases in which the dataset curation cycle involves the publishing of snapshots at regular intervals of time that are intended to be identifiable as different "versions". However, if the dataset is subject to small changes that are wanted to go exposed as they come, a unit of lower granularity is desirable. In that second case, the most straightforward decision is that of using the minimum "retrieval and addressing" unit, i.e. those resources that have a unique, dereferenceable unit.

A convention for publishing datasets in IPFS would be requiring a VoID object [1] resolved from a human-readable IPNS address. At the time of this writing, you can only publish a single entry per IPFS node (nodeId) using IPNS, but this is likely to change in the future. In any case, the node of the publisher should provide a way to access the VoID file in a mutable way via IPNS. A convention may be that of publishing it associated to the "root" folder, under the subfolder /.well-known/void, imitating the IANA registered well-known

---

[3] https://www.w3.org/TR/void/.

URI convention[4]. It should be noted that VoID descriptions can also link to provenance information following the WC3 PROV proposed specification [16].

Once the VoID description is available, the two approaches mentioned above have some differences and the next section discusses the details.

A related approach to differentiating types of dataset publishing is described in [5], concretely two main approaches to temporal annotation for Linked Data are discussed: document-centric and sentence-centric. The former refers to annotating "whole RDF documents" which may be interpreted as a whole dataset or the RDF graph of a dereferenceable element. The latter is for a more fine-grained temporal annotation. Here we follow an archival model that encompasses versioning instead of temporal annotation, and takes as units the usual data curation units: entire datasets or entities that have are intended for as independent units of information (which is decided by the publisher implicitly by publishing it with a separate URI).

3.2 Documents and Entry Points

In the *graph as an object* approach, the VoID file can be simply published together with a compressed version of the dataset in the same folder. This way, the snapshot archived is self-described. The VoID description allows for timestamping the snapshot using dcterms:modified. Other Dublin Core terms as dcterms:relation can be used to point to the previous version, i.e. linking to the IPFS address of the previous snapshot.

In the case of finer granularity, as mentioned above, a common organization for LOD is that of using dereferenceable URIs for each entity of interest in the dataset. For example, http://dbpedia.org/page/Berlin provides the description of the city of Berlin using triples. Content negotiation can be used to get the information in different formats (e.g. different RDF mappings, XML, JSON).

In this case, each of the documents that has a URI could be added to IPFS independently. This would make them permanent and uniquely identifiable independently from the others. It should be noted that the directory and block structure of IPFS is inherently deduplicating files or fragments that are identical, as they are addressed by hashes of the contents.

However, in this case two important conditions must be met in order to guarantee an appropriate accessibility *inside* the addressing system of IPFS (we will call it IPFS linking):

1. All elements should be reachable from entry points available in the description of the full dataset, e.g. as made available in the VoID file with void:rootResource. This imposes the requirement of knowing some roots of trees or DAGs (i.e. publishing the roots of a forest structure).
2. Links as IPFS addresses (not the original, "normal" links) should be included *whenever* IPFS files available.

---

[4] http://www.iana.org/assignments/well-known-uris/well-known-uris.xhtml.

The recent[5] IPLD specification[6] formalizes the concept of content-addressed links. The specification in its current form assumes graphs are DAGs, so the discussion about links presented here is applicable.

It should be noted that IPFS can still be used without converting the RDF links to the corresponding IPFS addresses. Such a simpler model is discussed as an example in Sect. 4.2.

The second condition establishes that if an independent IPFS address for the resource exists, it must be used. But this may not be the case in many situations, notably when the links are to other datasets, that may have not (yet) been moved to IPFS. For intra-dataset links, this requires a bootstrapping step in which all the links are changed to the IPFS counterparts. The problem with this is that it entails two issues:

- If the RDF graph is a directed acyclic graph (DAG), the process of adding the documents to IPFS should start from the leaves of the graph and move up to nodes pointing to them. This may require mechanisms for dealing with the graphs *out of core* for large datasets.
- The approach does not work for non-DAG cases, as the IPFS address (the hash) of the document changes with even the smallest change in it.

The latter restriction calls for maintaining the links in a separate index system. While this at first sight conflicts with the usual idea of having links embedded in the documents, it is a model used in the early hypermedia models, and it is also used as a representation in scalable parallel graph architectures as Apache Spark [6].

3.3 Graph Evolution

The approach described so far that involves IPFS linking assumes a static graph, which fits well with the IPFS property of immutable objects. However, many datasets evolve naturally, being DBPedia a prominent example. This is sometimes known as dynamic datasets [5].

A practical approach for this kind of evolution is summarized in the following principles:

1. Deletions are simply impossible due to the permanent nature of IPFS. This fits well with the IPFS approach, and avoids "broken links".
2. New versions of a document entail submitting a new version and having thus a new IPFS address.

The problem with this update is that maintaining a record of the latest changes requires some mechanisms for applications to be aware of them. Several options are available. A possibility is that of using backlinks to previous versions,

---

[5] Note that this paper was writing before that specification was published in Github, so some of the ideas on this paper may need re-working to be fully IPLD compatible.
[6] https://github.com/ipld/specs/tree/master/ipld.

in the style of `owl:DeprecatedClass` that allows to point to a substitute as `owl:equivalentClass`. The problem is that this needs to be implemented from the new to the previous version (since the previous one is immutable), and has problems of accessibility (how applications could find out the most recent version) and of evolution (the links to the older version should be changed).

These problems point out again to the need to adopt a separation of IPFS links and resources. A simple mechanism could be that of implementing releases or snapshots as index files. This can be done by maintaining an index for each version of the dataset with all the IPFS addresses of the resources for the dataset and the given timestamp, and a separate file with all the links (arcs in the graph). Having complete transformation of the links could be done by specifying links as triples `<ipfs-addr-src, ipfs-addr-dest, URI>` with the source and destination addresses in IPFS, and the normal URI of the link, so that the triple(s) with the original URIs can be re-interpreted by applications by simple substitution. The IPLD format can be adapted to fit under this structure, since in its current form links are embedded in the source document, and there would be a need to adapt to non-embedded links.

A deletion can be achieved by simply making the document inaccessible from entry points. Changes could be marked as substitutions, i.e. pairs of IPFS addresses `<ipfs-addr-old, ipfs-addr-new>`, or simply substituting the old by the new (this complicates versioning back). The IPFS links could be maintained in the same or in a separate but linked file, so that retrieving the full graph entail retrieving the link file and the master index, but the retrieval of the documents with the descriptions can be done separately, delayed or lazily, depending on the needs of the application.

4 Example Implementation

A proof of concept design prototype was built using IPFS version 0.4.3. We have selected two representative cases to illustrate how the above presented approach can be deployed.

4.1 Example Snapshot Dataset

The Ordnance Survey of the UK government makes available their 1:50 000 Scale Gazetteer in the form of LOD[7]. This dataset contains at the time of this writing 258,404 different named places (including farms, antiquities and hills among other types) that are described with only 9 predicates in a total of 2,362,412 triples. Gazetteers are complements to maps that historians use to locate the places, mainly for places that no longer exist and names that are no longer used or whose spelling has significantly altered. Due to the nature of gazetteers, the frequency of update is expected to be low.

---

[7] http://data.ordnancesurvey.co.uk/datasets/50k-gazetteer.

It should be noted that the Linked Data Cloud diagram[8] does not consider this particular dataset as independent but integrated into the *"Ordnance Survey Linked Data"*[9] dataset, even though the dataset is independently described by their publishers also, in a `VoID` description available in the URI pointed to the `foaf:homepage` of the dataset description in the master file.

The `VoID` description of the dataset is available[10] for all the datasets, but also the fragment related only to the gazetteer. The approach thus to detect changes in the dataset can be that of looking at the `dct:modified` predicate that contains the latest modification date. It is also possible to monitor other elements in the `VoID` description that entail changes as `void:triples` but modification time seems the most sensible alternative.

The approach to publish and update the dataset should then be that of:

- Monitoring changes in the `VoID` description as described above.
- When a change is detected, download the new snapshot pointed to by the `void:dataDump` predicate.
- Include the `VoID` file within the folder of the dataset.
- Add the resulting structure in *path* to IPFS with `ipfs add -r path`.
- The resulting IPFS URI generated from the content of the data and its dataset metadata is retained in a file for enabling location, browsing (e.g. `ipfs ls`) and broader indexing capabilities (as commented below).

This simple approach is sufficient for low-frequency updates of LOD datasets, and it could be used to move that part of the LOD cloud into IPFS provided that the datasets conform to the minimal conventions on using `VoID` that are used for the detection of changes. Of course, this creates full copies of the graphs, which only makes sense in some cases. Also, this simple approach does not convert RDF links to IPFS links.

We have tested that approach using a simple script using `rdflib` in Python, that could be applied to any dataset following similar conventions. Once the data is downloaded, a SPARQL engine can be setup using the same libraries by importing all the `.nt` files downloaded. It should be noted that this approach does not store copies of ontology or vocabulary versions, which should be done separately if the full semantics are to be made permanent with the copy of the data.

In the domain of open data, a problem of P2P systems is that they do not provide built-in capabilities for discovery of datasets, which require building an index of static URIs on top of IPFS itself. At the time of this writing this is still a debated topic, but some prototypes as *Noetic*[11] already exist. In any case, a master file of the snapshots of the dataset indexed using IPNS could be a viable alternative to implement search tools by following common URI standards.

---

[8] http://lod-cloud.net/.

[9] https://datahub.io/dataset/ordnance-survey-linked-data.

[10] http://data.ordnancesurvey.co.uk/.well-known/void.

[11] https://github.com/doesntgolf/ipfs-search.

4.2   Example URI-Based Dataset

As a second example, we have chosen Europeana, the European digital library that provides an access point to millions of cultural objects (paintings, books, etc.) that have been digitized throughout Europe, gathered from hundreds of individual cultural institutions. This represents an aggregation, that is typically maintained by using the standardized OAI-PMH protocol. The Europeana LOD pilot [7] currently implements a SPARQL endpoint to a regularly updated copy of the Europeana database.

In this case, replicating the entire dataset for a single change would be inefficient, and maintaining snapshots do not appear to be a good candidate, as record updates are determined by the providers, which are independent institutions as museums or archives. The most sensible approach here is that of following the OAI-PMH approach, that works with temporal deltas of the datasets, so that in each harvesting cycle, only the changed or deleted records are collected.

Also, as the collection is diverse and large, it is unlikely that users retrieve it in full but only some particular section of the records using some content criteria. This matches well the idea in P2P networks of a possible content-clustering of data copies. Figure 1 shows the structure of a record, including the entity referring to the real physical object and the description of the view of the provider (`ore:Proxy`) about it. That structure is the logical unit of transfer, so it makes sense to be the unit to be added to IPFS as independent objects.

The prototype implementation used the `Sickle` Python library[12] was used to write a simple OAI-PMH client. The initial harvesting produced the first version of the dataset in this case by iterating the records returned by the `ListRecords` verb request and submitting them to IPFS via `ipfs add`. This generates a list of initial IPFS files which URIs are listed in the initial index file for the dataset with the corresponding OAI-PMH identifiers. The timestamp used in the harvesting

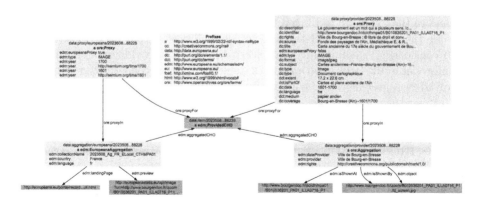

Fig. 1. Example structure of a Europeana CHO

---

[12] https://pypi.python.org/pypi/Sickle.

request is also included in the index file. The VoID file for the dataset is currently not available, but its URI could also be included in the index file.

Subsequent harvesting cycles create a new version of the index file and of the modified records, so that the IPFS addresses of the modified records are changed to the newly created files. A deletion involves simply removing the address of the record from the file. A back link to the previous index is also provided that allows for a retrieval of the history of revisions.

This approach maintains permanent storage of records and the modification granularity is that of a record, which appears a reasonable approach, since this is likely the unit of change at the provider's systems. However, it does not implement IPFS linking, as the URIs used internally are still the original ones.

Regarding linking, the original RDF links are maintained, but when adding or updating the records, they are inspected for the presence of intra-dataset links, i.e. those following the URI design pattern:

http://data.europeana.eu/item/collectionID/itemID

These are then used to generate a separate link file. Both the index and the files are added to IPFS in the same folder. Concretely, the current Web page of Europeana Linked Data states that "When applicable, the Europeana URIs for these [ProvidedCHO] objects also link, via owl:sameAs statements, to other linked data resources about the same object". This is a typical case for having a separate structure of links for the integration of different provider's metadata views on the same cultural objects.

Such approach could be expanded in the future for between-dataset links, e.g. once other datasets as DBPedia are eventually in IPFS.

5   Conclusions and Outlook

Decentralized P2P file systems as IPFS have the potential to remove technical barriers to the exposure of Linked Open Data (LOD) by providing a infrastructure for the publishing of data that detaches datasets from their institutions when considering sustainability. However, they do not provide explicit support for interoperability and good practices, which remain a concern that can only be solved via agreement and community curation.

In this paper, we have sketched a potential implementation pattern for common cases of sharing LOD datasets that require a limited deployment effort.

There are a number of additional important non-technical implications that should be stressed in adopting IPFS-like technology:

- Datasets become a property of the commons, as there is no way to revoke the publishing of datasets. This is a benefit from the perspective of openness but it would require careful consideration in licensing.
- Availability and performance become a feature of the P2P network, in which nodes decide to host some of the datasets, increasing global and local availability.

- Consequently, there is no upfront cost or investment to publish the dataset, but there is also a lack of control of the quality of service for a particular dataset. In any case, institutions may choose to IPFS-*"pin"* their datasets (or other's dataset that are considered of interest) to ensure local copies.
- P2P systems feature mechanisms to incentivize sharing, and this may result in a lack of neutrality in access that need to be addressed.
- Permanent storage requires versioning and linking among versions, which requires some sort of indexing layer to make dataset versions and derivatives easily findable, and eventually the separation of links and content.
- Semantics and interoperability remain a matter of consensus and adoption of good practices, as demonstrated in the transfer of LOD practices and ideas into the IPFS deployment.

It is still too early to value if the adoption of file systems as IPFS will become widespread and how they would tackle with the problem of distributing the storage responsibilities with some form of incentives to sharing. But in any case, they represent a new playground for experimenting with new ideas and approaches to open data that make it more transparent and independent from their original curators.

References

1. Alexander, K., Hausenblas, M.: Describing linked datasets-on the design and usage of VoID, the vocabulary of interlinked datasets. In: Linked Data on the Web Workshop (LDOW 2009), in Conjunction with 18th International World Wide Web Conference (WWW 2009) (2009)
2. Benet, J.: IPFS-Content Addressed, Versioned, P2P File System. arXiv preprint arXiv:1407.3561 (2014)
3. Bizer, C., Heath, T., Berners-Lee, T.: Linked data-the story so far. In: Emerging Concepts, Semantic Services, Interoperability and Web Applications, pp. 205–227 (2009)
4. Cabello, F., Franco, M.G., Haché, A.: The social web beyond 'Walled Gardens': interoperability, federation and the case of Lorea/n-1. PsychNology J. **11**(1), 43–65 (2013)
5. Fernández, J. D., Polleres, A., Umbrich, J.: Towards efficient archiving of dynamic linked open data. In: Proceedings of DIACHRON, pp. 34–49 (2015)
6. Gonzalez, J.E., Xin, R.S., Dave, A., Crankshaw, D., Franklin, M.J., Stoica, I.: Graph processing in a distributed dataflow framework. In: 11th USENIX Symposium on Operating Systems Design and Implementation (OSDI 2014), pp. 599–613 (2014)
7. Haslhofer, B., Isaac, A.: Data.europeana.eu: The Europeana Linked Open Data pilot. In International Conference on Dublin Core and Metadata Applications, pp. 94–104 (2011)
8. Hausenblas, M., Karnstedt, M.: Understanding linked open data as a web-scale database. In: 2010 Second International Conference on Advances in Databases Knowledge and Data Applications (DBKDA), pp. 56–61. IEEE (2010)
9. Janssen, M., Charalabidis, Y., Zuiderwijk, A.: Benefits, adoption barriers and myths of open data and open government. Inf. Syst. Manag. **29**(4), 258–268 (2012)

10. Johnson, J.A.: From open data to information justice. Ethics Inf. Technol. **16**(4), 263–274 (2014)
11. Khayyat, M., Bannister, F.: Open data licensing: more than meets the eye. Inf. Polity **20**(4), 231–252 (2015)
12. Konstantinou, N., Spanos, D.E.: Methodologies and software tools. In: Materializing the Web of Linked Data, pp. 51–71. Springer International Publishing, Switzerland (2015)
13. Langille, M.G., Eisen, J.A.: BioTorrents: a file sharing service for scientific data. PLoS One **5**(4), e10071 (2010)
14. Lanthaler, M., Gtl, C.: On using JSON-LD to create evolvable RESTful services. In: Proceedings of the Third International Workshop on RESTful Design, pp. 25–32. ACM (2012)
15. Liu, G., Shen, H., Ward, L.: An efficient and trustworthy P2P and social network integrated file sharing system. IEEE Trans. Comput. **64**(1), 54–70 (2015)
16. Missier, P., Belhajjame, K., Cheney, J.: The W3C PROV family of specifications for modelling provenance metadata. In: Proceedings of the 16th International Conference on Extending Database Technology, pp. 773–776. ACM (2013)
17. Rajabi, E., SanchezAlonso, S., Sicilia, M.A.: Analyzing broken links on the web of data: an experiment with DBpedia. J. Assoc. Inf. Sci. Technol. **65**(8), 1721–1727 (2014)
18. Schmachtenberg, M., Bizer, C., Paulheim, H.: Adoption of the linked data best practices in different topical domains. In: Mika, P., Tudorache, T., Bernstein, A., Welty, C., Knoblock, C., Vrandečić, D., Groth, P., Noy, N., Janowicz, K., Goble, C. (eds.) ISWC 2014. LNCS, vol. 8796, pp. 245–260. Springer, Heidelberg (2014). doi:10.1007/978-3-319-11964-9_16
19. Wong, A., Liu, V., Caelli, W., Sahama, T.: An architecture for trustworthy open data services. In: Jensen, C.D., Marsh, S., Dimitrakos, T., Murayama, Y. (eds.) Trust Management IX. IFIP, pp. 149–162. Springer International Publishing, Switzerland (2015)
20. Zaveri, A., Rula, A., Maurino, A., Pietrobon, R., Lehmann, J., Auer, S.: Quality assessment for linked data: a survey. Semant. Web **7**(1), 63–93 (2015)
21. Zuiderwijk, A., Janssen, M.: Open data policies, their implementation and impact: a framework for comparison. Govern. Inf. Q. **31**(1), 17–29 (2014)

MORe: A Micro-service Oriented Aggregator

Dimitris Gavrilis[1], Vangelis Nomikos[1], Konstantinos Kravvaritis[1], Stavros Angelis[1],
Christos Papatheodorou[1,2(✉)], and Panos Constantopoulos[1,3]

[1] Digital Curation Unit – IMIS, Athena Research Center, Athens, Greece
{d.gavrilis,v.nomikos,k.kravvaritis,s.angelis,c.papatheodorou,
p.constantopoulos}@dcu.gr
[2] Department of Archives, Library Science and Museology,
Ionian University, Corfu, Greece
[3] Department of Informatics, Athens University of Economics and Business,
Athens, Greece

Abstract. Metadata aggregation is a task increasingly encountered in many projects involving data repositories. The small number of specialized software for this task indicates that in most cases customized software is used to perform aggregation, which in turn relates to the highly complex tasks and architectures involved. In this paper, the metadata and object repository aggregator (MORe) is presented, which has been effectively used in numerous projects and provides an easy and flexible way of aggregating metadata from multiple sources and in multiple formats. Its flexible and scalable architecture exploits cloud technologies and allows storing content into different storage systems, defining workflows dynamically and extending the system with external services. One of the most important aspects of MORe is its curation/enrichment services which allow curation managers to automatically apply and execute enrichment plans employing enrichment micro-services in order to aggregated data.

Keywords: Metadata aggregation · Metadata interoperability · Enrichment · Scalable architectures · Micro-services · Metadata quality

1 Introduction

In many research, as well as industrial, data management applications an aggregation step is performed whereby data (or metadata) are aggregated from multiple sources into one database/system and from multiple formats into a unified/common one. The plethora of sources and formats can be explained by the diversity of technologies and requirements that exist. Metadata aggregation is the special case where the main resources being aggregated comprise metadata (not data). In the past years organizations like the European Library (Europeana) or the Digital Public Library of America (DPLA) have aggregated large amounts of content (measured in many tens of millions of metadata records) from different formats [1] and a large number of different sources into one format (EDM [2] or DPLA respectively). The handling of multiple formats (metadata schemas) that

© Springer International Publishing AG 2016
E. Garoufallou et al. (Eds.): MTSR 2016, CCIS 672, pp. 15–26, 2016.
DOI: 10.1007/978-3-319-49157-8_2

are either custom or based on standards but used in a custom manner requires significant effort in order to:

- map them properly to the target schema (e.g. EDM);
- validate the incoming content (structural validation, schema validation, link checking, etc.);
- curate and enrich content with poor quality.

Moreover, efficiency presents another crucial challenge as in many cases millions of records are aggregated periodically and in a short amount of time. It is clear that traditional monolithic approaches would not work against the above challenges whereas a scalable and elastic architecture could stand better chances.

One of the systems used by organizations like Europeana for the aggregation task is the Metadata & Object Repository (MORe)[1] developed by the Digital Curation Unit/ IMIS – Athena Research Centre. The large number of different projects, formats and applications, such as CARARE, LoCloud, 3DIcons, ARIADNE[2], that a single instance of MORe has proved capable of serving is an indication that the system, featuring an innovative architecture to deal with several complexity and efficiency issues, has addressed the challenges of the task in a cost-effective way. This paper presents the architectural modules of MORe aggregator and its functionalities for performing aggregation workflows and curating information in accordance with commonly accepted standards and conventions of the aggregation workflows.

The next section briefly presents the state of the art for the aggregation process and the existing systems, while Sect. 3 presents the MORe architectural modules. In Sect. 4 the information enrichment functionalities are presented and Sect. 5 presents some results from the usage of MORe. Finally Sect. 6 concludes the main results of the paper.

2   Related Work

The traditional approach to aggregate metadata and links to digital resources involves an aggregator [3], which implements a crosswalk to transform original metadata records to records following a common output schema. Three architectural views are relevant to the aggregator system discussed in this paper:

- a scalable, distributed and elastic architecture
- a micro-services oriented architecture
- a pluggable enrichment services architecture

The above distinct architectural features have been explored in the literature with the majority of papers focusing on curation micro-services. Enrichment micro-services can be considered as a category of curation micro-services.

---

[1] http://more.dcu.gr/.

[2] CARARE: http://www.carare.eu/, LoCloud: http://www.locloud.eu/, 3DIcons: http://3dicons-project.eu/, ARIADNE: http://www.ariadne-infrastructure.eu/.

In [4], news items are automatically enriched with information from Linked Open Data (LOD) Datasets and use an ontology-based browser to demonstrate the advantages of LOD enabled navigation. In [5] the authors use an annotation tool to help users annotate records with information drawn from LOD thesauri. Regarding the micro-services approach, this is demonstrated in [6] where the authors propose and present a curation micro-services infrastructure in order to demonstrate the powerful characteristics and flexibility of such an approach. In [7] a micro-services architecture is presented which focuses on digital curation and preservation. Curation micro-services have also been used for enriching content. In [8] curation micro-services were used in a thematic aggregator to improve the quality and information of content.

MORe upgrades the current state of the art by integrating the traditional aggregation workflow with curation functionalities. It is a metadata aggregator that integrates several services for supporting metadata managers to (a) perform and monitor complex workflows, (b) handle huge volumes of million metadata datasets (c) validate and curate metadata with flexibility, i.e. according to enrichment plans that reflect different needs of metadata curation and (d) publish metadata in various schemas.

3   MORe Architecture

The architecture of MORe is based on established cloud technologies and focuses on three main requirements: (i) *scalability* that refers to both storage and services, (ii) *elasticity*, that refers to handling efficiently high bursts of requests and (iii) *flexibility* that refers to addressing different requirements with the use of services that are deployed in a distributed manner and are applied per case dynamically. All of the above requirements are discussed extensively in the following sections.

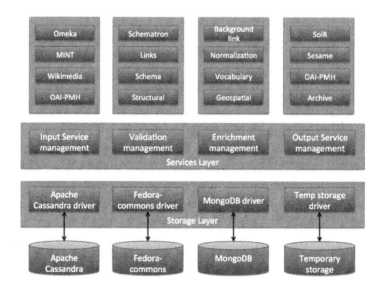

Fig. 1. Overall architecture

The overall architecture of MORe (Fig. 1) includes the following major components: (i) a storage layer, (ii) a services layer that provides a set of core services and (iii) a set of micro-services (Table 1).

Table 1. Curation/Enrichment micro-services integrated in MORe

Geo-normalization	A geo-normalization micro-service.
Geo-coding	A geo-coding micro-service based on Geo-names.
Reverse geo-coding	A reverse geo-coding micro-service based on Geo-names.
Rule based thematic enrichment	A subject collections micro-service that allows the user to create thematic collections of concepts from 27 standard vocabularies (encoded in SKOS).
Automatic thematic enrichment	An automatic vocabulary matching micro-service that identifies SKOS concepts from 27 standard vocabularies (based on title, descriptions and subject related information found in each metadata record).
Wikipedia & DBPedia automatic enrichment	A background links service that automatically identifies Wikipedia and DBPedia entries (based on title, descriptions and subject related information found in each metadata record).
Language identification	A language identification service, which identifies languages based on a title or description employing Apache Tika.
Historic place names enrichment	A historic place names micro-service.
Thesauri mappings	The thesauri mappings service allows loading and managing SKOS concepts mappings from SKOSified subject terms to a target SKOS thesaurus.

The storage layer provides an API that allows attaching virtually any "create, read, update and delete" (CRUD) based storage technology. For each storage technology a driver implementation is required and currently the Apache Cassandra, Fedora-commons and Temporary storage have been implemented.

The services layer consists of a number of core services. Some of them, such as enrichment services, follow the micro-services approach, increasing the flexibility of certain tasks:

- Harvest: harvest content from multiple sources.
- Ingest: ingest content into the appropriate storage.
- Validation: validate content.
- Indexing: index specific elements.
- Quality: measure metadata quality.
- Transform: transform content from one schema to another.

– Enrichment: enrich content using specific enrichment micro-services.
– Publish: publish aggregated content to a specific target.

Inter-service communication involves the communication between micro-services. For that, JMS Queues are used (see section below) which provide elasticity, routing and scalability. Each core service and micro-service consumes a separate queue thus enabling multiple instances to operate without any race conditions. The messages published to queues describe Jobs. A job (e.g. a transformation from one format to another) may contain additional information (e.g. the XSLT document that should be used for the transformation). This information can be part of the message. Core services (e.g. the enrichment management service) usually have to streamline tasks to micro-services and this process involves some kind of business logic (or workflow). This business logic is handled by the core service itself. The generation of jobs among core services (this essentially constitutes the workflow) is handled by a Dispatcher (or workflow management service) which is responsible for interpreting the user's input and enforcing the appropriate workflow.

3.1   Information Organization

Content aggregation is inherently a data driven task. This raises the importance of the content model, which needs to be robust, flexible and most important: domain agnostic. The latter is necessary in order to be able to aggregate information for different domains, schemas and for different purposes.

In order to address the above requirements and be able to cope with multiple users, content providers, metadata schemas and aggregation projects, information is organized in a simple hierarchical structure which can be seen from the data management perspective, as well as from the administrative perspective.

Regarding data management, all incoming information is organized in datasets. Each dataset always falls under an aggregation project. Hence a metadata provider, who participates to one or more aggregation projects, provides one or more datasets in an aggregation project. Each dataset contains one or more items which all belong to one metadata schema. MORe represents dataset items as complex items that comprise versionable datastreams. An item comprises seven datastreams:

1. The administrative metadata stream, which contains information about the provider, package, and general the history of the item.
2. The technical metadata, which contains technical metadata regarding the contents of the item.
3. The native metadata, which contains the source representation (e.g. the native metadata as they were initial harvested).
4. The enriched native metadata, which contains a representation of the enriched version of the native metadata.
5. The target metadata which contains the representation to the target schema
6. The enriched target metadata, which contains a representation of the enriched version of the target metadata.

7. Preservation metadata, which is a log of events of the PREMIS [9] metadata standard.

Regarding the administrative perspective, users are divided into four distinct roles:

1. Administrators that can setup the aggregation flows, define schemas, and system parameters.
2. Project managers that can have project scope access (e.g. see all information aggregated from different providers into a project).
3. Content providers that can initiate harvests and run the entire aggregation flow for their own organizations horizontally across projects.
4. Developers that can use MORe's RESTful API and deploy enrichment micro-services.

Each user can participate in different projects and assume different roles in each one.

3.2 Aggregation Workflows

One of the most important design considerations of MORe concerns workflows that can be adapted to particular aggregation scenaria. This is essential in order to be able to cope with the diverse needs that are found in large aggregation projects. The main aggregation workflow can be seen in Fig. 2; as indicated in the figure, some stages incorporate validation and indexing services.

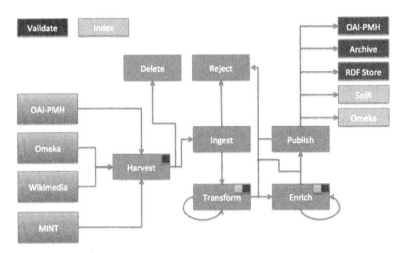

Fig. 2. Aggregation workflow

The user starts by initiating a harvest from one of the available input sources e.g. an OAI-PMH server or other intermediate aggregators, such as MINT, Omeka [10], etc. After a harvest is completed, the incoming dataset is validated and then ingested into the system. After ingest it is transformed into a common schema (e.g. EDM) and, if needed, it can be enriched using various enrichment micro-services (see next section).

After each transformation and enrichment operation, the dataset is validated and indexed. Finally, if the dataset is accepted for publication, the publishing service can publish it to one or more targets.

The overall content aggregation process involves a number of systems besides the aggregator itself, such as the content providers' native repositories and the publish targets. In order to provide interoperability so that these systems can be directly linked to the aggregation infrastructure, a RESTful API is provided alongside the human interface. The human interface is provided through a responsive Bootstrap based template that provides an improved user experience and intuitiveness.

3.3   Validation

When aggregating large amounts of content, validation and metadata quality measurement is essential in order to allow the aggregation manager to make informed decisions and ensure that the publication will be accepted. The validation of datasets in MORe is handled by the validation service that comprises four distinct micro-services:

- Schema validation (based on a given XSD)
- Structural checking of an XML record
- Link checking (broken links)
- Schematron rule based validation

The above four types of validation services are streamlined into validation schemes allowing for different schemes to be configured for different projects/sources or providers. This enables skipping heavy duty tasks such as link checking for very big packages that come from reliable sources.

3.4   Metadata Quality

Alongside with the validation results, a metadata quality service evaluates the quality of a metadata record (after each transformation or enrichment action) and produces quality related information that currently includes:

- Completeness per element for each schema
- Completeness per element set (mandatory and recommended element sets are used)

In order for the aggregation manager to make informed decisions on whether and how to enrich an information package, indexes (except metadata completeness) concerning spatial, thematic, temporal and rights information, are computed and presented by the system. These indexes are configured per schema, similarly to the metadata completeness index. In the case of spatial information indexing, a small map widget is used to project the data directly on the map thus offering a better, more intuitive user experience.

In the cases of thematic, temporal and rights information, a small widget presents the entries for each index on a list view and the number of their occurrences in the metadata records (items) per dataset.

3.5 Publish Targets

A dataset can be published after it has been transformed into a predefined metadata format, according to the specific project target schema, and it has passed validation and has been enriched. Traditionally in most aggregator infrastructures, publication either means exposing the published items through an OAI-PMH provider or pushing them to a SolR index server. In order to provide more flexibility MORe's publishing service supports a number of different publish targets, which can be extended to include more publish targets as they also follow a micro-service architecture. Currently, the publish service supports the following targets:

- Internal OAI-PMH repository (published under a specific Set per provider and project)
- Sesame RDF store (if the target schema is RDF formatted)
- OpenLink Virtuoso RDF store
- Elastic Search index server
- Archive (dump in tar.gz format for all published items)

Furthermore, it is possible for a project to have multiple targets, for example to publish all content through an OAI-PMH provider and also download them through an XML archive. The administrator can define these targets for each project separately and the aggregation manager can choose in which of the available targets to publish each dataset. This implies that the same content can be published simultaneously in multiple targets.

Some publish targets require the content to be in a format different than that of the project's target schema. For example, when aggregating content in XML and the goal is to transform into a common XML schema (e.g. EDM) the transformed EDM representations of each item can be directly published to the internal OAI-PMH provider or to an archive, but they cannot be published to an RDF store as this requires an RDF format. Similarly, Elastic Search accepts JSON format and, although it can automatically convert from XML to JSON in most cases it is more practical and efficient to encode information in specific JSON format.

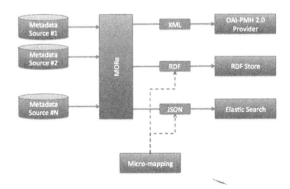

Fig. 3. Publishing to multiple targets using micro-mappings

In order to enable publishing to multiple targets simultaneously, a micro-mapping mechanism is employed which allows mapping directly and on the fly during the publication process to the target's desired output (Fig. 3). Micro-mappings are realized using XSLT transformations and the parameters are defined per project and partner. For example, in the case of Elastic Search, the server URL, credentials and index name are provided along the micro-mapping XSLT document.

3.6 Elasticity and Scalability

Elasticity is a characteristic very common to big data architectures (along with scalability) as bursts of requests that need not to be processed in real time can occur. An elastic architecture can ensure that all requests can be received, acknowledged and dispatched when a worker node becomes available. MORe, provides an elastic architecture which offers all of the above and is based on message queues.

Scalability is a critical characteristic of an aggregation process as it is resource consuming both in terms of storage and processing power. These two aspects are addressed in MORe in the following two ways:

- At the storage level by using a cloud based storage (like Apache Cassandra) that can scale out using a clustered architecture.
- At the data processing level by adopting scalable services architecture that allows services to scale out in a clustered environment (such as Apache Storm).

4   Curation

One of the most important aspects of MORe is that it is curation aware. This means that apart from simple XSLT based transformations from a native schema to a target schema, MORe employs a number of curation/enrichment micro-services that perform various curation/enrichment actions on the metadata. Examples of such micro-services that have been integrated/developed are listed in Table 1.

It is apparent that the above micro-services are heterogeneous from several aspects:

- they have been developed using different programming languages and frameworks;
- they require and produce different information (e.g. spatial coordinates, links, language codes, etc);
- they are encoded in different ways (e.g. json, xml, etc);

Some of them are self-sustained, some others, such as the Geo-names gazetteer rely on external databases and services.

4.1   Micro-services de-Coupling/Micro-schemas

The heterogeneity of micro-services presents a challenge for the system to be extensible with a minimum amount of effort and to take advantage of the richness of innovative services that are freely available. To this end, two main methodologies/technologies were employed:

(a) a service oriented architecture (SOA) relying on HTTP REST to facilitate communication; and

(b) an abstraction layer that de-couples the logic of the enrichment services from that of the aggregator.

The communication through REST enables simple and efficient remote invocation of micro-services while retaining the ability to scale them.

The abstraction layer enables to dynamically map parts of the target schema (e.g. part of metadata that provide only geographical data and coordinates) to the inputs of each enrichment service (e.g. Geo-coding micro-service). This technique employs *micro-schemas* to perform the mappings dynamically and thus enabling MORe to apply the same enrichment micro-service to multiple target schemas without having to adapt/ code. This technique is illustrated in Fig. 4.

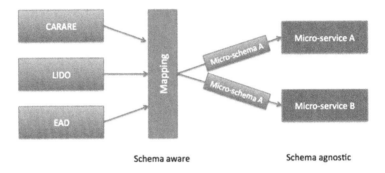

Fig. 4. Use of micro-schemas in enrichment micro-services

4.2 Streamlining Enrichment Micro-services: Enrichment Planning

In order for these services to be integrated and put into effective use in the aggregation workflow, service orchestration is employed through the *Enrichment Management Service*. This service is responsible for enriching a dataset through streamlining the execution of enrichment micro-services. The tasks it performs are:

– Iterating through all the valid items of a dataset.
– Executing a set of enrichment micro-services (called *Enrichment Plan* – see next section) for each one of the valid items by feeding the output of an enrichment micro-service to the input of the next enrichment micro-service.
– Handling errors in the enrichment process.
– Applying specific project/provider or run-time parameters to each enrichment service.
– Dispatching and monitoring the health of each enrichment micro-service.
– Compiling a report on the output of the enrichment.

All of the above tasks are provided through an API so that configuration and new micro-services integration tasks are provided easily.

Enrichment planning is an important and innovative feature of MORe as it allows each content provider or aggregation manager to easily create complex and powerful enrichment workflows by combining simple enrichment micro-services through an intuitive graphical interface.

Each enrichment plan operates on a single metadata schema and apart from streamlining the execution of enrichment micro-services, it defines configuration parameters for each one of them (if and when available). After the execution of an enrichment plan, a report is compiled and presented to the user so that he/she can see which items were enriched by which service etc.

5   Experimental Results

MORe manages huge volumes of data and provides a diversity of services to its users. Therefore it is obvious that the system performance depends on two parameters: (a) the number of micro-services that are processing data concurrently and (b) the number of concurrent users that call and apply micro-services on datasets. Regarding the second parameter it was observed that when the number of workers increase from 1 to 8, the response time decreases by 63.47 %.

Therefore, in order to evaluate the performance of the micro-service architecture, experimental results have been carried out using a fixed number of datasets coming from the CARARE project. The datasets included 1.3 million records approximately encoded in the CARARE schema [1]. The same datasets were used in the different experiments as well as the same environment (debian linux server).

The experiments investigated two scenarios: in the first scenario, the core service package contains all micro-services whereas in the second one the micro-services are completely distributed (running in different machines on the same subnet). It is clear

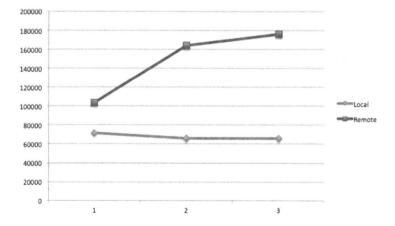

Fig. 5. Execution time in ms (vertical) versus the number of enrichment micro-services executed (horizontal) for an enrichment task. The blue line (circle) refers to a scenario where all the micro-services are packaged in the same worker. The red line (square) refers to a scenario where the micro-services are distributed. (Color figure online)

that when the enrichment plan becomes more complex and contains more micro-services the response time greatly decreases in the first case. In the example shown in Fig. 5, for 3 micro-services this increase reaches 62.42 %.

6   Conclusions

In this paper, the innovative architecture of the MORe metadata aggregator has been presented. MORe addresses the complexities found in metadata aggregation tasks through a micro-service oriented architecture that provides elasticity, flexibility and scalability. MORe has been effectively used to aggregated millions of records in various projects in different domains involving different formats and targets. It is accessible both as a Web application and through a RESTful API and it allows developers to extend it with curation/enrichment micro-services easily. The performance evaluation experimental results were encouraging since they indicate that MORe is a stable system even when it manages huge volumes of datasets.

References

1. Papatheodorou, C., Dallas, C., Ertmann-Christiansen, C., Fernie, K., Gavrilis, D., Masci, M.E., Constantopoulos, P., Angelis, S.: A new architecture and approach to asset representation for europeana aggregation: the CARARE way. In: García-Barriocanal, E., Cebeci, Z., Okur, M.C., Öztürk, A. (eds.) MTSR 2011. CCIS, vol. 240, pp. 412–423. Springer, Heidelberg (2011). doi:10.1007/978-3-642-24731-6_41
2. Isaac, A.: Europeana data model primer. Europeana Project (2011)
3. Reis, D., Freire, N., Manguinhas, H., Pedrosa, G.: REPOX – a framework for metadata interchange. In: Agosti, M., Borbinha, J., Kapidakis, S., Papatheodorou, C., Tsakonas, G. (eds.) ECDL 2009. LNCS, vol. 5714, pp. 479–480. Springer, Heidelberg (2009). doi: 10.1007/978-3-642-04346-8_65
4. Mannens, E., Troncy, R., Braeckman, K., Van Deursen, D., Van Lancker, W., De Sutter, R., Van de Walle, R.: Automatic metadata enrichment in news production. In:10th Workshop on Image Analysis for Multimedia Interactive Services (WIAMIS 2009), pp. 61–64. IEEE Press (2009)
5. Rainer, S., Haslhofer, B., Jung, J.: Annotations tags and linked data. Metadata enrichment in online map collections through Volunteer-Contributed Information. e-Perimetron **6**, 129–137 (2011)
6. Clair, K.: Metadata for a micro-services-based digital curation system. In: International Conference on Dublin Core and Metadata Applications, pp. 58–62 (2011)
7. Abrams, S., Kunze, J., Loy, D.: An emergent micro-services approach to digital curation infrastructure. Int. J. Digit. Curation **5**, 172–186 (2010)
8. Gavrilis, D., Dallas, C., Angelis, S.: A curation-oriented thematic aggregator. In: Aalberg, T., Papatheodorou, C., Dobreva, M., Tsakonas, G., Farrugia, C.J. (eds.) TPDL 2013. LNCS, vol. 8092, pp. 132–137. Springer, Heidelberg (2013). doi:10.1007/978-3-642-40501-3_13
9. Gavrilis, D., Angelis, C., Papatheodorou, C.: MOPSEUS: a digital repository system with semantically enhanced preservation services. In: 7th International Conference on Preservation of Digital Objects (iPRES 2010), pp. 135–143 (2010)
10. Kucsma, J., Reiss, K., Sidman, A.: Using Omeka to build digital collections: the METRO case study. D-Lib Mag. **16**(3–4) (2010)

Linkset Quality Assessment for the Thesaurus Framework LusTRE

Riccardo Albertoni, Monica De Martino, and Paola Podestà(✉)

Istituto di Matematica Applicata e Tecnologie Informatiche,
Consiglio Nazionale delle Ricerche, Via de Marini, 6, 16149 Genova, Italy
{albertoni,demartino,podesta}@ge.imati.cnr.it

Abstract. Recently a great number of controlled vocabularies (e.g., thesauri) covering several domains and shared by different communities, have been published and interlinked using the Linked Data paradigm. Remarkable efforts have been spent from data producers to make their thesauri compliant with Linked Data requirements both for the content encoding and for the connections (aka, linkset) with others thesauri. Also in our experience in the creation of the framework of multilingual linked thesauri for the environment (LusTRE), within the EU funded project eENVplus, the development of the interlinking among thesauri, have required significant efforts, thus, the evaluation of their quality in term of usefulness and enrichment of information became a critical issue. In this paper, to support our claim, we discuss the results of the quality evaluation of several linksets created in LusTRE. To this purpose, we consider two quality measures, the *average linkset reachability* and the *average linkset importing*, able to quantify the linkset-accessible information.

Keywords: Linkset quality · SKOS · Linked data · Environmental thesauri · Metadata

1 Introduction

The continued expansion of the Web as a medium for the exchange of interoperable data and the sustained growth of the Linked Open Data Cloud[1], represent important factors for the Linked Data paradigm success. In this context, where data sharing and consumption are accessible to a large number of actors, the quality of the exposed data become one of the most critical issues, since as widely known and accepted, data is only worth its quality [16].

Linked Data paradigm [10] is based on two core characteristics: data sources and connections of information belonging to different sources through the linksets. Thus, not only data but also the connections among data are essential to keep the Linked Data promise to "evolve the current web data into a Global Data Space" [10]. In fact, through a link a consumer can navigate

---

[1] http://lod-cloud.net/.

© Springer International Publishing AG 2016
E. Garoufallou et al. (Eds.): MTSR 2016, CCIS 672, pp. 27–39, 2016.
DOI: 10.1007/978-3-319-49157-8_3

in a seamless way between objects belonging to different datasets, possibly of different domains, accessing to richer and more complete information than the data at hand. The quality of connections (in the following, *linkset quality*) should be as important as the quality of data, but, since now, the research on Linked Data quality has been mainly focused on datasets [16].

During the EU funded project eENVplus (CIP-ICT-PSP grant No. 325232) supporting the Infrastructure for Spatial Information in the European Community (INSPIRE)[2] directive implementation, we have developed a multilingual linked thesaurus framework for the environment (LusTRE)[3]. The framework aims to provide shared standard and scientific terms for a common understanding of environmental data among the different communities operating in different fields of the Environment, It supports better cross-domain and multilingual metadata compilation and information discovery. In order to construct the framework, we have analysed several environmental thesauri exposed on the Web according to the Simple Knowledge Organization System (SKOS)[4] Vocabulary [4]. Then, we have spent considerable efforts to publish in Linked Data some of them (ThiST, EARTh [3]) and to interlink each other GEMET, AGROVOC [8], EUROVOC, TheSoz, DBpedia, ThiST, EARTh; but, until now, there is no way to calculate the information reachable navigating the interlinks.

In this paper, we evaluate the quality of the connections created in Lus-TRE. We rely on the notion of dataset and linkset provided in the Vocabulary of Interlinked Datasets (VoID)[5]. Thus, a *dataset* is a set of Resource Description Framework (RDF)[6] triples published, maintained or aggregated by a single provider; a *linkset*, is a special kind of dataset containing only RDF links between two datasets, the *subject*, and the *object* of the linkset. We focus on the `skos:exactMatch` linksets (i.e., linkset composed only by `skos:exactMatch` mapping relation) developed in LusTRE and we consider two measures: the *average linkset importing*, based on a measure presented in [5], and a new measure, the *average linkset reachability*. Using the average linkset importing, we evaluate LusTRE linksets assessing the multilingualism enrichment in terms of new translated labels reachable in the object through a linkset. Average linkset importing can help to address the incomplete language coverage issue, that is, when `skos:prefLabel` and `skos:altLabel` are provided in all the expected languages only for a subset of the thesaurus concepts (in [14]), affecting many popular SKOS thesauri. Average linkset reachability evaluates the number of new concepts reached by crossing a linkset. It can be exploited to evaluate the potential enrichment of the space of concepts that can be browsed (aka, the thesaurus browsing space). The application of both measures to the linksets of LusTRE shows that, in average, linksets bring an effective enrichment in terms of multilingualism and browsing space.

---

[2] http://inspire.ec.europa.eu/.

[3] http://linkeddata.ge.imati.cnr.it/.

[4] https://www.w3.org/TR/skos-reference/.

[5] https://www.w3.org/TR/void/.

[6] https://www.w3.org/RDF/.

The paper is organized as follows. Section 2 presents the related work on quality of Linked Data, while, Sect. 3 presents the LusTRE framework. *Average linkset importing* and *average linkset reachability* measures, are presented in Sect. 4 and the evaluation of LusTRE linksets is discussed in Sect. 5. Finally, Sect. 6 illustrates conclusions and future work.

2   Related Work

A recent systematic review of quality assessment for Linked Data can be found in the SWJ submission [16]. This paper reviews quality dimensions traditionally considered in data quality (e.g., availability, timeliness, completeness, relevancy, availability, consistency) and Linked Data specific dimensions, such as licensing and interlinking, considering, for the latter, the framework LINK-QA [9] and the works [2,15]. LINK-QA defines two network measures specifically designed for Linked Data (i.e., Open SameAs chains, and Description Richness) and three classic network measures (i.e., degree, centrality, clustering coefficient) for determining whether a set of links improves the overall quality of Linked Data. Whilst, [2,15] detect the quality of interlinking via crowd-sourcing. Recently, logic detection of invalid SameAs has been proposed in [12]. The main differences with respect to the two linkset measures deployed in this paper are: (i) [2,9,12,15] work on links independently from the fact that links are part or not of the same linksets; (ii) [2,9,12,15] address the correctness of links, and not the gain in terms of multilingualism or browsing space. A set of scoring functions measuring the gain obtained when complementing a dataset with its `owl:sameAs` linksets is proposed in our previous work [6], that, we extend here deploying two new measures based on `skos:exactMatch` linkset among environmental SKOS thesauri. A set of quality measures specific for SKOS thesauri have been proposed in [14]. The paper summarizes a set of 26 quality issues for SKOS thesauri and shows how these can be detected and improved by deploying qSKOS [11], PoolParty checker, and Skosify [13]. Unfortunately, an analysis on linksets among thesauri is not included in [14]: missing out-links and in-links are adopted as indicators of SKOS thesaurus quality, but, their potential in terms of reachability of new concepts or importing of `skos:prefLabel` and `skos:altLabel` values is not considered. A first attempt to evaluate linkset quality is the linkset importing measure presented in our previous work [5]. Average linkset importing differs from linkset importing, since the former considers the absolute number, while the latter considers the percentage of "new values" of an RDF property accessed through a linkset.

3   LusTRE Framework

Multilingual linked thesaurus framework for the environment (LusTRE) is an interesting example of the exploitation of Linked Data to support metadata compilation and information discovery, for describing and for finding INSPIRE data and INSPIRE data services (or, environmental geodata in general). The main

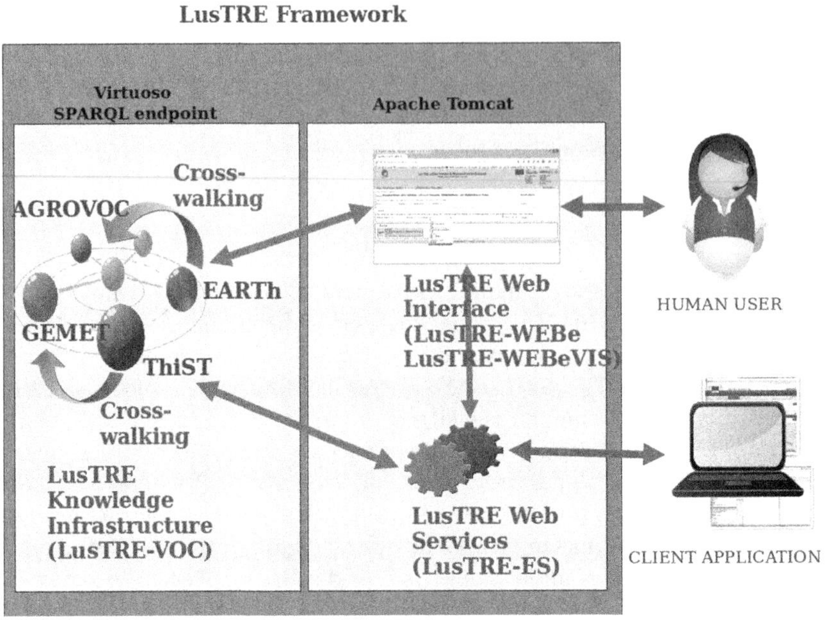

LusTRE Framework

Fig. 1. LusTRE components and their interactions.

goal of having such a framework is to be able to preserve and retrieve the information based on the semantic definitions, rather than just lexical keywords. This would guarantee the uniformity of the persisted metadata information, as well as discovery of metadata based on the semantic meanings even if metadata include diverse and dissimilar keywords [1].

The main components of LusTRE and their interactions, illustrated in Fig. 1, are the following:

- **LusTRE knowledge infrastructure (LusTRE-VOC)** contains different environmental thematic vocabularies, the interlinking among them, and the access to them as one virtual integrated Linked Data source. It is deployed on Virtuoso server and accessible by SPARQL endpoint.
- **LusTRE Exploitation Services (LusTRE-ES)** is a set of end-user oriented web services with a REST interface. It allows to exploit the knowledge contained in the LusTRE-VOC for improving client applications such as a metadata editor or a geodata portal. In particular, LusTRE-ES supports the automatic navigation among vocabulary (that is, *the cross-walking* explained in the following).
- **LusTRE web interface** provides a human-accessible interface to manually search and navigate the interlinked knowledge infrastructure using a textual (LusTRE-WEBe) or a visual browsing (LusTRE-WEBeVIS).

To be concrete, as shown in Fig. 1, there are two kinds of usage provided by LusTRE: (i) for the direct manual interaction, LusTRE offers to the end user a number of Web GUI elements for browsing, inspecting, searching and translating thesaurus concepts and for visualizing the knowledge structures hidden in the interlinked thesauri; and (ii) for the transparent access, it offers the possibility to extend the functionalities of existing third-party tools by accessing the web services. Currently, LusTRE web services are being integrated in client applications such as the EUOSME[7] and QSPHERE[8] metadata editors and the under development version of the INSPIRE Geoportal[9].

A relevant feature provided by LusTRE is the *cross-walking*: the possibility to automatically navigate among matching concepts belonging to different linked thesauri. It supports easier working beyond the scope and limitations of a single thesaurus, possibly enriching data at hand, and, thus, improving user satisfaction in data consuming process. Considering on one hand the important efforts spent to construct all the linksets in LusTRE, and on the other hand the potentiality of the exploitation of the linksets through the cross-walking, we decide to investigate the quality for LusTRE linksets. In particular, we focus on: (i) the multilingualism improvement and (ii) the widening of the browsing space. We consider in LusTRE the following SKOS thesauri, with the number of concepts and languages indicated within brackets: ThiST (34150 concepts, 2 languages), AGROVOC (32310 concepts, 24 languages), EARTh (14350 concepts, 2 languages), EUROVOC (6883 concepts, 23 languages), GEMET (5223 concepts, 32 languages). Concerning linksets, we consider the twenty `skos:exactMatch` linksets, presented in Table 1, describing the couple of thesauri involved and the number of links in each linkset (that is, linkset cardinality). Linksets have been created by working out the transitive closure on existing `skos:exactMatch` and applying specific linkset discovery tasks. Every linkset discovery task has followed a two-steps process: firstly, SILK[10] has been applied to discover new links, then the link correctness has been validated by some domain experts. Linkset completeness is reasonably ensured by having applied different and not very restrictive matching functions during the discovery task. SKOS entailments have been materialized to support clients with limited processing power. As a consequence of the such materialization and the `skos:exactMatch` symmetry, reciprocal linksets (e.g., EARTH2GEMET and GEMET2EARTh or EUROVOC2AGROVOC and AGROVOC2EUROVOC in Table 1) have exactly the same links but inverted.

---

[7] http://showcase.eenvplus.eu/client/editor.htm.

[8] http://plugins.qgis.org/plugins/qsphere/.

[9] http://showcase.eenvplus.eu/client/geoportal.htm.

[10] http://silkframework.org/.

Table 1. Linksets in LusTRE.

Linkset name	Subject thesaurus	Object thesaurus	Linkset cardinality
AGROVOC2EARTH	AGROVOC	EARTH	1438
AGROVOC2EUROVOC		EUROVOC	1269
AGROVOC2GEMET		GEMET	1188
AGROVOC2THIST		THIST	1695
EARTH2AGROVOC	EARTH	AGROVOC	1438
EARTh2EUROVOC		EUROVOC	1346
EARTH2GEMET		GEMET	4365
EARTH2THIST		THIST	1140
EUROVOC2AGROVOC	EUROVOC	AGROVOC	1269
EUROVOC2EARTh		EARTH	1346
EUROVOC2GEMET		GEMET	1683
EUROVOC2THIST		THIST	733
GEMET2AGROVOC	GEMET	AGROVOC	1188
GEMET2EARTH		EARTH	4365
GEMET2EUROVOC		EUROVOC	1683
GEMET2THIST		THIST	792
THIST2AGROVOC	THIST	AGROVOC	1695
THIST2EARTH		EARTH	1140
THIST2EUROVOC		EUROVOC	733
THIST2GEMET		GEMET	792

4   Quality Measures for Linkset

In this section, we present the two linkset quality measures, evaluating the information accessed cross-walking the linksets of LusTRE. To this purpose, we identify in the following with L the linkset between the subject thesaurus T_s and the object thesaurus T_o, the cardinality of a thesaurus T with $|T|$. The two linkset measures address different aspects of linkset quality: (i) the average linkset reachability, that estimates the enrichment of thesaurus browsing space. It evaluates, for each linkset, the average number for link of concepts in the object thesaurus browsable starting from the concepts involved in the linkset; (ii) the average linkset importing, that focuses on the average number for link of new values of a certain RDF property, reachable by the linkset in the object thesaurus. In this paper, the average linkset importing evaluates the average number of new skos:prefLabel and skos:altLabel reachable through the linkset, and it can help in addressing the incomplete language coverage issue, which affects many popular SKOS thesauri [14].

The application of the two measures requires the correctness of thesauri. Note that correctness is not the focus of our measures, in fact, our objective is to evaluate the additional information collected by the subject SKOS thesaurus from different object SKOS thesauri through different linksets. For reachability, we assume also *completeness*, described before, for skos:exactMatch linkset, that is, *any concept in the subject thesaurus having an exact equivalent concept in the object thesaurus must be involved in a skos:exactMatch link and of course in the linkset.* Nevertheless, if correctness and completeness are not satisfied our measures might take into account duplicated information and the final evaluation might differ from the real one. In any way, these assumptions seem reasonable since: (i) currently, all applications consuming Linked Data implicitly assumes at least correctness (trusting on publisher reliability); (ii) there are several tools (SILK, LIMES and qSKOS) that can help to reach linkset correctness and completeness. Finally, we consider the set of SKOS properties for average linkset importing $SKOSlabel = \{$skos:prefLabel,skos:altLabel$\}$ and for average linkset reachability $SKOSrel = \{$skos:narrower, skos:related, skos:broader$\}$.

4.1   Average Linkset Importing Measure

The average linkset importing measure evaluates the average number of "new values" of a RDF property p, accessible through a link. Generalising to the entire linkset, "new values" are those not already present in the subject thesaurus T_s, but reachable through the linkset L in object thesaurus T_o.

Definition 1 *(Average linkset importing).* *Let ln be an ISO language tag or _ (which stands for all languages), l be a link in L of the form (t_s, skos:exactMatch, t_o) and p be in SKOSlabel. Let TsVal(p,l,ln) be the values of p for concept t_s and ToVal(p,l,ln) be the values of p for concept t_o. The average linkset importing measure can be defined as follows:*

$$ALI^p_{ln} = \frac{1}{|L|} \sum_{l \in L} |ToVal(p,l,ln) - TsVal(p,l,ln)|$$

Example 1. Considering the thesauri T_s, T_o and linkset L in Fig. 2, and p $=$ skos:prefLabel, TsVal(skos:prefLabel,l2,en) $=$ {Dog@en}, whilst TsVal(skos:prefLabel,l2,_) $=$ {Dog@en, Perro@es}, since in the latter there is no constraint on the language tag. Considering y_3, the skos:exactMatch-linked concept for x_3, all the translations for the skos:prefLabel in the object thesaurus are ToVal(skos:prefLabel,l2,_) $=$ {Dog@en, Cane@it}, while for the

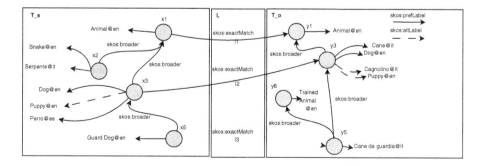

Fig. 2. Example of RDF/SKOS thesauri and skos:exactMatch linkset.

subject the translations are TsVal(skos:prefLabel,l2,_) = {Dog@en,Perro@es}. Thus, $ALI_-^{skos:prefLabel} = \frac{1}{3} * (0 + 1 + 1) = \frac{2}{3} = 0.67$. A value of 0.67 represents the average number of new skos:prefLabel values, in any language, accessible through each link of the linkset. The result means that every three links, two new translations of skos:prefLabel are accessible.

4.2   Average Linkset Reachability Measure

The *average linkset reachability* evaluates the average number of concepts reachable when cross-walking a link of the linkset and exploring the object thesaurus T_o until a certain depth, identified with the number of hops. Concepts in the T_o directly involved in the linkset are not considered.

Definition 2 *(Average linkset reachability)*. *Let L_o be the set of object concepts of each link in L, $ConceptsT_o(k,SKOSrel,L_o)$ be the set of concepts in T_o, from the concepts in L_o, through the relations in SKOSrel in a number of hops $<= k$. The average linkset reachability is defined as follows:*

$$ALR_k^{SKOSrel} = \frac{|ConceptsT_o(k, SKOSrel, L) - L_o|}{|L|}$$

Example 2. Considering the thesauri T_s and T_o and linkset L in Fig. 2, L_o is represented by $\{y_1, y_3, y_5\}$, while considering a number of hop k equal to 2 and p=skos:broader, $ConceptsT_o(2, \text{skos:broader}, L_o) = \{y_1, y_3, y_5, y_6\}$. The $ALR_2^{SKOSrel} = \frac{|\{y_1,y_3,y_5,y_6\}\setminus\{y_1,y_3,y_5\}|}{3} = \frac{1}{3} = 0.33$. The 0.33 represents the average number of concepts in object thesaurus reachable, in 4 hops, to each link in the linkset, which implies one new concept every three links.

5   Linksets Quality Evaluation of LusTRE

We have developed a prototype in JAVA/JENA that implements both average linkset importing and reachability, and applied it to all the linksets developed among within SKOS thesauri of LusTRE (see Table 1 for details). LusTRE linksets satisfy both requirements of correctness and completeness, since they have been validated by domain experts. As already discussed, the considerable efforts spent during the project eENVplus to create and validate LusTRE linksets, suggest that an evaluation of the linkset in terms of usefulness and information enrichment is important. In this application case, the *average linkset importing* measure evaluates the average number of new values for skos:prefLabel and skos:altLabel, in different languages, reachable through each link, in the object thesaurus. While, the *average linkset reachability* represents the average number of concepts accessible in the object thesaurus, through each link, in a number of hops less or equal to 4, considering the SKOS relations skos:narrower, the skos:broader and skos:related. We chose k = 4, since it seems a reasonable number of steps that a user should performs, for example,

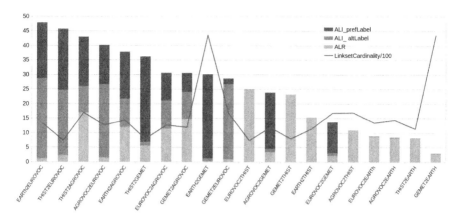

Fig. 3. Average linkset measures results (histograms) and cardinality/100 (line) of LusTRE linksets.

during the search of a specialization/generalization of a SKOS concept in a thesaurus. The results obtained by both measures allow to analyse the quality of the linkset at different levels of detail. An overall evaluation of the linksets is shown in Fig. 3, where the x axis shows the linksets and the y axis shows the average linkset reachability and the average linkset importing. In each bar in the graph, we have piled the results of importing for skos:altLabel and skos:prefLabel, and of reachability for number of hops k = 4. All the linksets provide a set of new values in terms of skos:prefLabel, skos:altLabel or new concepts. Thus, globally we can say that the efforts spent in the development of LusTRE linksets have paid off, in fact, the multilingualism and/or the browsing space of a single thesaurus are enriched by its linksets. After that, we can also observe another important fact: the substantial independence between the quantity of new label translations/concepts accessed and the cardinality (divided by 100) of the linkset shown as a line in Fig. 3. Often, the quality of a linkset is identified with its cardinality, more link means more quality, but, as shown in the graph the two facts are not related. In particular, the linkset EUROVOC2GEMET has about 1600 links, and it provides for each link a total of about 15 new translations and concepts, while the linkset EARTH2EUROVOC has about 1300 links, but it brings a total greater than 45 of new translations and concepts for each link, almost three times EUROVOC2GEMET. As a consequence, it seems clear that the quality of a linkset is substantially independent and more complex than its cardinality.

More in details, we analyse the result of each measure for a single link. We notice that, as obvious, the two measures captures different aspects of the linkset quality, thus, a linkset can have two different evaluation in the two measures. Considering Fig. 4, for example, we notice that linksets EARTH2EUROVOC and THIST2EUROVOC are "good" for average linkset importing and "poor" for average linkset reachability, and vice versa GEMET2THIST and EUROVOC2THIST

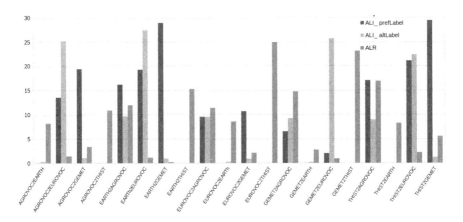

Fig. 4. Average linkset importing and reachability in details.

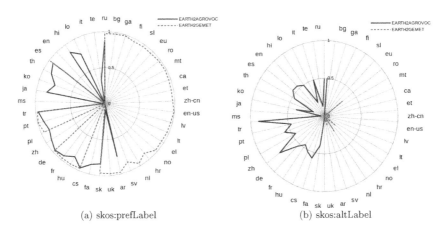

(a) skos:prefLabel   (b) skos:altLabel

Fig. 5. Average linkset importing for EARTH2AGROVOC and EARTH2GEMET.

are "good" for average linkset reachability and "poor" for average linkset importing. These results can be used, also, to formulate some hypothesis concerning the content of the involved thesauri, for example, Fig. 4 shows that EARTH thesaurus should not have to many translations or it should have but in languages different from GEMET, since it adds, in average, about 25 new `skos:prefLabel` for each link.

Finally, it is possible to deepen the analysis of the average link importing focusing on the average number of values for each distinct languages accessible for each link of the linksets. As an example, we focus on EARTH2AGROVOC and EARTH2GEMET, and we report the result in the radial graph in Fig. 5, where radial axes include one axis for each language (41 different languages are considered) representing the average importing for link. In Fig. 5, we observe that, both linksets import the same set of 10 languages (i.e., ar, ru, es, tr, pt, pl, de, fr, hu, cs). Besides, 19 languages (e.g., bg, ga, fi, sl, eu, ro) can be imported only through EARTH2GEMET, and 9 only through EARTH2AGROVOC. Discussing the graphs in detail, we have two opposite situations: (i) for `skos:prefLabel`, Fig. 5(a), it is evident that both the average importing and the number of languages of the linkset EARTH2GEMET are higher than in EARTH2AGROVOC; (ii) for `skos:altLabel`, Fig. 5(b), we have exactly the opposite situation. In fact EARTH2AGROVOC imports more `skos:altLabel` and also more different languages than EARTH2GEMET. In fact, we import about 20 languages from EARTH2AGROVOC and only 4 from EARTH2GEMET.

As a synthesis, we can say that in average the linksets developed in LusTRE bring an advantage in term of multiligualism and browsing space, as a consequence all the actors (human or software) of the data consuming process can benefit from this information enrichment, exploiting the cross-walking feature provided by LusTRE. Focusing on the linkset quality issue, it is evident that, the choice of the best linkset is not straightforward, but it depends on the specific point of view of the analysis. In our case, we have demonstrated the ability of a linkset to enrich the thesaurus multilingualism and browsing space. However, the linkset quality is a very complex issue, thus, its full characterizations requires the definition of further measures capturing different aspects. On the other hand, it is evident that considering only the number of links is not enough.

6 Conclusions and Future Work

This paper presents the evaluation of the linksets developed in the multilingual linked thesaurus framework for the environment LusTRE. Our purpose, in the assessment of linkset quality of LusTRE, is to evaluate if the efforts spent in the creation of such connections results in some improvement in consuming LusTRE data. This is a quite new point of view in Linked Data quality assessment since almost all the existing methods focus on datasets. Our approach is based on two considerations: (i) linkset is as important as data in the evolution of the Web of Data into the Global Data Space; (ii) the creation of linksets require efforts, thus, the evaluation of their quality is pivotal. To this purpose we consider two measures evaluating the information accessible through a linkset: average linkset importing and average linkset reachability. Average linkset importing can be exploited to evaluate the multilingual enrichment, in terms of translated labels, while average linkset reachability assesses the browsing space extension in terms of new concepts accessible through the linkset. Both measures are used to evaluate the `skos:exactMatch` linksets developed in LusTRE. The results show that the efforts spent in the creation of the linksets are paid off, in fact, all the linksets provide an increase of the multilingualism and of the browsing space. As future work, we plan to extend the quality framework adding other measures able to capture other aspects of linkset quality. Moreover we foresee to encode the quality results according to the Data Quality Vocabulary (DQV) [7], developed by the Data on the Web Best Practices Working Group, and to face the challenge to apply the measures "into the wild", that is to the `skos:exactMatch` linksets exposed in the LOD Cloud.

Acknowledgements. The paper activity has been carried out within the EU funded project eENVplus (CIP-ICT-PSP grant No. 325232). The authors would like to thank all partners and, in particular, Paolo Plini (IIA-CNR) for the important collaboration, and the team of the European Commission's Joint Research Centre (Italy) for the valuable contribution.

References

1. Abecker, A., Wössner, R., Schnitter, K., Albertoni, R., De Martino, M., Podestà, P.: Latest developments of the linked thesaurus framework for the environment (LusTRE). In: 29th EnviroInfo and 3rd ICT4S Conference 2015, 7–9 September 2015, Copenhagen, Denmark (2015)
2. Acosta, M., Zaveri, A., Simperl, E., Kontokostas, D., Auer, S., Lehmann, J.: Crowdsourcing linked data quality assessment. In: Alani, H., et al. (eds.) ISWC 2013. LNCS, vol. 8219, pp. 260–276. Springer, Heidelberg (2013). doi:10.1007/978-3-642-41338-4_17
3. Albertoni, R., De Martino, M., Di Franco, S., De Santis, V., Plini, P.: EARTh: an environmental application reference thesaurus in the linked open data cloud. Semant. Web **5**(2), 165–171 (2014)
4. Albertoni, R., De Martino, M., Podestà, P.: Environmental thesauri under the lens of reusability. In: Kő, A., Francesconi, E. (eds.) EGOVIS 2014. LNCS, vol. 8650, pp. 222–236. Springer, Heidelberg (2014). doi:10.1007/978-3-319-10178-1_18
5. Albertoni, R., De Martino, M., Podestà, P.: A linkset quality metric measuring multilingual gain in SKOS thesauri. In: Rula, A., Zaveri, A., Knuth, M., Kontokostas, D. (eds.) Proceedings of the 2nd Workshop on Linked Data Quality Co-located with 12th Extended Semantic Web Conference (ESWC 2015), 1 June 2015, Portorož, Slovenia, vol. 1376. CEUR Workshop Proceedings. CEUR-WS.org (2015)
6. Albertoni, R., Gómez-Pérez, A.: Assessing linkset quality for complementing third-party datasets. In: Guerrini, G. (ed.) EDBT/ICDT Workshops, pp. 52–59. ACM (2013)
7. Albertoni, R., Isaac, A., Debattista, J., Dekkers, M., Guret, C., Lee, D., Mihindukulasooriya, N., Zaveri, A.: Data on the web best practices: data quality vocabulary (2016), W3C Working Draft. http://www.w3.org/TR/vocab-dqv/. Accessed 28 July 2016
8. Caracciolo, C., Stellato, A., Morshed, A., Johannsen, G., Rajbhandari, S., Jaques, Y., Keizer, J.: The AGROVOC linked dataset. Semant. Web **4**(3), 341–348 (2013)
9. Guéret, C., Groth, P., Stadler, C., Lehmann, J.: Assessing linked data mappings using network measures. In: Simperl, E., Cimiano, P., Polleres, A., Corcho, O., Presutti, V. (eds.) ESWC 2012. LNCS, vol. 7295, pp. 87–102. Springer, Heidelberg (2012). doi:10.1007/978-3-642-30284-8_13
10. Heath, T., Bizer, C.: linked data: evolving the web into a global data space. Synthesis Lectures on the Semantic Web. Morgan & Claypool Publishers (2011)
11. Mader, C., Haslhofer, B., Isaac, A.: Finding quality issues in SKOS vocabularies. In: Zaphiris, P., Buchanan, G., Rasmussen, E., Loizides, F. (eds.) TPDL 2012. LNCS, vol. 7489, pp. 222–233. Springer, Heidelberg (2012). doi:10.1007/978-3-642-33290-6_25

12. Papaleo, L., Pernelle, N., Saïs, F., Dumont, C.: Logical detection of invalid SameAs statements in RDF data. In: Janowicz, K., Schlobach, S., Lambrix, P., Hyvönen, E. (eds.) EKAW 2014. LNCS (LNAI), vol. 8876, pp. 373–384. Springer, Heidelberg (2014). doi:10.1007/978-3-319-13704-9_29
13. Suominen, O., Hyvönen, E.: Improving the quality of SKOS vocabularies with skosify. In: Teije, A., Völker, J., Handschuh, S., Stuckenschmidt, H., d'Acquin, M., Nikolov, A., Aussenac-Gilles, N., Hernandez, N. (eds.) EKAW 2012. LNCS (LNAI), vol. 7603, pp. 383–397. Springer, Heidelberg (2012). doi:10.1007/978-3-642-33876-2_34
14. Suominen, O., Mader, C.: Assessing and improving the quality of skos vocabularies. J. Data Semant. **3**(1), 47–73 (2014)
15. Zaveri, A., Kontokostas, D., Sherif, M.A., Bühmann, L., Morsey, M., Auer, S., Lehmann, J.: User-driven quality evaluation of DBpedia. In: Sabou, M., Blomqvist, E., Noia, T.D., Sack, H., Pellegrini, T. (eds.) I-SEMANTICS 2013, 4–6 September 2013, Graz, Austria, pp. 97–104. ACM (2013)
16. Zaveri, A., Rula, A., Maurino, A., Pietrobon, R., Lehmann, J., Auer, S.: Quality assessment for linked open data: a survey. Semant. Web **7**(1), 63–93 (2016)

General Session: Information Extraction and Retrieval

GaiusT 2.0: Evolution of a Framework for Annotating Legal Documents

Nicola Zeni[1], Luisa Mich[2(✉)], and John Mylopoulos[3]

[1] Department of Information Engineering and Computer Science,
University of Trento, Trento, Italy
nicola.zeni@unitn.it
[2] Department of Industrial Engineering, University of Trento, Trento, Italy
luisa.mich@unitn.it
[3] School of Electrical Engineering and Computer Science, University of Ottawa,
Ottawa, CA, Canada
jmylopou@eecs.uottawa.ca

Abstract. Semantic annotation technologies support the extraction of legal concepts, for example rights and obligations, from legal documents. For software engineers, the final goal is to identify compliance requirements a software system has to fulfill in order to comply with a law or regulation. That implies analyzing and annotating legal documents in prescriptive natural language, still an open problem for research in the field. In this paper we describe GaiusT 2.0, a system for extracting requirements from legal documents. GaiusT 2.0 is the result of the evolution of GaiusT, and has been designed and implemented as a web-based system intended to semi-automate the extraction process. Results of the application of GaiusT 2.0 show that the new version improves performance of the extraction process and also makes the tool more usable.

Keywords: Semantic annotation · Legal documents · Legal requirements · Natural language processing · Linguistic resources · User needs

1 Introduction

Recent trends in technologies, as well as social, economic and political issues, have greatly increased the importance of ensuring compliance of software systems with regulations, laws, policies and other types of legal documents. Autonomous vehicles, the Internet of Things, cloud services, augmented reality, videogames, online auctions, e-commerce and e- related sectors, and social networking platforms are only some of the many technologies [1] challenging the elicitation of legal requirements [2] from a variety of legal documents. That in turn has driven researchers to design and develop tools and systems to support requirements analysts in eliciting compliance requirements from regulatory documents [3, 4]. Existing systems range from editors for manually annotating (markup) legal documents to frameworks integrating linguistic tools and natural language processing (NLP) features. NLP systems constitute essential components, as they fully or partially automate requirements

© Springer International Publishing AG 2016
E. Garoufallou et al. (Eds.): MTSR 2016, CCIS 672, pp. 43–54, 2016.
DOI: 10.1007/978-3-319-49157-8_4

extraction, demanding an in-depth semantic and pragmatic analysis, an objective that is far from being reached [5]. In this context, the first version of GaiusT represented a multi-phase framework to support the different tasks in which legal requirements extraction can be decomposed [6]. The framework was implemented as a modular system with a semantic annotation system core [7] coded in TXL, a structural transformation system [8].

The applications and experiments with GaiusT exposed a number of issues that required a deep reengineering of the system. The reengineering project started in 2013 and the first steps included (a) an analysis of existing tools and linguistic resources that could be adopted to address the high level requirement for an improved version of GaiusT, but still preserving its nature of a lightweight system [9]; (b) an investigation of usability issues for a tool supporting the extraction of requirements from legal documents [10].

In this paper we describe the architecture of the new version of GaiusT, called GaiusT 2.0. The evolution project moved from the output of those steps. In addition to the above, the project aimed to address new challenges that are: (a) how to deal with legal documents in different domains and cultures; (b) how to support user intervention in all the steps of the requirements extraction process. GaiusT 2.0 is a completely revised system supporting requirements extraction (SSRE) where all the modules have been re-implemented and new ones have been added. To evaluate the new system, preliminary results obtained with GaiusT 2.0 are reported and compared with those obtained with GaiusT for the HIPAA (Health Insurance Portability and Accountability Act) [6, 11].

This paper is structured as follows. Section 2 describes the problems to be addressed by a system supporting requirements elicitation annotating legal documents and the requirements for such systems. Section 3 illustrates the architecture of GaiusT 2.0 comparing it with that of GaiusT. Section 4 presents the main results of the evaluation of GaiusT 2.0. Section 5 illustrates the related work and finally, the conclusions are drawn in Sect. 6.

2   Extracting Legal Requirements from Textual Documents

2.1   Supporting Requirements Elicitation for Regulatory Compliance

The core task a SSRE for regulatory compliance has to support is the identification of deontic concepts described in legal documents, from compound concepts such as rights and obligations to requirements concepts, such as goal, actor, action, resource. A detailed description of the characteristics of an SSRE is given in [6]. Here we review the issues most relevant to the evolution of our system described in the next section. First of all, automatic annotation and processing of legal documents is particularly difficult because they are written in "legalese", a specialized prescriptive language. Moreover, from a linguistic point of view an SSRE from legal documents has to deal with knowledge in different domains, or rather the specialized language and knowledge of laws for specific domains. A SSRE has also to take into account the structure of legal documents which generally respect a multi-level hierarchical structure. Such structure constitutes the basis for internal and external references [12]; for example, describing a right, the involved actors may be found defined in another document or section; exceptions can refer to actions defined in another subsection, etc. As a result, at least two models are required for annotating a given regulation: (a) a conceptual model, including the deontic concepts that represent prescribed behaviors – rights and obligations (descriptive metadata) – and (b) a structural model of the text of the regulation (structural metadata). The second model is particularly important to delimit the scope of concepts identified in the text according to the conceptual model. Still, a comprehensive SSRE had to support many other activities including the very first steps in which (a) the conceptual model for a law is designed, or adapted from an existing one, for a given regulation and (b) indicators corresponding to the concepts are defined in an annotation schema. The variety of tasks and elements involved also requires the creation and management of a large knowledge base; multi-language annotation is also in place in some projects. Existing systems usually focus only on one of the tasks described in this section that is the annotation of deontic concepts, or the definition of a conceptual model for a given law [13], or of a structural model. The goal of the new version of GaiusT is to deal with all these tasks, integrating and consolidating the results of previous researches but also adding new modules.

As regards the implementation of a SSRE, a variety of programming languages and technical frameworks could be used and it is quite common that some of the linguistic activities (e.g. parsing, word frequency analysis) or editing features (both on texts and on the graphical diagrams for the conceptual models) are developed by customizing open source programs or looking for existing libraries. These modules have then to be integrated into the framework taking into account transportability and inter-operability issues. All these issues were critical to the success of the reengineering project for GaiusT 2.0; the complexity of the system implied a high risk of worsen its performances relative to the less automatic process supported by GaiusT.

2.2   Requirements for a System Supporting Requirements Extraction from Legal Documents

There are a number of users involved in an SSRE - requirements analysts, business analysts, lawyers, experts on standards, experts on linguistics, developers, domain experts (e.g. health experts for laws like HIPAA) to name but a few. Users of the SSRE can be classified into three main categories: researchers, developers and end users. Each of these classes have specific needs and requirements that an SSRE from legal documents has to satisfy and are described in the following adding to those given in [10].

Researchers. SSRE from legal documents are often developed by researchers working on large-scale international research projects (e.g. European projects involve research groups in different countries) which often last years. This is indeed the case with GaiusT 2.0. The system evolved from a research project to design and implement semantic annotation tools where the final framework, called Cerno, was a large-scale semantic annotation system [7]. The Cerno project started about ten years ago [14]; from its inception the researchers were scattered in different organizations and countries. The application of Cerno to legal documents required a re-design of some of its modules and the development of new ones and resulted in the first version of GaiusT [6]. These evolutions revealed further high-level requirements to be satisfied by a new version of GaiusT:

- to deal with conceptual models other than the ones used in GaiusT, describing laws at a different abstraction level [10], for example the model used in the Nómos project [15] (for another model see [16]);
- to allow researchers in different fields - compliance with legal documents implies a multi-disciplinary approach - to use the modules of the system;
- additional features for choosing the concepts to be annotated in legal documents for a given research project;
- management of multi-accesses, spaces and tools for different projects;
- support while testing the impact of changes in the annotation schema on the output of the annotation steps;
- storing all the artifacts of a project for traceability;
- interfaces to the core modules to support the definition of annotation rules for new conceptual models;
- modules for the analysis and comparison of output produced in all the steps of the requirements extraction process.

Developers. Developers of an SSRE have not only to implement solutions for a variety of tasks but they also have to choose from a large number of platforms, technologies, libraries, programming languages and standards. The most critical decisions made for the GaiusT 2.0 project are related to the need to integrate modules, libraries and new software and, whenever possible, to use open source resources. Mashing them up is often more challenging than expected. For example, available off-the-shelf modules frequently turn out to be incompatible, or have low performance levels, or do not implement all the needed features. An analysis of the available tools and resources (and of their adaption needs) is given in [9]. The main developers' requirements identified for GaiusT 2.0 are the following:

- to address the limitations of TXL, which had resulted less scalable than expected to process long documents, and to support an evolutionary approach to the implementation of prototypes for the modules of GaiusT 2.0;
- to adopt the most suitable technologies to implement GaiusT 2.0 as a full-fledged web-based system, a high level requirement shared by all the user classes;
- to port available open source linguistic tools and resources that often are in Java [17–19] in C#, the language adopted for the Gaius project.

End users. Prospective users of GaiusT 2.0 are requirements engineers or analysts, and lawyers. A particular class of users of SSRE are the participants in experiments, often students whose role and needs are similar to those of (junior) requirements analysts. For these users, needs are also related to the support of the manual analysis and annotation of legal texts (necessary to create a gold reference for comparison with the GaiusT 2.0 results). For end users, requirements are mainly related to:

- ease of use and learnability, which are hampered by the trade-off between the need of the developers to implement new solutions in an effective and efficient process, often with throwaway prototypes, and the need of the researchers to test their analysis and annotation approaches without the help of a developer;
- usability of the annotation schema and engine modules (Table 1), that turned out to be critical also to allow researchers and developers to gather users' feedback and data to compare different solutions to a specific problem;

Table 1. GaiusT component comparison matrix.

Component/Module	GaiusT	GaiusT Version 2.0
Annotation schema generator		
Conceptual model parser	–	Parses XMI models
Word frequency list (WFL) generator	Generates WFLs	+ Generates n-grams
Lexical database manager	Uses WordNet to validate WFL entries	+ Uses Google n-grams to validate n-grams
Part of Speech (PoS) manager	External (TreeTagger)	Embedded (porting Java libraries in C#)
Template Library	–	Generates templates for different types of laws (domains) and projects
Annotation engine		
Text extractor	Extracts text from different file formats	+ Parallel processing and web downloader
Normalizer	Clear texts and puts 1 sentence per line	+ Parallel processing
Document structure analyzer	TXL rules	Regular expressions
Annotation rules generator	External (TXL rules)	Embedded (regular expressions)
Annotation generator	Annotates concepts	Annotates concepts and relationships
Repository		
Relational database	Annotated items	Structured artifacts
Evaluation module	Provides data for basic statistics	Provides HTML statistical reports, graphical traceability, document browsing
Graphical model generator		
Graph generator	–	Generates graphical models of annotations
Graphical User Interface		
Interface	Client application	Web application (responsive, multi-user)

- support to group work as requirements analysts and lawyers are sometime working in different places and have to cooperate on interdisciplinary projects;
- to annotate legal documents at different granularity levels, that is associating texts of different size - ranging from single words to the entire document - to a given concept;
- a specific requirement for lawyers is that of having legal citations for external references according to the Bluebook [12].

3   The Framework

3.1   GaiusT

The explanation of GaiusT given here is abbreviated to what is necessary to understand this paper. A full description can be found in [6]. GaiusT was designed as a multi-phase framework to semi-automatically identify deontic concepts in legal documents. It has been successfully applied to the annotation of documents both in English and in Italian. Starting from a meta-model of legal concepts - a conceptual model - GaiusT identifies instances of complex deontic concepts, such as rights and obligations. From the meta-model one can derive an annotation schema which specifies the rules for identifying the instances of legal concepts in a document. For each concept, the annotation schema specifies its identifiers with their syntactic roles, and patterns are used to represent complex concepts. Indicators can be single words, phrases, or references to previously parsed basic entities while patterns are collection of concepts related by regular expressions. For each legal document, the output of GaiusT is an XML file that lists all the annotations generated by the system.

The first version of GaiusT was about 50 k lines of code and more than 130 MB in size.

3.2   GaiusT 2.0

Based on the requirements gathered for the different classes of users described in Sect. 2.2 and the limitations identified from the experiments and applications of GaiusT, the SSRE evolved into GaiusT 2.0, a web-based system with new and improved modules. Most of the requirements are related to the need to deal with different domains (privacy for health data, accessibility and ICT, etc.) and with different legal systems where general principles determine different conceptual models (descriptive metadata) and different document structure (structural metadata). The architecture of GaiusT 2.0 includes a number of components and is represented in Table 1, where each of them is compared to the corresponding module of GaiusT.

The Annotation schema generator supports the process to semi-automatically create annotation schemas i.e. to associate each concept in a conceptual model to a list of indicators to be used to identify the concepts in a legal document. For this component, two of the modules – the WFL generator and the Lexical database manager – have been fully reengineered. In particular, the statistical analyses of the input documents provided by the WFL generator module have been integrated with the calculation of TF/IDF (term frequency/inverse document frequency). The Lexical Database Manager has been extended by adding Google n-grams [20] (consisting of 1-grams to 5-grams; the last available version contains more than 346 million bigrams and more than 2,616 million trigrams) and the WordNet ontology [21]. The part of speech (PoS) module has been replaced by porting in C# sharpNLP (in Java [22]), and expanding it, as it only covers part of the English language and does not deal with Italian. A new Template library module has been added to support the creation of annotation templates and the reusability of items (concepts, indicators, patterns, structural items). This module helps the users in the definition of annotation schemas according to the legal documents to be analysed. For example, a user can define a template for the semantic annotation of American laws, including structural patterns, legal concepts and patterns to annotate any American legal document; as a result compliance requirements extraction from a specific American law does not have to start from scratch as items defined in the template can be reused. Besides, to support different users groups and annotation works, a project template was added.

The Annotation engine supports a number of steps in order to: (1) extract text from different file formats (Text extractor), (2) normalize the input document by removing leading and unprintable characters and trailing spaces to produce a text document where each line represents a phrase (Normalizer); (3) annotate text units with tags for structure and cross references identification (Document structure analyzer); (4) generate rules for annotating concepts; (5) identify concepts, applying the annotation schema i.e. indicators and patterns defined for the conceptual model; find relationships between identified concepts by using the heuristic patterns (Annotation generator). In GaiusT, the Annotation rules generator was in TXL [8]; to overcome its limitations, GaiusT 2.0 uses regular expressions. These expressions allow to effectively capture grammar and syntactic rules for the chosen languages. The design and implementation of this component has been the most demanding in the entire project.

The redesign of the system included the expansion of the Database component, a relational database that increased from 10 to 71 tables. A new module deals with unstructured documents and artifacts of the annotation process.

A new component, the Graphical model generator, has been added to create graphical models with the instances of the deontic concepts extracted from the annotated texts (a specialized module for the conceptual model used for Nómos is described in [23]).

The GUI component is the result of the migration of GaiusT to a web-based system. An explorative web prototype was developed using NodeJs [24]. Later, for maintenance and integrability reasons, it was migrated to ASP.NET MVC [25]. Mobile and multi-users accesses are also supported.

The actual version of GaiusT 2.0 is about 120 k lines of code, a more than two-fold increase, and 415 MB in size of libraries.

4   Evaluation

GaiusT 2.0 satisfies most of the requirements described in 2.2. It is a full-fledged web-based system and it is used by researchers working in groups and on different projects. The system deals with multi-accesses, storing artefacts for projects which adopt different annotation models and schemas. Researchers and end users utilize it for new experiments and applications. Besides HIPAA, the results of which are reported in the next subsection, a new project is now applying GaiusT 2.0 to the English version of German [26] and Italian privacy law [27] i.e. laws whose documents have a different structure and are in a different domain. For the developers, technologies deployed for implementing GaiusT 2.0 offer better support to extend and adapt the system. Although quantitative results are not yet available, the experiences gained so far confirm that GaiusT 2.0 usability has largely improved thanks to the web interface that has reduced the time needed to start using the system. Requirements not yet implemented regard the future work outlined in the conclusion.

To evaluate the new version of GaiusT we applied it to the HIPAA (with the same assessment methodology described in [11]) in particular to Sect. 164.524 and Sect. 164.526, thus allowing us to compare the results obtained from GaiusT 2.0 with those reported in [6]. We adopted experts' manual annotation as gold standard as an upper bound of what automatic annotation can do. Moreover, the performance of the tool was calibrated with the degree of disagreement among experts in the experiment (an average of 23 %). We used standard information retrieval metrics[1] - precision, recall, fallout, accuracy, error and F-measure – to compare the performance of the tool to that of humans. The results of the experiment are presented in Table 2.

Table 2. Evaluation rates of experts and the two versions of GaiusT annotating HIPAA.

	Human	GaiusT	GaiusT 2.0
Precision	0.83	0.84	0.86
Recall	0.92	0.87	0.90
Fallout	0.49	0.42	0.73
Accuracy	0.78	0.78	0.74
Error	0.22	0.22	0.28
F-measure	0.86	0.85	0.87

---

[1]
- Recall is a measure of how well the tool performs in finding relevant items $TP/(TP + FN)$;
- Precision is a measure of how well the tool performs in not returning irrelevant items $TP/(TP + FP)$;
- Fallout is a measure of how quickly precision drops as recall is increased $FP/(FP + TN)$;
- Accuracy is a measure of how well the tool identifies relevant items and rejects irrelevant ones $(TP + TN)/N$;
- Error is a measure of how much the tool is prone to accept irrelevant items and rejects relevant ones $(FP + FN)/N$;
- F-measure is a harmonic mean of recall and precision $2 \times Recall \times Precision/(Recall + Precision)$

where TP is the number of items correctly assigned to the category; FP is the number of items incorrectly assigned to the category; FN is the number of items incorrectly rejected from the category; TN is the number of items correctly rejected from the category; and N is the total number of items $N = TP + FP + FN + TN$.

The new system has improved recall without diminishing precision. It should be noted that, taking into account the disagreement among experts, GaiusT 2.0 does a little bit better with respect to precision and F-measure than human annotators.

GaiusT 2.0 has also improved the process of structure identification and the identification of structure elements is performed with a recall of 100 %. The tool is able to correctly identify all cross-references but, for example, the cross reference "… in paragraphs (b)(1) (ii)(A) or (B)" is partially matched as "paragraphs (b)(1)" because in the input text there is an extra-space between "(b)(1)" and "(ii)(A)" so that the regular expression fails to catch the entire occurrence.

Further experiments need to be carried out to evaluate the productivity of the new system. To this end the effort to adapt it to the analysis of new laws had to be taken into account: by providing a larger support to the activities in the annotation process, GaiusT 2.0 should help to reduce the time needed for the annotation. Real time support (or nearly real time) would be useful to analyze online documents; for example, for moderating posts according to given social networking rules.

5   Related Work

The extraction of requirements from legal documents is a challenging task and several approaches have been presented to tackle the problem, ranging from manual analysis to the use of NLP techniques to semi-automate the process.

Heuristic rules for extracting rights and obligations from regulations are given in [28]; in this field Breaux proposed a framework to acquire legal requirements using a systematic frame-based requirements analysis methodology. The framework uses an upper ontology (which describes general concepts that are the same across all knowledge domains) to classify regulatory statements, for context-free mark-up and to handle the structural organization of regulatory documents. Recently a Hit (Human Intelligence Tasks) was called for on Amazon Mturk [29] to have people apply their approach to the annotation of legal documents.

To automatically support the most critical steps of the semantic annotation process some researchers propose the application of NLP tools and machine learning approaches to achieve full, domain-independent analysis of legal documents. The purpose of these approaches includes activities like the extraction of a particular kind of concept and the identification of text fragments relevant for a given goal or for document classification. In [30] authors propose a name entity recognition methodology for the automatic identification of actors relevant for the analysis of a regulation or a law. Lesmo et al. [31] present TULSI, a system for the extraction of semantic annotations from legal documents, a core task of the GaiusT 2.0 architecture. GATE (General Architecture for Text engineering) [32] provides a rule-based language to build annotation patterns with an approach similar to that used by GaiusT 2.0, however GATE is not web–based and none of its modules could be integrated into it.

As regards the analysis of the structure of legal documents and the management of cross references - seemingly easy tasks but a complete, affordable, robust system is still lacking - it is worth citing two recent papers [33, 34]. Both propose a systematic approach based on regular expressions and linguistic techniques to identify cross references in legal text: a necessary step, but unfortunately insufficient to fully address this task. GaiusT 2.0 uses regular expressions to capture formal partitions (the document structure) and cross-references, achieving good performances as an evolution of the approach described in [35].

Other authors have focused on the problem of managing knowledge in legal documents, to support lawyers in searching for legal cases and sentences related to their professional activities. Among the tools used to manage legal knowledge, EuNomos [36] supports the identification and classification of legal documents through an ontology-based process. Documents are analyzed and annotated in XML format, according to the Akoma Ntoso standard [37], using a combination of linguistic and machine learning techniques (text classification, text similarity and pattern matching) to extract concepts and structure from the legal texts. GaiusT 2.0 also uses conceptual and structural models - small ontologies intended to capture domain semantics - to identify main concepts and relevant clues in the structure of the texts. Another knowledge management tool for lawyers is SALEM [38], which uses linguistics technologies and machine learning (a technique similar to SMV, support vector machine) to extract provisions from legal documents. An interesting result is reported in [39] where authors compare machine learning classification of legal sentences versus pattern based classification and highlight how a pattern based classifier resulted more robust in the categorization of legal documents than a SVM classifier. Gaius T 2.0 is not directly comparable with any of the existing tools, even though the techniques used for semantic annotation are shared by many of the existing approaches. GaiusT 2.0 differs from all them as it is a comprehensive large scale system which includes modules to deal with most of the tasks related to requirements extraction from legal documents. It uses a semi-automatic pattern-based approach to analyze legal documents dealing with the needs related to the application of the system to different domains and legal systems. Worth citing are the (complementary) approaches addressing the problem the other way around, that is starting from software requirements and checking if they comply with a given law; see for example [40]. A methodology for reasoning and modeling of obligations is described in Hashmi [41]. Reasoning would be useful to check if a given event breaks a law (rule) and is one of the goals of Nómos [15].

6   Conclusions and Future Work

Regulatory compliance for software systems is a complex problem. In this paper we described GaiusT 2.0, a large-scale system supporting the extraction of legal requirements from textual documents. The design of the system started in 2013 and evolved from GaiusT by improving and including new modules. The design of the modules was based on the users' needs gathered in different projects and on the analysis of available linguistic resources and development technologies. GaiusT 2.0 addresses most of the requirements described in Sect. 2.2. However there are a number of issues that have to be further investigated and features to be added. The following are prospective points for future study: (1) add reasoning mechanism to improve the annotation step; (2) adopt one of the standard mark-up languages for legal documents for example, LegalRuleML [42]; (3) add a layer to transform legal concepts into legal requirements, a step that none of the existing systems support yet.

Acknowledgements. Nicola Zeni's work was supported in part by the ERC advanced grant 267856 'Lucretius: Foundations for Software Evolution'. Our thanks go to the experts and users who made themselves available to test GaiusT 2.0.

References

1. IEEE: IEEE Computer society predicts Top 9 technology trends (2016). http://www.computer.org/web/pressroom/Technology-Trends-2016
2. Systems Engineering Body of Knowledge (SEBoK) v.1.5.1. (2015). http://sebokwiki.org/wiki/Guide_to_the_Systems_Engineering_Body_of_Knowledge_(SEBoK)
3. RELAW workshops. http://gaius.isri.cmu.edu/relaw
4. Jurix conferences. http://jurix.nl/proceedings
5. Davis, E.: The singularity and the state of the art in artificial intelligence: the technological singularity. In: Ubiquity symposium (Ubiquity 2014), 12 pages (2014)
6. Zeni, N., Kiyavitskaya, N., Cordy, J.R., Mich, L., Mylopoulos, J.: GaiusT: supporting the extraction of rights and obligations for regulatory compliance. Req. Eng. **20**(1), 1–22 (2015). (online 2013)
7. Kiyavitskaya, N., Zeni, N., Cordy, J.R., Mich, L., Mylopoulos, J.: Cerno: light-weight tool support for semantic annotation of textual documents. Data Knowl. Eng. **68**(12), 1470–1492 (2009)
8. Cordy, J.R.: The TXL source transformation language. Sci. Comput. Program. **61**(3), 190–210 (2006)
9. Zeni, N., Mich, L., Mylopoulos, J., Cordy, J.R.: Applying GaiusT for extracting requirements from legal documents. In: 6th International Workshop on Requirements Engineering and Law (RELAW 2013), pp. 65–68. IEEE (2013)
10. Zeni, N., Mich, L.: Usability issues for systems supporting requirements extraction from legal documents. In: 7th International Workshop on Requirements Engineering and Law (RELAW 2014), pp. 35–38. IEEE (2014)
11. Health Insurance Portability and Accountability Act – HIPAA. http://www.hhs.gov/ocr/privacy/hipaa/understanding
12. Bluebook. https://www.legalbluebook.com
13. Souza, V., Zeni, N., Kiyavitskaya, N., A, P., Mich, L., Mylopoulos, J.: Automating the generation of semantic annotation tools using a clustering technique. In: Kapetanios, E., Sugumaran, V., Spiliopoulou, M. (eds.) NLDB 2008. LNCS, vol. 5039, pp. 91–96. Springer, Heidelberg (2008). doi:10.1007/978-3-540-69858-6_10
14. Kiyavitskaya, N., Zeni, N., Cordy, J.R., Mich, L., Mylopoulos, J.: Applying software analysis technology to lightweight semantic markup of document text. In: Singh, S., Singh, M., Apte, C., Perner, P. (eds.) ICAPR 2005. LNCS, vol. 3686, pp. 590–600. Springer, Heidelberg (2005). doi:10.1007/11551188_65
15. Siena, A., Jureta, I., Ingolfo, S., Susi, A., Perini, A., Mylopoulos, J.: Capturing variability of law with *Nómos* 2. In: Atzeni, P., Cheung, D., Ram, S. (eds.) ER 2012. LNCS, vol. 7532, pp. 383–396. Springer, Heidelberg (2012). doi:10.1007/978-3-642-34002-4_30
16. Winkels, R., Hoekstra, R.: Automatic extraction of legal concepts and definitions, pp. 157–166. IOS (2012)

17. OpenNLP. https://opennlp.apache.org
18. Stanford CoreNLP. http://nlp.stanford.edu/software
19. NLP toolkit. http://www.nltk.org
20. Google n-grams. http://storage.googleapis.com/books/ngrams/books/datasetsv2.html
21. WordNet. https://wordnet.princeton.edu
22. SharpNLP. https://sharpnlp.codeplex.com
23. Zeni, N., Seid, E.A., Engiel, P., Ingolfo, S., Mylopoulos, J.: Building large models of laws with NómosT. In: 35th International Conference on Conceptual Modeling (ER 2016). Springer (2016, to appear)
24. NodeJs. http://nodejs.org
25. ASP.NET MVC. www.asp.net/mvc
26. Germany's Federal Data Protection Act - Bundesdatenschutzgesetz – BDSG. https://www.loc.gov/law/help/online-privacy-law/germany.php
27. Data Protection Code - Legislative Decree no. 196/2003. http://194.242.234.211/documents/10160/2012405/DataProtectionCode-2003.pdf
28. Breaux, T.D., Antón, A.I.: Analyzing regulatory rules for privacy and security requirements. IEEE Trans. on SW. Eng. **34**(1), 5–20 (2008)
29. Amazon Mechanical Turk. https://www.mturk.com
30. Dozier, C., Kondadadi, R., Light, M., Vachher, A., Veeramachaneni, S., Wudali, R.: Named entity recognition and resolution in legal text. In: Francesconi, E., Montemagni, S., Peters, W., Tiscornia, D. (eds.). LNCS (LNAI), vol. 6036, pp. 27–43Springer, Heidelberg (2010). doi:10.1007/978-3-642-12837-0_2
31. Lesmo, L., Mazzei, A., Palmirani, M., Radicioni, D.P.: TULSI: an NLP system for extracting legal modificatory provisions. Art. Intell. Law **21**(2), 139–172 (2013)
32. GATE. https://gat.ac.uk
33. Sannier, N., Adedjouma, M., Sabetzadeh, M., Briand, L.: An automated framework for detection and resolution of cross references in legal texts. Req. Eng., 1–23 (2015). http://link.springer.com/journal/766/onlineFirst/page/1
34. Oanh Thi, T., Bach Xuan, N., Le Minh, N., Akira, S.: Automated reference resolution in legal texts. Artif. Intell. Law **22**(1), 29–60 (2014)
35. Zeni, N., Kiyavitskaya, N., Mich, L., Mylopoulos, J., Cordy, J.R.: A lightweight approach to semantic annotation of research papers. In: Kedad, Z., Lammari, N., Métais, E., Meziane, F., Rezgui, Y. (eds.) NLDB 2007. LNCS, vol. 4592, pp. 61–72. Springer, Heidelberg (2007). doi:10.1007/978-3-540-73351-5_6
36. Boella, G., di Caro, L., Humphreys, L., Robaldo, L., van der Torre, L.: NLP challenges for EuNomos a tool to build and manage legal knowledge. In: 8th International Conference on Language Resources and Evaluation (LREC 2012). European Language Resources Ass., Istanbul (2012)
37. Akoma Ntoso standard. http://www.akomantoso.org
38. Soria, C., Bartolini, R., Lenci, A., Montemagni, S., Pirrelli, V.: Automatic extraction of semantics in law documents. In: 5th Legislative XML Workshop (2007)
39. de Maat, E., Krabben, K., Winkels, R.: Machine learning versus knowledge based classification of legal texts. In: 23rd Conference on Legal KW and IS (JURIX 2010), pp. 87–96. IOS Press (2010)
40. Massey, K.: Legal requirements metrics for compliance analysis. Ph.D. dissertation, North Carolina State University (2012)
41. Ingolfo, S., Siena, A., Mylopoulos, J., Susi, A., Perini, A.: Arguing regulatory compliance of software requirements. Data Knowl. Eng. **87**, 279–296 (2013)
42. LegalRuleML. https://www.oasis-open.org/committees/legalruleml

Knowledge Extraction from Audio Content Service Providers' API Descriptions

Damir Juric[✉] and György Fazekas

Queen Mary University of London, London, UK
{d.juric,g.fazekas}@qmul.ac.uk

Abstract. Creating an ecosystem that will tie together the content, technologies and tools in the field of digital music and audio is possible if all the entities of the ecosystem share the same vocabulary and high quality metadata. Creation of such metadata will allow the creative industries to retrieve and reuse the content of Creative Commons audio in innovative new ways. In this paper we present a highly automated method capable of exploiting already existing API (Application Programming Interface) descriptions about audio content and turning it into a knowledge base that can be used as a building block for ontologies describing audio related entities and services.

Keywords: Metadata · Audio content · Ontologies · Natural language processing · Knowledge extraction

1 Introduction

The field of digital music or more general digital audio content (content that does not include only songs, but sounds or soundscapes) is very propulsive one, but still creatives who work in the industries that are using digital audio content in their daily work face with some basic problems. One of those problems is a lack of technologies for accessing and easily incorporating audio content directly into a creative workflow[1]. In this paper we will not deal with problems of accessing the data or workflow enhancements but with a first and very important step that will eventually allow us to tackle those problems in the future. That first step is conducting the task of knowledge and metadata extraction for potentially very large amount of unstructured data that already exists in this domain. In the particular case of sound and music, a huge amount of audio materials like sound samples, soundscapes and music pieces, is available online and released under Creative Commons licenses. That data is coming from both amateur and professional content creators. We refer to this content as the *Audio Commons*. Because the Audio Commons is large and very diverse (and with time that dataset will increase considerably) it is problematic for creatives to grasp the potential benefits of such a dataset, exactly because of those characteristics. There is a need for tools and methods that will allow the

---

[1] Deliverable D2.1: Requirements Report and Use Cases: http://www.audiocommons.org/materials/.

© Springer International Publishing AG 2016
E. Garoufallou et al. (Eds.): MTSR 2016, CCIS 672, pp. 55–66, 2016.
DOI: 10.1007/978-3-319-49157-8_5

assessments and extraction of Audio Commons data from its existing poorly structured and fragmented form into something more formal. A perfect candidate for that would be an ontology. However, building the Audio Commons Ontology (and ontology in general) requires extensive knowledge about the domain that the ontology will describe. Knowledge about the domain includes the vocabulary of the domain and knowledge about the workflows (processes) that are being carried out by various roles involved in them. This paper will describe the method for gathering the formalized knowledge (knowledge database) from unstructured data of audio content providing services that already exist. This should be only the first step in a much bigger european project that will try to offer creative users more integrated ways of searching for and using audio content. There is a lack of globally unique and interoperable identifiers that creators could easily get familiar with and use it in their creative process. The aim of the Audio Commons project[2] is to create an ecosystem of content[3], technologies and tools to bring the Audio Commons to the creative industries, enabling creation, access, retrieval and reuse of Creative Commons audio content in innovative ways that fit the requirements of the use cases considered (e.g., audio-visual, music and video games production). Currently creative users can access various audio content by using existing APIs for programmatically accessing content from sites like Jamendo[4], Freesound[5], Europeana[6], etc. Despite the number of providers and large and their libraries are extensive users face with problems such as limited access to data due to the lack of high quality and unified metadata [1], there is no unified access mechanism that will connect the different APIs (APIs have different specifications), retrieval tools are inadequate and that all leads to the fact that audio commons content is not frequently used in the professional environment. The biggest challenge left to solve is to define the metadata requirements in creative applications, to design the appropriate ontologies for data representation and finally to provide reliable metadata to facilitate access to Audio Commons content. In this paper, we will present a highly automated method for harvesting the API descriptions of audio content providers to build the knowledge database and vocabulary that will be used as a basic building block for the Audio Commons ontology. In Sect. 2 we will mention the work that has been already done on the creation of music and audio related ontologies. In Sect. 3 we will describe the music API dataset and present the method that will conduct machine reading task on the dataset and the creation of knowledge database (implemented as a graph database). Finally, we evaluate the method on an audio content provider dataset in Sect. 4 and conclude in Sect. 5.

2   Related Work

Ontology construction is normally carried out manually but in recent years automated approaches have emerged. Most of these approaches deal with raw text, but some also

---

[2] Audio Commons Project - http://www.audiocommons.org/.

[3] The Audio Commons Ecosystem (ACE) referred to as ACE in the rest of the paper.

[4] Jamendo - https://developer.jamendo.com/v3.0.

[5] Freesound - https://www.freesound.org/help/developers/.

[6] Europeana - http://www.europeana.eu/portal/.

use other sources such as Wikipedia pages and HTML (HyperText Mark-Up Language) forms. Manually developed general ontologies are still the most widely used type [2]. The construction of such ontologies is a very expensive and time-consuming process. Moreover, the process of acquiring new knowledge is always needed and it requires ongoing work by human experts, even after the ontology has been released. In order to solve the problem of reliance on a cumbersome manual construction, some techniques propose broader collaboration during the ontology construction process, as in the case of Semantic Wikipedia [3], where facts are created and incorporated into an ontology by many volunteers. As for automated approaches, Zhou [4] gives a typical scenario of an ontology learning process (which can either be manual or automated) and it consists of: creating concepts, creating relations, ontology population and ontology evaluation. Wiszniewski [5] introduces a metamodel for ontology learning from text and presents an extensive survey of ontology learning models. As for the audio and music domain, there have been research carried out on the construction of music related ontologies and metadata. The Music Ontology [6] allows for describing the music production workflow from composition to delivery, while the Studio Ontology is for capturing the nuances of record production by providing an explicit, application and situation independent conceptualisation of the studio environment [7]. Both are presented as a modular framework of ontologies using core elements that allow for the representation of time-based events (using the Event and Timeline ontologies), and the workflow of music production in an editorial context subsumed under broader terms defined by the Functional Requirements for Bibliographic Records (FRBR) [8]. Expressed as a layered entity-relationship (ER) model, FRBR distinguishes three types of things, entities, attributes and relationships. An entity can be anything from a physical object to an abstract concept, relationships specify interactions between entities and attributes are properties or characteristics of entities or relationships. A particular group of entities represent the products of intellectual or artistic works are of specific interest. In the music domain *Work* may represent a certain musical composition such as a Beethoven violin sonata. A respective *Expression* may be the recording by Itzhak Perlman, a *Manifestation* may refer to the CD release by the record label Naxos and finally an *Item* represents a specific physical CD copy of this release. The Audio Feature Ontology [9] provides a descriptive framework for expressing different conceptualisations of the audio feature extraction domain and enables designing linked data formats for representing feature data. There is an ongoing work on Europeana Data Model for sounds[7].

3 Music API Descriptions

3.1 Ontology and Web Services

The Audio Commons project intends to include different Web Services providing music related metadata into its ecosystem. The project started with well-established Web Services for music data retrieval like Freesound offering its content. Freesound is a sound

---

[7] Europeana profile for sound - http://pro.europeana.eu/get-involved/europeana-tech/european-atech-task-forces/edm-profile-for-sound.

sharing site with more than 300 000 sound samples (including sound effects, instrument samples and field recordings). The content is released under several types of licences. The service has built its own API that is available for users (Fig. 1). From the aspect of a creative industry user there is an ongoing problem in accessing that content (or content from similar services) because there is no unifying ontology that is describing metadata from different services in the music domain, making it difficult to query these services consistently using unified terms. Building such ontology can be time demanding and cumbersome task for domain experts but it is a necessary task that will be used later on as a basis for building semantic web services and orchestrating user queries.

Name	Type	Description
id	number	The sound's unique identifier.
url	URI	The URI for this sound on the Freesound website.
name	string	The name user gave to the sound.
tags	array[strings]	An array of tags the user gave to the sound.
description	string	The description the user gave to the sound.
geotag	string	Latitude and longitude of the geotag separated by spaces (e.g. "41.00823;
created	string	The date when the sound was uploaded (e.g. "2014-04-16T20:07:11.145")
license	string	The license under which the sound is available to you.
type	string	The type of sound (wav, aif, aiff, mp3, or flac).
channels	number	The number of channels.
filesize	number	The size of the file in bytes.
bitrate	number	The bit rate of the sound in kbps.
bitdepth	number	The bit depth of the sound.
duration	number	The duration of the sound in seconds.
samplerate	number	The samplerate of the sound.

Fig. 1. Example of the Freesound API metadata and service descriptions

One of the first challenges that needs to be overcome is the problem of knowledge acquisition. The ontology should be capable of describing entities and actions that are already defined in the data models of various service providers and that number can be potentially large. This is the reason why it is necessary to provide support to knowledge engineers at an early stage of the knowledge acquisition task. Following well defined methodology that describes the process of ontology building can be helpful and there is considerable amount of work addressing this issue [10]. These methodologies tend to be generic and they often can't be scaled for the specific problem. In a project like the one described in this paper there is a need for an automated or semi-automated tool that can be used by a knowledge engineer to analyse different API dictionaries (including the one that will be joining the AC ecosystem in the future).

3.2 Music API Dictionaries

Music service APIs allow users to browse, search, and retrieve information about other users interacting with the service, but most importantly the collections of sounds and music pieces or particular sounds from their extensive databases. For example, it is possible to retrieve similar sounds to a given target (based on content analysis) or retrieve automatically extracted features from audio files. The Freesound API allows users to perform advanced queries combining content analysis features and various metadata. The API also allows different actions for manipulating the music data like upload action, writing comments, rating, etc.

When looking into descriptions for each of the parameter in the API dictionary it is possible to distinguish different types of information that are implied in the text. For example, parameter named *channels* is described using a simple description: *the number of channels*. On the other hand, parameter like *tags* is described as: *an array of tags the user gave to the sound* or parameter *descriptions* described as: *the description the user gave to the sound* are describing parameter and its context (caused by a certain action). These descriptions bring valuable insight into a set of actions that are being carried out by various entities connected with the service. Having the capability to identify different contexts where various entities can take different (or the same) roles can help us to model the ontology that will describe specific contexts in a more general way. We can say that Web Service API parameters and their descriptions could serve as basis for a bottom-up approach of building the ontology. These descriptions are an important source of knowledge about the entities existing in the service for a knowledge engineer with a task of building the ontology that will exist on top of those services.

3.3   Method Overview

As mentioned in the previous section each service that allows its content to be queried (using REST (REpresentational State Transfer) API) needs to build the dictionary that will be made of parameters names, parameter types and parameter descriptions. Those dictionaries are exposed publicly on the Internet for users or developers for retrieving the content they want by calling specific parameters.

We are proposing a set of tasks that will allow the knowledge engineer to better understand the concepts and relationships between entities that exist in a web service data model that will potentially be mapped to an ontology as classes and properties. The method is using various natural language processing techniques to analyse the parameter descriptions and create a collection of facts that will be represented in a graph database. The pipeline for the proposed method can be seen on Fig. 2. It consists of the following tasks:

1. Web Scraping (Harvesting) – technique of extracting information from websites
2. Repository – of extracted parameter descriptions
3. Information Extraction - extraction of structured relation triples from plain text
4. Semantic Role Labelling - detection of the semantic arguments associated with the predicate or verb of a sentence and their classification into their specific roles
5. Visualisation and Manipulation – one of the possible usages would be annotation of WDSL (Web Services Description Language) or OWL-S (Semantic Markup for Web Services) service descriptions.

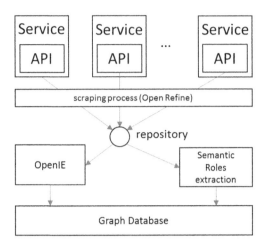

Fig. 2. Pipeline overview

3.4   Information Extraction

Information extraction systems are often used for tasks like question answering, relation extraction, and information retrieval, because they can produce relation triples from unstructured text. IE systems search for a collection of patterns over either the surface form or dependency tree of a sentence. Although a small set of patterns cover most simple sentences (e.g., subject verb object constructions), relevant relations are often spread across clauses or presented in a non-canonical form [11]. For example, the parameter *id* described as: *the sound's unique identifier* will be transformed into a triplet: *sound, have, unique identifier*.

All produced triplets have high confidence $C = 1.0$ but the produced statements are not canonical. For example, entity URI (Uniform Resource Identifier) came in four different variants: *(1.0, URI, point to, complete analysis result sound)*, *(1.0, URI, point to, complete analysis result)*, *(1.0, URI, point to, analysis result sound)*, *(1.0, URI, point to, analysis result)*. A solution to this problem is to implement a simple algorithm that will keep the statement with minimum numbers of terms constituting an object (Fig. 3) as a candidate for ontology. Since our goal is not to create an ontology in a completely automated fashion (manual refinement by domain expert is important) we decided to include all statements as a candidate for the ontology and decide which one to keep in a later stage (visualisation).

```
for each triple T(S, V, O)
    for triple T(S, V, O) == T'(S, V, O')
        numObjectTerms = count(split(O'))
        minObjectTerms = numObjecTerms
        if numObjectTerms < minObjectTerm
            candinateT = T'(S, V, O)
```

Fig. 3. Choosing a statement

3.5   Semantic Role Labelling

As a parallel task with information extraction we use an NLP technique called dependency parsing to analyse music service API descriptions. The Stanford dependencies provide a representation of grammatical relations between words in a sentence that are designed to be easily understood and effectively used by people who want to extract textual relations. Stanford dependencies (SD) are triplets: name of the relation, governor and dependent [12]. This approach was adopted as it is generally accepted as the best way forward when one does not know what is being looked for *a priori*. Entities extracted from textual descriptions should correspond to the lexical pattern shown in Fig. 4. The number of identified patterns depends on a number of identified Subject-Verb-Object (S-V-O) patterns – also one activity can contain more than one S-V-O pattern (e.g., *If the sound is part of a pack, this URI points to that pack's API resource* contains two S-V-O patterns). Since the result of the information extraction task described in previous section produces triples in S-V-O form we will include those triples to the semantic role labelling process to get the ontological pattern that correspond to the lexical pattern shown on Fig. 4. Our decision to create an ontological representation shown in Fig. 4 is

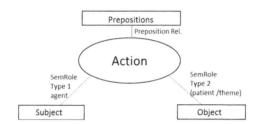

Fig. 4. Lexical pattern

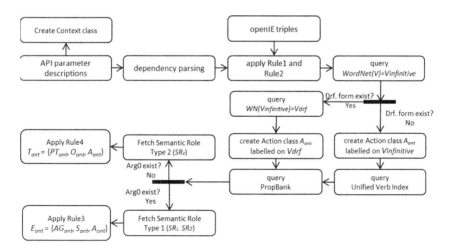

Fig. 5. Algorithm for automated creation of ontological patterns from API descriptions

influenced by the Provenance Ontology[8] and the Media Value Chain Ontology (MVCO)[9]. Both the Provenance (PROV) ontology and the MVCO ontology use action/activity entity that is connected with role and object entities through relationships as shown on Fig. 6. This kind of representation will be reproduced with the algorithm described below.

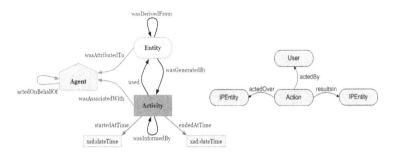

Fig. 6. Activities in PROVO and MVCO ontologies

We can describe the algorithm as follows (Fig. 5): If A is a set of elements describing one particular action A then, for every A, we have A_N, $I_1,...,I_n$, $O_1,...,O_k$, R where N is a string containing a description of an action, I_n is a set of strings describing elements that are part of the action. Set O_k is a set of strings describing the entities that are created as a result of the action, and R is an entity that started the action A (subject).

The following rules are implemented:

- Rule1: For every action A implied in the parameter description there should be a corresponding class A_{ont} in the ontology.
- Rule2: For every parameter description P_{verb} containing verb V_1, $V_2,..., V_n$ create A_{ont1}, $A_{ont2},..., A_{ontn}$ class by splitting P_{verb} into a set $\{P_{verbpart1},..., P_{verbpartN}\}$ (e.g., *If the sound is part of a pack, this URI points to that pack's API resource* should create two action classes $A_{ont1} = being$ and $A_{ont2} = pointer$.
- Rule3: For A_{ont} class there is a set of elements $E_{ont} = \{AG_{ont}, S_{ont}, A_{ont}\}$. Relationship S_{ont} between AG_{ont} and A_{ont} classes is called agent relationship. For example: $E_{ont} = \{URI, pointer, pointingAction\}$.
- Rule4: For A_{ont} class there is a set of elements $T_{ont} = \{PT_{ont}, O_{ont}, A_{ont}\}$. Relationship O_{ont} between PT_{ont} and A_{ont} classes is called patient\theme relationship. For example: $T_{ont} = \{API\_Resource, thing\_pointed, pointingAction\}$.
- Rule5: For some parameter descriptions P_{verb} containing verb V_n there is a set of elements $Z_{ont} = \{A_{ont}, V_{ont}, PO_{ont}\}$ where V_{ont} is a relationship between action class and the object of a preposition. For example: $Z_{ont} = \{giveAction, to, sound\}$.

We use the Stanford dependency parser [12] to uncover the dependency tree for each parameter description. Our method is focusing on the following dependency relations:

---

[8] Provenance ontology - https://www.w3.org/TR/prov-o/.

[9] Media Value Chain ontology: http://dmag.ac.upc.edu/ontologies/mvco/

.

- *nsubj*(V, S) - a nominal subject is a nominal phrase which is the syntactic subject and the proto-agent of a clause. (Implies subject-verb relationship.)
- *nsubjpass*(V, S) - a passive nominal subject is a noun phrase which is the syntactic subject of a passive clause. (Implies verb-object relationship.)
- *dobj*(V, O) - the direct object of a verb is the second most core argument of a verb after the subject. (Implies verb-object relationship.)
- *nn*(N, S/O) - a noun compound modifier of an NP is any noun that serves to modify the head noun. This pattern is used to expand the subject or object (example: *information descriptors* instead *descriptors*).
- *prep*(A, S/O) - A prepositional modifier of a verb, adjective, or noun is any prepositional phrase that serves to modify the meaning of the verb, adjective, noun, or even another preposition.

To automate the process of creation of the ontological classes described by the rules method uses various lexical repositories. The Unified Verb Index[10] is large list of English language verbs and a system that merges links and Web pages from four different natural language processing projects that are providing lexical information about verbs. One of those projects is PropBank [13] - a corpus of text annotated with information about basic semantic propositions. Verbs in the PropBank corpus can have a semantic role, also called argument, associated to them, which connects the verb with the subject (the agent) and the object called patient/theme (some authors differentiate between patient and theme but the Penn Treebank regards it as a single patient/theme argument). Each argument is given a number. The agent and the patient are always given the argument numbers 0 (Arg0) and 1 (Arg1), respectively. In some cases Arg0 does not exist so the role of the agent is given by Arg1. The next lexical resource we use is WordNet [2]. WordNet is a large lexical database for English language and it's often used in the NLP domain. WordNet is used here to transform the verb found in the parameter description into its infinitive form (c). The main reason for conducting this task is the fact that a verb denoting one meaning or implying one specific action can appear in different representations. Also, acquiring the infinitive form of the verb allows the method to query for the derivationally related form of the verb so the action that the verb is implying can be labelled more naturally (example: $V =$ gave, $V_{infinitive} =$ give, $V_{drf} =$ giving). Since a verb can have a large number of different senses of derivationally related forms, gloss or dictionary definition is searched for predefined clues (a list of words with meanings similar to words *act* or *event*).

For example, verb *gave* have three derivationally rated forms in the WordNet dictionary (*giver, giving, giving*) but only the third one is describing the act (gloss: *the act of giving*). The Unified Verb Index is used to find mappings between the $V_{infinitive}$ and the representation or the verb in PropBank (containing arguments Arg_0 and Arg_1). The values of arguments Arg_0 and Arg_1 are used to label the agent and the patient roles that are representing S_{ont} and O_{ont} in triples $E_{ont} = \{AG_{ont}, S_{ont}, A_{ont}\}$ and $T_{ont} = \{PT_{ont}, O_{ont}, A_{ont}\}$. Ontological patterns created from Music API descriptions should give insight into

---

what kind of actions we can expect to deal with as well as objects (or possible IP entities) and roles (users of the system).

4   Use Case: The Freesound API

To assess the method discussed in Sect. 3 we deployed it over the Freesound API. We chose a graph database[11] as a repository for storing the ontological patterns extracted with the pipeline described in previous sections. In graph database every graph is a collection of two elements: vertices and edges, or it's a set of nodes and the relationships that connects them. Graphs represent entities as nodes and the ways in which those entities relate to the world as relationships. We are using the popular variant of a graph model called the *property graph*. The property graph is made up of nodes, relationships and properties. Nodes can contain properties. Nodes are entities that can store properties in the form of arbitrary key-value pairs. The keys are strings and the values are arbitrary data types. Relationships act as connectors between nodes. Having a properly labelled relationship is very important because it adds semantic clarity to the nodes structure. The graph database is used not just for storing the ontological patterns but is also a tool for visualising them. It supports Cypher a SPARQL like query language that can be used to manipulate the patterns and views of the patterns. Our information extraction system has been running over the collection of Freesound API parameter descriptions. The result of this action is a set of facts produced by the system with a certain confidence C (Table 1).

Table 1. Facts produced by information extraction

	Freesound
nr. of param. desc.	81
nr. of facts	75
nr. of dist. fact	25

The algorithm took as an input 81 parameters (each parameter is coming with a short parameter description). The information extraction system suggested 75 facts (S-V-O patterns) from the input data. Since the IE system can produce variants of the same fact we used the algorithm from the Fig. 3 to keep only distinct facts and discarded the rest. Figure 7a is showing how certain entities are connected with identical object through different roles (contexts). For example, URI entity is used as a bookmark for a sound, as a pointer to a comment about the sound and simply as a link for downloading the sound. On the same input data, the semantic role labelling algorithm described on Fig. 5. created 10 action nodes (*download, rate, part, give, comment, retrieve, contain, point* and *upload*). The action nodes occupy the centre of the pattern that is shown on Fig. 4 (Table 2).

---

[11] Graph database - https://neo4j.com/.

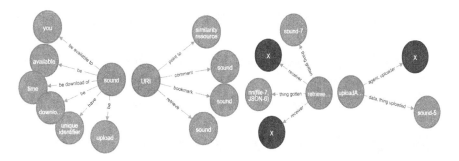

Fig. 7. a. Subjects pointing to different objects, **b.** Action node with semantic roles

Table 2. Semantic role labelling

	Freesound
nr. of actions/events	10
nr. of agent/themes	31/20
nr. of semantic roles	51

The example on Fig. 7b is showing action nodes representing *retrieve* and *upload* actions that were identified in the parameter descriptions. Each action that denotes uploading of something (like *sound* in our example) will be merged with one unique node *uploadAction* and object/theme node of the action will be connected with a relationship *data, thing uploaded*. Relationship between *uploadAction* and agent/subject node is labelled as *agent, uploader*. The agent/subject node is labelled as *X* because there was not enough data in the parameter description to extract the actual label for that node (this can be done manually in the knowledge-base).

5 Conclusion and the Future Work

In this paper we presented an approach for extracting ontological patterns from music service API descriptions. Web service API descriptions represent a valuable source of knowledge about the capabilities of the service. Analysing these descriptions can give us insights about the actions/activities that are being carried out by different roles (usually users of the service). We presented an approach that is highly automated. While the construction of ontologies may never be completely automated, there is a need for methods proposed in this paper, that will ease the cumbersome task of ontology creation from scratch. Additionally, considering the amount of different services that the Audio Commons ecosystem can potentially have in the future, the "ontology bottleneck" could become a serious problem. Ontological patterns extracted from service API descriptions will be used by knowledge engineers as the next step towards the creation of the Audio Commons ontology (an ontology that will describe various music and audio content entities and services). Our method proved successful in extracting ontological patterns from the Freesound API, thus it presents a promising direction. Future work will involve the creation of an interface that allows the knowledge engineer to build on the patterns extracted in an automated fashion providing an easy and convenient way of improving the quality of labels or filing the labels that could not be extracted from the descriptions. Recognizing important entities and actions from music service API descriptions will also assist in future tasks of the Audio Commons project that involves building semantic web services and orchestrating user queries.

References

1. Aslam, N., Ullah, I., Rohullah, B.S., Akram, T., Shabir, M.: Tracking the progression of multimedia semantics: from text based retrieval to semantic based retrieval. World Appl. Sci. J. **20**(4), 549–553 (2012)
2. Fellbaum, C. (ed.): WordNet: An Electronic Lexical Database. MIT Press, Cambridge (1998)
3. Krotzsch, M., Vrandecic, D., Volkel, M., Haller, H., Studer, R.: Semantic Wikipedia. J. Web Semant. **5**(4), 251–261 (2007)
4. Zhou, L.: Ontology learning: state of the art and open issues. Inf. Technol. Manag. **8**(3), 241–252 (2007)
5. Wisniewski, M.: Metamodel of ontology learning from text. In: Badr, Y., Chbeir, R., Abraham, A., Hassanien, A.-E. (eds.) Emergent Web Intelligence: Advanced Semantic Technologies. Advanced Information and Knowledge Processing, pp. 245–276. Springer, London (2010)
6. Raimond, Y., Abdallah, S., Sandler, M., Giasson, F.: The music ontology. In: International Society for Music Information Retrieval Conference, pp. 417–422 (2007)
7. Fazekas, G., Sandler, M.B.: The studio ontology framework. In: 12th International Society for Music Information Retrieval Conference (2011)
8. Saur., K.G.: Functional requirements for bibliographic records: final report, vol. 19, 136 p. (1998). UBCIM Publications, ISBN: 978-3-598-11382-6
9. Allik, A., Fazekas, G., Sandler, M.B.: An ontology for audio features. In: 17th International Society for Music Information Retrieval Conference (2016)
10. De Nicola, A., Missikoff, M., Navigli, R.: A software engineering approach to ontology building. Inf. Syst. **34**(2), 258–275 (2009)
11. Angeli, G., Premkumar, M.J., Manning., C.D.: Leveraging linguistic structure for open domain information extraction. In: Proceedings of the Association of Computational Linguistics (ACL) (2015)
12. de Marneffe, C.-M., MacCartney, B., Manning, C.D.: Generating typed dependency parses from phrase structure parses. In: Proceedings of LREC 2006, pp. 449–454 (2006)
13. Palmer, M., Gildea, D., Kingsbury, P.: The Proposition Bank: an annotated corpus of semantic roles. Computat. Linguist. **31**(1), 71–106 (2005)

Association Rule Mining with Context Ontologies: An Application to Mobile Sensing of Water Quality

Eliot Bytyçi[1](✉), Lule Ahmedi[1], and Arianit Kurti[2,3]

[1] University of Prishtina, Prishtina, Kosovo
{eliot.bytyci,lule.ahmedi}@uni-pr.edu
[2] Linnaeus University, Växjö, Sweden
arianit.kurti@lnu.se
[3] Interactive Institute Swedish ICT, Norrköping, Sweden
arianit.kurti@tii.se

Abstract. Internet of Things (IoT) applications by means of wireless sensor networks (WSN) produce large amounts of raw data. These data might formally be defined by following a semantic IoT model that covers data, meta-data, as well as their relations, or might simply be stored in a database without any formal specification. In both cases, using association rules as a data mining technique may result into inferring interesting relations between data and/or metadata. In this paper we argue that the context has not been used extensively for added value to the mining process. Therefore, we propose a different approach when it comes to association rule mining by enriching it with a context-aware ontology. The approach is demonstrated by hand of an application to WSNs for water quality monitoring. Initially, new ontology, its concepts and relationships are introduced to model water quality monitoring through mobile sensors. Consequently, the ontology is populated with quality data generated by sensors, and enriched afterwards with context. Finally, the evaluation results of our approach of including context ontology in the mining process are promising: new association rules have been derived, providing thus new knowledge not inferable when applying association rule mining simply over raw data.

Keywords: Association rules · Ontology · Context · Wireless sensor networks · Internet of Things

1 Introduction

In the era of the Internet of Things, wireless sensor networks (WSN) play a major part, due to their affordability. The problem, however arises in collection, exchange and analysis of the data produced. These data, referred to as raw data initially, are collected directly from the source and streamed towards data repositories for further handling. However, this leads to problems we usually encounter with raw data in terms of data quality, heterogeneity and the space needed for their preservation.

© Springer International Publishing AG 2016
E. Garoufallou et al. (Eds.): MTSR 2016, CCIS 672, pp. 67–78, 2016.
DOI: 10.1007/978-3-319-49157-8_6

Nevertheless, those problems have been mitigated using different scenarios, in order to minimize the effect of the challenges obtained. But, some new challenges have come forward in the last few years. One of them is also the semantic integration of data. Meaning of a concept and their description is basis of the semantic study. That challenge might be resolved by using the ontologies for description of concepts and their relations [7]. Furthermore, new relations can be drawn from the ontology representation of initial concepts, thus improving not only the quality of the data but checking also possible inconsistencies occurred during data integration.

Lastly, data needs to be analyzed in order to discover useful information, which is the reason why those data have been initially gathered and processed. The analysis can be descriptive, using data to summarize important components, and predictive, using data to predict further relations and discover new knowledge. Knowledge discovery in databases (KDD) is the field concerned with developing techniques and methods for making sense of data [8]. One of the stages of the process is data mining, which besides different tasks involves also association rule mining. Association rule mining represents a powerful method for discovering relations between data [9]. By using it jointly with ontology, we consider that further interesting relations can be drawn.

The relations drawn, can vary from one situation to another. Furthermore, in order to entirely understand them, one needs to know the circumstances that create that situation. That represents the context upon which rules are formed. In [1], authors showed that context could provide accuracy and efficacy to data mining outcomes used in medical applications.

In our case, we tend to use association rule mining with context ontologies in surface water quality monitoring. The monitoring is performed through mobile sensing devices, which measure several parameters and forward them to a repository. This component is part of a bigger system, which involves also static monitoring stations for water quality monitoring. Such application can be further extended for usage in other domains.

The paper is organized as follows: in the next section we provide insight on related work, while in Sect. 3 we describe data preprocessing process. In Sect. 4, our ontology modeling for mobile sensing of water quality is described, with the context inference module included and in Sect. 5, the results from association rule mining on data and on context-aware ontology data are presented. Conclusions, challenges and future contribution are covered in Sect. 6.

2   Related Work

The authors of [1] have introduced a framework of representing context in ontology, firstly captured during data mining process and then adapting it accordingly. They have used a classification tree to predict accurately the patient's heart attack risk. In the end, their framework showed that use of the context factor had increased effectiveness in data mining.

Authors in [13] have addressed the challenge of mining knowledge encoded in domain ontologies. They have demonstrated the usefulness of their approach by mining biological data, showing major improvements and advantages.

In [14], authors have mined with association rules over RDF and OWL data repositories. The appropriate transactions were derived from ontology through schema knowledge for further mining through association rules algorithms. Their initial experiments have proved usefulness and efficiency of the approach.

The concept of "mining configurations" has been introduced by authors in [15], allowing mining of RDF data at different levels. Among configurations is the one describing relation among the subjects and the objects in the RDF triples through basket analysis. Authors at the same time call for further research in the field of association rule mining over RDF by combining configurations and different use cases.

The discussed related approaches are characterized by the combination of ontologies and data mining, but none of them have used association rules, and generated context and ontology to then further advance with data mining of the Semantic Web. That has served us as a motivation for further work that related these three concepts: association rule mining, ontologies, and context.

3   Background and Hypothesis

Internet of Things relies on sensors to monitor the environment. Sensors produce data that should be further processed and analyzed in order to infer new knowledge. New knowledge should help us on differing between a usual or unusual process happening in the environment. That can be used to respond to the environment with possible actuators.

3.1   Problem Definition – Water Quality Case Study

Let us consider an example where sensors are used to measure values of water quality parameters in a river. The measurements are performed for several parameters such as pH, water temperature, dissolved oxygen and conductivity. In [2], it is acknowledged that during the night the values of dissolved oxygen will fall sharply, mainly due to the process of photosynthesis that occurs only during the day. Another parameter related to that is temperature, which during the night is lower due to the deprivation of sun. Related to both parameters, an elevated turbidity can increase the water temperature and will lower the dissolved oxygen (DO), imitating the process of photosynthesis as a regular process that can happen during the night. Furthermore, as presented in [3], a direct variation exists as a correlation between pH and temperature. Therefore, hypothetically, if the dissolved oxygen falls sharply and temperature is lower as well, resulting in lower pH, we can observe this as a regular process if it occurs during the night. But, if the same process is happening during the day, we can treat it as something out of ordinary as a possible ongoing pollution. Hence, it depends if it is day or night, in order to make the substantial difference of the process.

3.2   The InWaterSense Project

InWaterSense project [5], which contributed with a wireless sensor network deployed in river Sitnica, has a static component consisting of several sensors, in order to measure the water quality parameters. In addition to the static part, the deployed system consists also of a mobile component, with the aim to discover other possible polluted water locations for the deployment of the static system in the future. The mobile component measured 4 parameters: pH, temperature, dissolved oxygen and conductivity. Using an open source wireless platform, with sensors attached to it, measurement data, herein after raw data, was sent in real time to a specific remote server for storage to a database, with timestamp data attached.

After several measurements, data were manually analyzed, where several anomalies were observed, which back up the need for simulated data. For example, in several cases, the temperature was measured below 0 or in some cases pH values were −1. After considerable measurements, it was concluded that those outlier data could have been result of several conditions:

– sensors have made measurements before entering to water: this due to the fact that all the measurements where done from the bridge (sensors where lowered to the water) due to impossible approach to the specific locations from nearby river.
– sensors were not calibrated: a periodic calibration of the sensors is mandatory, or
– sensor damage: at least in one case sensor was damaged due to fast water streams.

Another prevailing factor that determined usage of simulated data was the amount of data stored, less than 1000 records. Therefore, with a generator created as part of the web portal in project [16], data was simulated in large amounts. Besides that, the generator was used carefully in order to control values of data, so to back up claims by the water experts on their correlation. Furthermore, backing our initial hypothesis, that due to photosynthesis during the night the value of DO will be lower, we have created a constraint in generation of DO to be lower or equal to DO during day hours. If there is a rapid fall of DO during the day, then the reason should be searched with turbidity or direct pollution, according to the experts. Using the data generator, more than 100000 records were generated comprising of timestamp and sensor values.

After the generation of the data and before starting the process of data evaluation, a process of data pre-processing was concluded. The first step of preprocessing the division of the timestamp into several parts including: year, month, date, hour and minutes for the purposes of finding the day or night interval. For experimental purposes, day was described from 6 h in the morning until 18 h in the evening and in contrary the night was described as from 18 h in the evening until 6 h in the morning. After that, data were discretized or divided into several bins, each maintaining data with similar characteristics. A number of three bins were chosen while dividing parameters: a bin that holds parameters with lower values, those with medium values and lastly the ones with higher values. The automatic process of bin division was performed with help of a support tool - WEKA [11]. After that, we have removed some data that were seen unnecessary such as year, day, minutes, since they were repetitive and therefore not significant for the process. In the end, a final set of data was obtained for further processing and analyzing.

Similar preprocessing was conducted over Ontology, were additionally, in the beginning of the process, a data cleaning and preprocessing was conducted. That meant removing some of the relations between ontology concepts and concepts descriptions, due to the purpose of the experiment purely on data. Affirmatively, the process of ontology population was performed before hand, a process that will be described in the next section.

The final obtained set, both data and ontology data, was ready for the experiment of data mining. Data mining techniques that have been widely used to find patterns of mining data are Apriori and FPgrowth. Association rule mining may return interesting relations between values, in our case values obtained through water quality measurement performed by sensors, thus identifying new rules that describe correlations between data and have a certain percentage of trust [10]. Apriori has been used in databases containing transactions in order to find correlations between the items on such transaction. The values obtained from water quality measurements through sensors, can be viewed as such transactions. A transaction is comprised of timestamp and parameter values. Thus, using association rule mining, one can explore correlations between values. That would result in generation of new rules between sensors obtained values. Such rules would acknowledge a possible failure in the system or a possible dangerous situation, in regards to water quality monitoring and pollution. Besides that, backing our initial assumption, one could claim that by using Ontology, we can extract even more rules, which would result in new knowledge being revealed. The input from the ontologies would be in the context provided and therefore enhancing the knowledge base.

Therefore, in this paper we aim that by using simulated data, that imitate increase or decrease of specific sensor values during night, in conjunction with association rule mining and ontology, we would be able to derive new rules. To achieve that, ontology should be created and populated with generated data, before providing the necessary context for the inference of new knowledge.

4   Context Ontology

In order to formally define the entities of the mobile water quality monitoring sensor system, a lightweight ontology is introduced. The ontology is populated with generated sensor data. As known from the literature [7], ontology describes the overall agents involved in the system and their relations. Besides them, i.e., the mobile sensing component and the water quality related data it generates through measurements, the context is also covered by our newly introduced ontology. That motivated in our example domain by known facts, e.g., that due to photosynthesis during night, certain observed water quality parameters take different values when compared to their values during day. Thus, the day/night context enables modeling these domain specific context behaviors and their implications, as will be made explicit in the examples to follow.

Figure 1 depicts the proposed lightweight mobile context-aware ontology named LMINWS[1]. Whereas authors in [4] modeled an ontology (InWaterSense) that covers an arbitrary wireless sensor network for water quality monitoring, this lightweight ontology aims to cover modeling the rather more rich-in-context but simple mobile portable sensors of a wireless sensor network. The classes of our lightweight ontology are depicted in grey color, in Fig. 1, while the ones in white represent imports from other

---

[1] http://inwatersense.uni-pr.edu/ontologies/LMINWS.owl.

ontologies. The Time[2] ontology, its class DateTimeDescription has been extended with addition of two new subclasses in order to represent the day and night context. The DayDescription subclass expresses the time between 6 o'clock in the morning and 18 o'clock. The rest of time, from 18 o'clock until 6 o'clock in the morning is modeled as NightDescription. Another additional core class in our ontology is the MobileEquipment, representing only the mobile sensing part of the WSN system for water quality monitoring, and specializes as such the InWaterSense ontology introduced in [4]. An important concept introduced, which is aims to serve for future work, is the Activity class with its two subclasses CalibrationActivity and MeassuringActivity. It helps on context implementation when related to the user who performs the activity: an Engineer or a Technician. Both later concepts are introduced in the ontology as well belonging to the class Person of the FOAF[3] ontology, in order to determine by whom exactly the given activity is conducted, and whether data can be reliable or not. Another concepts used is Place[4], with two other subclasses introduced InDoor and OutDoor.

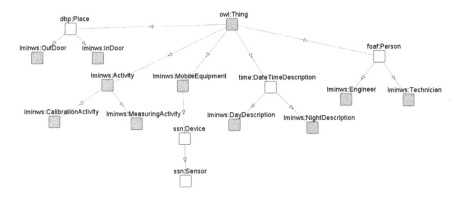

Fig. 1. Lightweight mobile ontology (LMINWS)

4.1   Populating the Ontology

Once sensors generate the data, the modeled ontology is populated right away. Since the tool used [12] for populating the ontology through mapping required the specific format of data representation, before populating it we have converted data to the requested specific format. Then a mapping file, partially presented in Fig. 2, was created in order to convert database data into an RDF/XML format. Subsequently data was added to the existing ontology as a repository.

---

2 https://www.w3.org/TR/owl-time/.

3 http://xmlns.com/foaf/spec/.

4 http://dbpedia.org/ontology/.

```
@prefix rdf: <http://www.w3.org/1999/02/22-rdf-syntax-ns#> .
@prefix rdfs: <http://www.w3.org/2000/01/rdf-schema#> .
@prefix xsd: <http://www.w3.org/2001/XMLSchema#> .
@prefix d2rq: <http://www.wiwiss.fu-berlin.de/suhl/bizer/D2RQ/0.1#> .
@prefix jdbc: <http://d2rq.org/terms/jdbc/> .

map:database a d2rq:Database;
    d2rq:jdbcDriver "com.mysql.jdbc.Driver";
    d2rq:jdbcDSN "jdbc:mysql://localhost/sensordata";
    d2rq:username "root";
    jdbc:autoReconnect "true";
    jdbc:zeroDateTimeBehavior "convertToNull";
    .

# Table data
map:data a d2rq:ClassMap;
    d2rq:dataStorage map:database;
    d2rq:uriPattern "data/@@data.Id@@";
    d2rq:class vocab:data;
    d2rq:classDefinitionLabel "data";
    .
map:data__label a d2rq:PropertyBridge;
    d2rq:belongsToClassMap map:data;
    d2rq:property rdfs:label;
    d2rq:pattern "data #@@data.Id@@";
```

Fig. 2. Partially described data mapping file

During the conversion process, several specific issues were encountered which needed manual intervention, especially when dealing with prefixes.

4.2 Context Inference

Once the ontology is populated with data, then by means of an ontology reasoner, new context-dependent data may get inferred from existing ontology data.

Let us consider again the running day/night context example, in DateTimeDescription class in the ontology (cf. Fig. 1), two subclasses are introduced: DayDescription and NightDescription. Initially, date and time data are assigned instances of DateTimeDescription. After applying the hour constraint, those instances are inferred as instances of either DayDescription subclass or the NightDescription subclass.

The constraint on DayDescription, subclass of DateTimeDescription, using the data property onProperty: hour, was introduced as follows:

$$hour\ some\ xsd:int[>= "6"^{\wedge\wedge}xsd:int, <= "17"^{\wedge\wedge}xsd:int]$$

It infers only instances of DateTimeDescription that belong to the day description, meaning only measurements performed during day hours.

A similar constraint is defined for NightDescription, by simply putting the negation over DayDescription instances as follows:

$$not\ (hour\ only\ xsd:int[>= "6"^{\wedge\wedge}xsd:int, <= "17"^{\wedge\wedge}xsd:int])$$

In the end, the lightweight ontology triplets were derived, describing ontology concepts and their relations. A snippet of these triplets may be seen in Fig. 3, where concepts such as e.g. MobileComponent can be seen standing in relation with NightDescription and the corresponding values of the water quality measurement sensors.

@type/1	@type/2	#hasConductivity	#hasDissolvedOxygen
#MobileEquipment	#NightDescription	\'(3383.333333-inf)\"	\'(100-200]\"
#MobileEquipment	#NightDescription	\'(3383.333333-inf)\"	\'(100-200]\"
#MobileEquipment	#NightDescription	\'(1766.666667-3383.333333]\"	\'(-inf-100]\"
#MobileEquipment	#NightDescription	\'(1766.666667-3383.333333]\"	\'(-inf-100]\"
#MobileEquipment	#NightDescription	\'(-inf-1766.666667]\"	\'(-inf-100]\"
#MobileEquipment	#NightDescription	\'(1766.666667-3383.333333]\"	\'(-inf-100]\"

Fig. 3. Portion of triplets generated from ontology

5 Association Rule Mining with Context Ontology

In our deployed wireless sensor network for water quality monitoring, its static and mobile sensors generate data, which are then sent to a remote server, as presented in Fig. 4. Those transactions include values of the sensor measurements on the water quality parameters and the timestamp when the measurement occurred. From the previous Section, the context ontology part of the architecture was explained, preceded by the description of the problem in Sect. 3. From the ontology, data were transformed into JSON format. That because the requested format for the tool used - WEKA [11] was CSV or Arff. After that, on preprocessed data, algorithms were used, which resulted in gaining new knowledge.

Fig. 4. The architecture of the system

Association rule mining is based on the market basket analysis, which analyses transaction repository [9]. In the repository, there exist sets of items, which once may describe by X and Y. Therefore, the association rule expressed as X => Y, denotes that in the transaction database, a number of transactions contain X, with a certain probability that the they will contain also Y.

In [6], authors have presented top ten most used algorithms in data mining. Amongst them is the Apriori algorithm [10], which is used for finding frequent subsets (item sets) from a transaction dataset and derive association rules. When there are no other subsets, the algorithm stops. It should be noted that one can limit the *support* threshold of the algorithm so that it generates only transactions that have a specific number of appearances. Besides that, also the *confidence* can be limited by user in order to find those transactions that contain both items of the rule. Best rules are those with the higher support and confidence.

An improvement over Apriori is the so-called FP-growth (frequent pattern growth) method that succeeds in eliminating candidate generation [6] used in Apriori.

We have conducted tests with both algorithms Apriori and FP Growth over our data, but have always obtained same results. Therefore, only results obtained when applying one of the algorithms, i.e. Apriori, will be presented next.

Data preparation. The sensor measurement data are preprocessed for the mining process by using several unsupervised methods such as normalization and discretization of data. Besides that, since we need to know the timing of data measured, we have split the timestamp into smaller pieces. One of those smaller pieces is the hour when the measurement occurred, which actually provides the context on whether the measurement happened during day or night. That may help on realizing which association rules generated are out of ordinary. Moreover, in order to find significant rules, data have first been discretized – a filter that allows distribution of data in separate bins. As previously mentioned, we have distributed our data into 3 bins proportionally depending on the values, with the help of a tool - WEKA [11]. For example, the temperature is divided into low temperature, mid temperature and high temperature.

Running example. The architecture presented in Fig. 4, depict the running example of the system in water quality monitoring. It explains the process and the solution according to our approach of a specific domain problem.

Results of using Apriori algorithm over sensor data enriched with the LMINWS context ontology are shown in Fig. 5.

Observing Fig. 5, one can see there exists a relation between temperature and day or night context. The association rule that during night, the temperature is lower, has bigger support and higher confidence. Further similar association rules exist, amongst others related to dissolved oxygen, which has lower values during night, as well.

Results and discussion. We consider that our approach of aiding association rule mining with context ontologies is more successful on providing inferred knowledge than existing approaches in the literature which proclaim using association rule mining over simply raw data [14, 16]. To back up that claim, we have performed an experiment only on raw data, i.e., without the context ontology, and compared the results to the previous example.

PATTERN	SUP	CONF
#hasConductivity/0/@value='\'(3383.333333-inf)\'" ==> #hasPh/0/@value='\'(9.666667-inf)\'"	36453	0.96271
#hasTemperature/0/@value='\'(-inf-10]\'" ==> #hasDissolvedOxygen/0/@value='\'(-inf-100]\'"	43024	0.961387
#hasConductivity/0/@value='\'(-inf-1766.666667]\'" ==> #hasPh/0/@value='\'(-inf-5.333333]\'"	36404	0.95737
#hasDissolvedOxygen/0/@value='\'(100-200]\'" ==> #hasTemperature/0/@value='\'(10-20]\'"	36150	0.953826
#hasDissolvedOxygen/0/@value='\'(-inf-100]\'" ==> @type/1=#MobileEquipment	45462	0.944567
#hasDissolvedOxygen/0/@value='\'(-inf-100]\'" ==> @type/2=#NightDescription	45462	0.944567
#hasDissolvedOxygen/0/@value='\'(-inf-100]\'" ==> @type/1=#MobileEquipment @type/2=#NightDescription	45462	0.944567
#hasDissolvedOxygen/0/@value='\'(-inf-100]\'" #hasTemperature/0/@value='\'(-inf-10]\'" ==> @type/1=#MobileEquipment	40402	0.939057
#hasDissolvedOxygen/0/@value='\'(-inf-100]\'" #hasTemperature/0/@value='\'(-inf-10]\'" ==> @type/2=#NightDescription	40402	0.939057
#hasDissolvedOxygen/0/@value='\'(-inf-100]\'" #hasTemperature/0/@value='\'(-inf-10]\'" ==> @type/1=#MobileEquipment @type/2=#NightDescription	40402	0.939057
#hasPh/0/@value='\'(-inf-5.333333]\'" ==> #hasConductivity/0/@value='\'(-inf-1766.666667]\'"	36404	0.93005
#hasTemperature/0/@value='\'(-inf-10]\'" ==> @type/1=#MobileEquipment	41182	0.920227
#hasTemperature/0/@value='\'(-inf-10]\'" ==> @type/2=#NightDescription	41182	0.920227
#hasTemperature/0/@value='\'(-inf-10]\'" ==> @type/1=#MobileEquipment @type/2=#NightDescription	41182	0.920227
#hasPh/0/@value='\'(9.666667-inf)\'" ==> #hasConductivity/0/@value='\'(3383.333333-inf)\'"	36453	0.919439
#hasTemperature/0/@value='\'(-inf-10]\'" ==> @type/1=#MobileEquipment #hasDissolvedOxygen/0/@value='\'(-inf-100]\'"	40402	0.902798
#hasTemperature/0/@value='\'(-inf-10]\'" ==> @type/2=#NightDescription #hasDissolvedOxygen/0/@value='\'(-inf-100]\'"	40402	0.902798
#hasTemperature/0/@value='\'(-inf-10]\'" ==> @type/1=#MobileEquipment @type/2=#NightDescription #hasDissolvedOxygen/0/@value='\'(-inf-100]\'"	40402	0.902798
#hasDissolvedOxygen/0/@value='\'(-inf-100]\'" ==> #hasTemperature/0/@value='\'(-inf-10]\'"	43024	0.893912
#hasDissolvedOxygen/0/@value='\'(-inf-100]\'" ==> @type/1=#MobileEquipment #hasTemperature/0/@value='\'(-inf-10]\'"	40402	0.839435
#hasDissolvedOxygen/0/@value='\'(-inf-100]\'" ==> @type/2=#NightDescription #hasTemperature/0/@value='\'(-inf-10]\'"	40402	0.839435
#hasDissolvedOxygen/0/@value='\'(-inf-100]\'" ==> @type/1=#MobileEquipment @type/2=#NightDescription #hasTemperature/0/@value='\'(-inf-10]\'"	40402	0.839435
#hour/0/@value='\'(15.333333-inf)\'" ==> @type/1=#MobileEquipment	36727	0.836473
#hour/0/@value='\'(15.333333-inf)\'" ==> @type/2=#NightDescription	36727	0.836473
#hour/0/@value='\'(15.333333-inf)\'" ==> @type/1=#MobileEquipment @type/2=#NightDescription	36727	0.836473
#hasTemperature/0/@value='\'(10-20]\'" ==> #hasDissolvedOxygen/0/@value='\'(100-200]\'"	36150	0.735055

Fig. 5. Results of association rule mining over sensor data with LMINWS context ontology

Thus, the Apriori algorithm over raw data has been applied to find all possible rules. Results are presented in Fig. 6.

PATTERN	SUP	CONF
Conductivity='\'(3383.333333-inf)\'" ==> pH='\'(9.666667-inf)\'"	36479	0.96271
Temperature='\'(-inf-10]\'" ==> DO='\'(-inf-100]\'"	43048	0.961386
Conductivity='\'(-inf-1766.666667]\'" ==> pH='\'(-inf-5.333333]\'"	36430	0.957374
DO='\'(100-200]\'" ==> Temperature='\'(10-20]\'"	36168	0.953823
pH='\'(-inf-5.333333]\'" ==> Conductivity='\'(-inf-1766.666667]\'"	36430	0.930001
pH='\'(9.666667-inf)\'" ==> Conductivity='\'(3383.333333-inf)\'"	36479	0.919399
DO='\'(-inf-100]\'" ==> Temperature='\'(-inf-10]\'"	43048	0.893891
Temperature='\'(10-20]\'" ==> Day='\'(12.666667-13.333333]\'"	38362	0.779541
Temperature='\'(10-20]\'" ==> DO='\'(100-200]\'"	36168	0.734958
Day='\'(12.666667-13.333333]\'" ==> Temperature='\'(10-20]\'"	38362	0.445174

Fig. 6. Apriori over raw data

At the first glance, we observe there are a set of rules derived with Apriori for the same running example on both cases, with and without the context ontology. As has been required, only the rules that have the support more than 30 % are provided on the result set. That was the same condition that we have put for both experimental cases. We observe that in both cases, obtained rules that relate parameters such as conductivity and pH, have high confidence. But, in the experiment with the LMINWS context ontology, we have derived new rules not derived when running the same experiment but without context ontology e.g. the rules that involve context. Therefore, context modeled through the ontology and considered while mining obviously makes the difference. Therefore, as expected in the beginning of running example, we have obtained results that relate certain parameters with the context, i.e., day or night context. This may assist experts in concluding whether certain inferred relations are confident or happening as a result of the natural process, such as photosynthesis. This in addition to the other fewer in number obtained rules when applied over raw data, which describe rather aid in inferring relations between parameters and are not related to the context.

6 Conclusion, Challenges and Future Work

In this paper we have presented an approach of using association rule mining over mobile component of wireless sensor network data, with context-aware ontology. Initially, the concepts and relationships to model a mobile water quality measurement sensing device through an ontology have been described. The ontology is also enriched by contextual concepts and restrictions. Following that, we have populated the ontology with the sensor measurement data. Only then, using association rule techniques, it is proved that the achieved results are richer, compared to the results obtained when association rule mining techniques are used over raw data and without context ontology. An increasing number of rules are obtained, in case of association rule mining with context ontology, i.e., with the LMINWS lightweight context ontology in our example domain. That was verified by the same experiment performed but over raw data.

We haven't been able to find a specific approach similar to ours on finding rules related to context by hand of a context-aware ontology. Furthermore, there is to the best of our knowledge no such a study in the domain of water quality monitoring domain and with wireless mobile sensors. Using the same approach, we aim to extend the experiments to other domains such as health in the future. Moreover, we will check with other data mining techniques to divide data into bins, since we believe that it could yield even better results.

References

1. Singh, S., Vajirkar, P., Lee, Y.: Context-based data mining using ontologies. In: Song, I.-Y., Liddle, S.W., Ling, T.-W., Scheuermann, P. (eds.) ER 2003. LNCS, vol. 2813, pp. 405–418. Springer, Heidelberg (2003). doi:10.1007/978-3-540-39648-2_32
2. "Water Quality" (2016). http://www.grc.nasa.gov/WWW/k-12/fenlewis/Waterquality.html. Accessed 24 June 2016
3. (2016) http://depts.alverno.edu/nsmt/archive/SagatClarkNathavong.htm. Accessed 24 June 2016
4. Jajaga, E., Ahmedi, L., Ahmedi, F.: An expert system for water quality monitoring based on ontology. In: Garoufallou, E., Hartley, R.J., Gaitanou, P. (eds.) MTSR 2015. CCIS, vol. 544, pp. 89–100. Springer, Heidelberg (2015). doi:10.1007/978-3-319-24129-6_8
5. Ahmedi, L., Jajaga, E., Ahmedi, F.: An ontology framework for water quality management. In: Proceedings of the 6th International Conference on Semantic Sensor Networks, vol. 1063, pp. 35–50. CEUR-WS (2013)
6. Wu, X., Kumar, V., Quinlan, J.R., Ghosh, J., Yang, Q., Motoda, H., McLachlan, G.J., Ng, A., Liu, B., Philip, S.Y., Zhou, Z.H.: Top 10 algorithms in data mining. Knowl. Inf. Syst. **14**(1), 1–37 (2008)

7. Gruber, T.R.: A translation approach to portable ontology specifications. Knowl. Acquisit. **5**(2), 199–220 (1993)
8. Fayyad, U., Piatetsky-Shapiro, G., Smyth, P.: From data mining to knowledge discovery in databases. AI Mag. **17**(3), 37 (1996)
9. Agrawal, R., Imieliński, T., Swami, A.: Mining association rules between sets of items in large databases. ACM SIGMOD Rec. **22**(2), 207–216 (1993)
10. Agrawal, R., Mannila, H., Srikant, R., Toivonen, H., Verkamo, A.I.: Fast discovery of association rules. Adv. Knowl. Discov. Data Mining **12**(1), 307–328 (1996)
11. Frank, E., Hall, M., Holmes, G., Kirkby, R., Pfahringer, B., Witten, I.H., Trigg, L.: Weka. In: Data Mining and Knowledge Discovery Handbook, pp. 1305–1314. Springer US (2005)
12. "D2rq". d2rq.org/. Accessed 24 June 2016
13. Lavrač, N., Vavpetič, A., Soldatova, L., Trajkovski, I., Novak, P.K.: Using ontologies in semantic data mining with SEGS and g-SEGS. In: Elomaa, T., Hollmén, J., Mannila, H. (eds.) DS 2011. LNCS (LNAI), vol. 6926, pp. 165–178. Springer, Heidelberg (2011). doi: 10.1007/978-3-642-24477-3_15
14. Nebot, V., Berlanga, R.: Finding association rules in semantic web data. Knowl. Based Syst. **25**(1), 51–62 (2012)
15. Abedjan, Z., Naumann, F.: Improving RDF data through association rule mining. Datenbank-Spektrum **13**(2), 111–120 (2013)
16. Ahmedi, L., Sejdiu, B., Bytyçi, E., Ahmedi, F.: An integrated web portal for water quality monitoring through wireless sensor networks. Int. J. Web Portals (IJWP) **7**(1), 28–46 (2015)

General Session: Synthesis of Semantic Models

Semantically Enhanced Virtual Learning Environments Using Sunflower

Daniel Elenius, Grit Denker$^{(\boxtimes)}$, and Minyoung Kim

SRI International, 333 Ravenswood Ave, Menlo Park, CA 94025, USA
grit.denker@sri.com
http://www.sri.com

Abstract. Teaching procedural skills is relevant for a broad range of applications, from IT administration to automotive repair to medical diagnostics. Virtual learning environments reduce the cost, time, and risk, and increase the availability of such training. We introduce ontologies and rules to characterize the objects in the learning environment, and the actions that the user can perform on them. These semantic models are used as the basis for automated reasoning about a student's actions and their effects, and guide automated assessment and feedback to the student. We describe our system and models in the context of weapon skills such as disassembling and assembling a rifle.

1 Introduction

Teaching procedural skills is relevant for a broad range of applications, from IT administration to automotive repair to medical diagnostics. While "learning by doing" approaches are highly effective because learners gain knowledge as they solve problems in the relevant environment, cost, time or risk often make it infeasible to provide learning systems in those environments.

Virtual environments (VEs) are a feasible solution that overcome these limitations while still providing "learning-by-doing"-type of user experiences. They also provide the added benefit of flexible delivery platforms that allow users to learn where and when they want.

To provide learning systems based on VEs, various capabilities and automated tools need to be implemented as part of the VE and provide functionality such as context-aware feedback, personalization to adapt learning content to a student's capabilities, or assessment. Such automated tools promise to make learning systems more effective for the individual student and they would both reduce the cost of using VEs for training and open the door to self-directed learning systems, in which users can acquire procedural skills at their own pace.

Traditional approaches to learning require direct observation by an instructor to provide functionality such as assessment, context-aware feedback or adaptation of learning content. Our approach uses semantic technologies to enable the automation of such functionalities. We have developed a framework called Semantically Enabled Automated Assessment in Virtual Environments (SAVE)

© Springer International Publishing AG 2016
E. Garoufallou et al. (Eds.): MTSR 2016, CCIS 672, pp. 81–93, 2016.
DOI: 10.1007/978-3-319-49157-8_7

which can observe learners operating within an instrumented VE, assess their performance, and provide helpful feedback to improve their skills.

At the core of SAVE is the capability to meaningfully understand what a student is doing in the VE, and the effects of those actions on the environment. Consider a VE for teaching a military student how to disassemble, clean and assemble a weapon. Knowledge that the student clicked on a given screen coordinate has very limited use for assessment. Instead, an understanding of the higher-level semantics of performing that mouse click, e.g., "the student released the charging handle while keeping the bolt catch pressed," is essential. This understanding extends beyond knowledge of what was done, and requires insight into important relationships (e.g., spatial, causal, functional) among the objects in the VE. With this level of characterization, the merits of a particular action can be understood: whether it is at all possible given the current state of the weapon, whether the action has the intended effect (e.g., removing an ammunition cartridge from the chamber), whether the student's actions satisfy the security protocol, and whether the action demonstrates specific domain knowledge.

Furthermore, the system should support exploration, where the student is free to choose among a wide range of actions. Emergent, rather than pre-programmed behavior, is key. We need to be able to base the assessment of students' performance on not only the ability to following exact procedures, but also on whether they achieve a given outcome – possibly in an unanticipated way. Furthermore, describing the behavior of objects in the environment (for example, an M4 rifle) is a task that requires domain expertise – not something that should be left to programmers.

1. Point weapon in safe direction.
2. Attempt to place the selector lever on SAFE. Note: If weapon is not cocked, lever can't be pointed toward safe.
3. Remove the magazine from the weapon, if present.
4. Lock the bolt open.
   (a) Pull the charging handle rearward.
   (b) Press the bottom of the bolt catch.
   (c) Move the bolt forward until it engages the bolt catch.
   (d) Return the charging handle to the forward position.
   (e) Ensure the receiver and chamber are free of ammo.
5. Place the selector lever on safe.
6. Press the upper portion of the bolt catch to allow the bolt to go forward.
7. Place the selector lever from SAFE to SEMI.
8. Squeeze trigger.
9. Pull the charging handle fully rearward and release it, allowing the bolt to return to the full forward position.
10. Place the selector lever on SAFE.

Fig. 1. A procedural task: clearing a rifle.

These considerations motivate an approach using declarative specifications that can be created, modified, re-used, and understood by domain experts – a *semantic* approach. In SAVE, we use ontologies and rules to provide semantic characterizations of objects and actions in the domain.

Though SAVE is applicable to any procedural skills, for the purpose of this paper, we discuss the semantic models and the reasoning for a military domain use case: disassembling, cleaning and reassembling a weapon. This task was of interest to our client and exhibited sufficient real-world complexity to challenge our system. Figure 1 shows the procedure, from [1], for clearing a weapon, which is part of a larger set of skills in this context. Note that, while this is a relatively straightforward procedure, in general, procedural skills can have different variants, optional parts, and so forth.

2   SAVE Overview

SAVE employs various components that generate or make use of semantic models. (1) The *Semantic 3D Annotation Editor* (S3D Editor) allows a 3D content author to associate objects in a 3D model with ontological concepts. The ontological concepts are part of a semantic model that is described in detail in Sect. 4. (2) The *Content Assembly Tool* allows a user to build the training-specific 3D scene. A 3D scene consists of various 3D assets, some with annotations (e.g., the objects with which the student will interact in their learning exercise) and others without annotations (e.g., background objects). (3) The *Exercise UI* serves two purposes. It is used by instructors to record a sequence of actions that will serve as a basis for solutions against which the student will be assessed. Because the VE objects were annotated with semantic classes using the S3D Editor, and the scene was assembled using semantically annotated 3D models, the VE can request actions for VE objects (and their components) from the underlying semantic reasoner and visualize them in the Exercise UI. Once the semantically enhanced virtual training environment has been set up, students use the Exercise UI to attempt to perform the intended tasks. The student sees the 3D objects (e.g., the M4 rifle and its components), and is able to apply generic actions (push, pull, etc.) to them. When the student does so, the action and its parameters (i.e., which components were selected) are communicated to the Flora reasoner, which has the M4 ontology loaded. The reasoner determines the effects of the given action, if any, and updates its KB (knowledge base) accordingly. It then communicates back to the Exercise UI the changes in the state of the environment. The UI uses this information to redraw the 3D components in their updated state. (4) The exercise solution editor shows the action traces to the instructors and allows them to add annotations to capture permissible generalizations to the solution. The generalized solution is the basis for SRI's assessment capability, which is designed to accommodate more open-ended procedural skills for which there can be a range of solutions with significant variations among them. (5) As the student interacts with the VE for a learning task, her actions are recorded as a semantically annotated action trace. The *Automated Assessment* component within SAVE analyzes semantic traces of learner actions against the generalized solution trace and provides contextually relevant feedback.

Details about the automated assessment and user studies for this use case are reported in [3], and the solution editor is described in [6]. This paper focuses on the semantic models used by the system, and the reasoning that happens in the VE at run-time, i.e., while the student is using the system for training.

3   Sunflower Overview

Existing languages like OWL, SWRL, and RIF, and associated editing and reasoning systems, do not support many of the features required for modeling virtual training environments. For example, SWRL does not support n-ary predicates, aggregation or higher order expressions, structured output (such as CSV or XML), or tracing or debugging of reasoning with rules. The Sunflower[1] suite is intended to fill this gap. Sunflower is a set of libraries and tools based on the Flora-2 language[2], which in turn is implemented as a layer on top of XSB[3].

Flora-2 is a highly expressive knowledge representation language and associated reasoning engine developed and maintained primarily by Michael Kifer at Coherent Knowledge Systems. While Flora-2 has its origins in the *logic programming* research community, OWL has its root in *description logics*. Flora supports, among other things, n-ary formulas, negation-as-failure, aggregation, higher-order predicates, functions, frame syntax for classes and instances, infix mathematical expressions, prioritized or default rules, and knowledge base update operators. Flora-2 integrates ontologies and rules in a powerful way.

On top of Flora-2, *Sunflower Foundation* is a library, implemented mostly in Java and partially in C/C++ and Flora itself, which provides many features that are essential to building applications based on Flora rules and ontologies. These features include a Flora parser that generates a detailed syntactic representation of Flora content in Java, syntactic manipulation of that representation, a higher-level ontology model, importers and exporters for other languages (RDF, OWL, SWRL, CSV, SQL, etc.), an interface to the Flora reasoner, a live RDF triple store connector, an explanation module that produces structured explanations of reasoning results to the user, and a natural language module that produces English paraphrases of reasoning results and explanations. The other main components of the Sunflower suite are *Sunflower Studio* – an Eclipse-based IDE for working with Flora-2 content, and *Sunflower Server* – a Web server that exposes much of the Sunflower Foundation functionalities over HTTP using REST APIs. More details on the Sunflower suite can be found in [2]. This paper describes how we use it in the SAVE system to represent and reason about actions in semantic VEs.

---

[1] http://sunflower.csl.sri.com.

[2] http://flora.sourceforge.net/.

[3] http://xsb.sourceforge.net/.

The Flora-2 Language. The authoritative documentation for Flora-2 is its user manual[4]. Here, we give a brief overview of only the features that we use elsewhere in this paper, without precisely defining syntax and semantics.

Terms. Flora identifiers can (optionally) use namespaces and namespace prefixes, as in RDF and OWL. We omit these for readability and space reasons here. There are the usual primitive data values like integers, strings, etc. Data values can be typed, e.g., `"Hello World"^^\string`. The boolean values are written `\true` and `\false`. Lists are written as `[1,2,3]`, optionally with a "tail" part, `[a|b]`. *Functional terms* are written `f(t1,...,tn)`, where the `ti` arguments are themselves terms.

Frames. `A : B` means `A` is an instance of `B`. `A :: B` means `A` is a sub-class of `B`. `A [ p -> V ]` means that `A` has value `V` for property `p` (i.e., this is a *subject, property, object* triple, in RDF terms). We call `[..]` an *instance frame*. `A [| p {m..n} => R |]` means that `A` has range `R` for property `p`, with min-cardinality `m` and max-cardinality `n` (the cardinality part is optional; `m` and `n` are non-negative integers, or `*` for "any"). We call `[|..|]` a *class frame*.

Formulas. Conjunctions of expressions are separated by comma (`,`). Several expressions can be grouped together into one statement, and frame expressions can be nested. For example, `a : A :: B [p -> V, q -> W [r -> Z]] [|p => R|]` is equivalent to `a : A, a :: B, a[p->V], a[q->W], W[r->Z], a[|p => R|]`.

Conjunction can also be written `\and`. Similarly, disjunction uses semi-colon (`;`) or `\or`. There are additional logical operators such as `\if..\then..\else`. There are several types of *negation*, including Prolog-style negation-as-failure, `\+` and Flora's well-founded negation `\naf`. Parentheses can be used to disambiguate operator precedence.

Statements. Flora statements are delimited by a period (`.`). *Rules* have the form `head :- body`, where `head` and `body` are flora expressions which may contain *variables*. Variables start with a question mark, e.g., `?x`, and may be typed using the `^^` notation. Rules may be preceded by a *rule id descriptor*, `@!{R}`, where `R` is a unique name for the rule.

An object-oriented-style *dot notation* can be used as a shortcut for property chains. For example, `a.b.c` refers to the value of `?x` in `a[b->?y[c->?x]]`.

Comments use the Java/C++ style: `//` for single-line comments, and `/* ... */` for multi-line comments.

Flora also has Prolog-style *predicates*, `p(t1,t2,t3)`. Predicates that have side effects are marked as *transactional* by prepending the name with a percent sign, e.g., `%p`.

Examples of operators that cause side effects are the *knowledge base update* operators, including `insert{p}` and `delete{p}`, for inserting and deleting the fact `p` to/from the knowledge base, respectively. The `writeln` predicate can be used to print to the console.

---

[4] http://flora.sourceforge.net/docs/floraManual.pdf.

4   Semantic Models

The main components of the semantic models for SAVE are: an ontology of
components (physical objects) that the student can interact with, rules for creating components (and their sub-components), an ontology of actions that the
student can perform, and rules for performing actions on components. We now
describe each of these in turn, followed by examples of querying these models.
These models and queries were tested by an in-house subject matter expert.

4.1   Component Ontology

In the SAVE scenario, we focused on procedural tasks around the M4 rifle.
Thus, we needed to model the components of this rifle, and their parts structure. Figure 2 shows an exploded component view of the lower half of the rifle.
We modeled the components to the level of detail necessary for the tasks we
were interested in (clearing the rifle, disassembly, cleaning, and assembly). For
other tasks, such as detailed gunsmithing work, a higher level of detail would be
required.

We created a simple ontology to capture the meronomy (parts hierarchy)
of physical objects, with properties like `hasDirectPart` and `hasRegion`. We also
introduced rules to introduce `hasPart` as the transitive closure of `hasDirectPart`,
so that we can reason about nested components.

Next, we introduced the specific classes for the M4's components. There
are about 80 of these classes in our ontology. Each class has sub-properties of
`hasDirectPart` and `hasDirectRegion` to support indication of the correct types

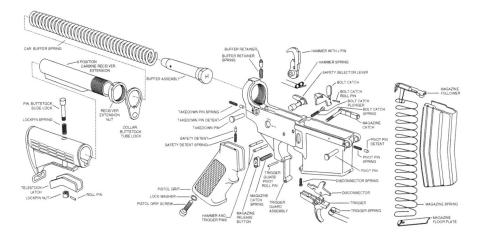

Fig. 2. M4 rifle parts diagram (lower half)

and cardinalities of its sub-components[5]. As an example, the definition of the "lower half" component is:

```
LowerHalf :: PhysicalObject [|
   selector {1..1} => Selector,
   magazine {0..1} => Magazine,
   magazineReleaseButton {1..1} => MagazineReleaseButton,
   hammer {1..1} => Hammer,
   trigger {1..1} => Trigger,
   pivotPin {1..1} => PivotPin,
   takedownPin {1..1} => TakedownPin,
   boltCatch {1..1} => BoltCatch,
   buttStock {1..1} => ButtStock,
   lowerReceiverExtension {1..1} => LowerReceiverExtension,
   bufferRetainer {1..1} => BufferRetainer
|].
```

Note that some of the components may have slightly different names in Fig. 2 due to differences in terminology. The figure shows many more components than the properties of our `LowerHalf` class have. This is primarily because, in our ontology, those components are found under nested sub-components.

The properties `selector`, `magazine`, and so on are all sub-properties of `hasDirectPart`. These all relate to further sub-components, like the `Selector`:

```
Selector :: Switch [| switchPosition {1..1} => SelectorMode |].
```

This component has no further sub-components. Instead, it illustrates another feature of our component classes: the ability to capture the current *state* of the component. The property `switchPosition` indicates the current position of the selector switch. The range class `SelectorMode` is essentially an enumeration of three possible values: `Safe`, `Semi`, and `Burst`. As we shall see, these state properties have essential importance when it comes to modeling the actions that one can perform on the components.

4.2   Component Creation Rules

In our SAVE framework, the student interacts with *instances* of the rifle and its components. Thus, we need to be able to create an instance hierarchy that corresponds to the class-level component hierarchy. Furthermore, we may need several copies of certain components, each with unique identifiers. Doing this manually (or in programming code) is tedious and error-prone. Instead, we define rules which allow us to create component instances, along with all their sub-components. These rules are made possible by Flora's support for *knowledge base update* primitives, which allow us to modify the KB at runtime. We call these rules *constructor rules*, since they are analogous to constructors in object-oriented programming languages. The constructor rule for the `LowerHalf` class is

---

[5] In OWL, one might instead use qualified cardinality restrictions. Other ways of modeling also exist in Flora.

```
@!{CreateLowerHalfRule}
%create(LowerHalf,?lower) :-
  %create(Selector,?selector), %create(Hammer,?hammer),
  %create(Trigger,?trigger), %create(PivotPin,?pivotPin),
  %create(TakedownPin,?takedownPin), %create(BoltCatch,?boltCatch),
  %create(Magazine,?magazine),
  %create(MagazineReleaseButton,?magreleasebutton),
  %create(ButtStock,?buttstock), %create(LowerReceiverExtension,?lre),
  %create(BufferRetainer,?bufferRetainer), %create_name(LowerHalf,?lower),
  insert{ ?lower : LowerHalf [
      selector -> ?selector, hammer -> ?hammer,
      trigger -> ?trigger, pivotPin -> ?pivotPin,
      takedownPin -> ?takedownPin, boltCatch -> ?boltCatch,
      magazine -> ?magazine, magazineReleaseButton -> ?magreleasebutton,
      buttStock -> ?buttstock, lowerReceiverExtension -> ?lre,
      bufferRetainer -> ?bufferRetainer ] }.
```

All the constructor rules use a common %create predicate, which takes two arguments: a component class, and a (resulting) instance object. The rule body has essentially three parts. First, we create all the child components. This step depends on the constructor rules for the sub-components. Secondly, we create a new name for our new component (using the %create_name predicate, which we define elsewhere). Finally, we insert into the KB facts which connect the sub-components to the new top-level component, and assert the type and initial state of the component. Now, we can issue a query, %create(LowerHalf,?x). This query will cause Flora to create a number of new instances, each connected in the appropriate way. The variable ?x will be tied to the top-level instance representing the lower half component itself. Normally, we create the whole rifle in one go, using the top-level M4 component as the first argument to %create.

4.3   Action Ontology

We built a high-level action ontology by adapting the taxonomy in [9] for our needs. The generic actions in our ontology are: Attach, Close, Detach, Extract, Insert, Inspect, Lift, Open, Point, Press, Pull, Push, and Release. Each of these is defined as a class, which is a subclass of the Action class. A specific action that occurs in space and time is considered to be an instance of the corresponding action class. Each action has a fixed set of parameters. These are defined on the action class. For example, the Insert class (here slightly simplified) is defined as:

```
Insert :: Action [
  description ->
    "Insert an object into another object"^^\string,
][|
  thingInserted {1..1} => PhysicalEntity,
  insertedInto {1..1} => PhysicalEntity
|].
```

The action takes two parameters, both of which are physical entities: the thing inserted, and the thing inserted into.

Modeling actions as instances presents us with a problem: We need to create a new instance, and related property assertions, for each individual action that the user takes. This is somewhat cumbersome, especially for testing purposes. Fortunately, Flora has some nice features that provide a solution to this problem. We can define a functional term pattern

```
insert(?_TI,?_II) : Insert [
  thingInserted -> ?_TI, insertedInto -> ?_II
].
```

This allows us to treat functional terms of the form `insert(?x,?y)` as terms, with property value `?x` for `thingInserted` and `?y` for `insertedInto`. We can use such terms directly in queries and rules, without having to explicitly declare a new instance first.

Next, we found that these generic actions were not quite sufficient to model all the intended tasks. At the same time, we did not want to pollute our generic task ontology with very specific tasks. Hence, we introduced a new ontology of "mechanics" actions: `PullAndHold`, `PushAndHold`, `TightenScrew`, `LoosenScrew`, and `SelectSwitchPosition`.

4.4 Action Rules

The final component of our semantic models is the set of *action rules*. These rules describe the preconditions and effects of the different actions, as applied to components of the M4 rifle. This is by far the largest part of our semantic models. As a simple example, we the rule for inserting a magazine is:

```
@!{InsertMagazineRule}
%do(?action^^Insert,?del,?add) :-
  // Action Parameters
  ?action [
    thingInserted -> ?mag^^Magazine,
    insertedInto -> ?lower^^LowerHalf
  ],
  // Preconditions
  ?lower [ magazine -> ?mag [ attached -> \false]],
  // Effects
  ?del = [ ${?mag [attached -> \false]} ],
  ?add = [ ${?mag [attached -> \true]} ],
  %kb_update(?del,?add).
```

Each action rule uses the predicate `%do`, which takes three arguments: the action instance, and two result arguments which we call the *delete-list* and the *add-list*. We will return to these lists shortly. The action variable is typed to the correct type of action (`Insert` in this case). The first part of the rule (*Action Parameters*) retrieves the parameters from the action instance, and checks the types of those arguments. In this case, the value of the `thingInserted` property must have type `Magazine`, and the `insertedInto` must be a `LowerHalf` (this is the part of the rifle that the magazine is inserted into). The second part of the rule is the *Preconditions* part. Here, we can check the state properties on the relevant components, to make sure the action is possible. In this case, we check that the magazine is not already inserted in the rifle. If the preconditions fail, the entire rule fails, and there is no change in the KB. Finally, in the *Effects* part of the rule, we perform the KB updates that represent the change in the world that the action performs. Typically, the KB update modifies the state properties of the components that are involved in the action. The KB updates are performed by a convenience predicate that we introduced (definition not shown here), called `%kb_update`. This predicate takes two arguments: a *delete-list* and an *add-list*. These lists contain the Flora formulas to delete from, and add to, the KB. In the current rule, we simply change the value of the `attached` property on the magazine. These two lists are also returned as result arguments of the entire `%do` predicate, in case the caller needs to know the rule's effects.

For each action rule, we also create a helper predicate that simplifies testing the rule. For example, for the action above:

```
@!{InsertMagazineHelperRule}
insert_magazine(?M4) :- %do(insert(?M4.lower.magazine,?M4.lower),?,?).
```

The action rules are very detailed and some of them get rather complex. Sometimes, the effects of an action are conditional, even after the preconditions have been satisfied. For example, to pull the trigger, the hammer must be cocked, and the selector must not be in the `SAFE` position. The effects of pulling the trigger depend on whether there is: (a) a round in the chamber, (b) a magazine in the magazine well, and (c) additional rounds in the magazine. Because these rules, like the component creation rules, utilize Flora's KB update operations, they are not expressible in less powerful languages such as OWL and SWRL.

4.5 Queries

As mentioned earlier, the action helper predicates can be useful in order to test our action rules. We can also create new predicates that represent sequences of actions, such as the "clearing a rifle" task in Fig. 1:

```
@!{ClearWeaponRule}
%clear_weapon(?M4) :-
  %point_weapon_at_target(?M4,ShootingBerm),
  \+%select_safe(?M4),
  %push_magazine_release_button(?M4),
  %pull_and_hold_charging_handle(?M4),
  %push_and_hold_bolt_catch_bottom(?M4),
  %release_charging_handle(?M4),
  %release_bolt_catch_bottom(?M4),
  %push_charging_handle(?M4),
```

```
%inspect_chamber(?M4),
%select_safe(?M4),
%push_bolt_catch_top(?M4),
%select_semi(?M4),
%pull_and_hold_trigger(?M4),
%pull_and_hold_charging_handle(?M4),
%release_charging_handle(?M4),
%select_safe(?M4).
```

Now, the query %create(M4,?m4), %clear_weapon(?m4) will succeed, and results in changes to the KB corresponding to the actions taken (i.e., the rifle is cleared and in a safe state).

We can also test an individual action rule and examine the add- and delete-lists that are returned. For example, we can execute a query to create a rifle, then load and fire it:

```
%create(M4,?m4), %insert_magazine(?m4),
%pull_and_hold_charging_handle(?m4),
%release_charging_handle(?m4),
%do(pull_and_hold(?M4.lower.trigger),?del,?add)
```

Note that pull_and_hold is a functional term defined using the technique described in Sect. 4.3, to avoid having to instantiate the action. The query results in the following value for ?del (recall that both the delete- and add-lists are lists of reified formulas):

```
[${Magazine_1 [rounds -> [Round_2, ..., Round_30]]},
 ${Round_2 [location -> Magazine_1]},
 ${Round_1 : Round}, ${Round_1 [location -> Chamber_1]},
 ${Round_1 [casing -> Casing_1]},
 ${Trigger_1 [pulled -> \false]}]
```

and ?add:

```
[${Magazine_1 [rounds -> [Round_3, ..., Round_30]]},
 ${Round_2 [location -> Chamber_1]},
 ${Casing_1 [location -> Outside]},
 ${Trigger_1 [pulled -> \true]}]
```

(We have abbreviated the long list of rounds in the magazine here). In other words: the round in the chamber; Round_1 is gone, its casing is in the Outside location (i.e., it is ejected from the rifle); the top round in the magazine, Round_2 is removed from the magazine and now located in the chamber; and the trigger is in the pulled state.

5   Related Work

In [5], the authors develop a "semantic-enabled assessment module" for a 3D environment, and [4] introduces a semantic approach to games, in order to enable more reusability and emergent gameplay. These projects each relate to different parts of the SAVE framework, but it is not clear what kind of semantic representations they use.

The approach of describing actions with preconditions and effects has a long history, dating back to the early days of AI planning systems [7]. These planning representations are typically focused on reasoning about achieving a certain goal state by chaining together a sequence of actions. Our present work, in contrast, *executes* actions selected by a user. More importantly, planning representations are typically specialized for a given domain, and are based on a less expressive logic. The action descriptions in our work have access to a full-featured ontology language.

In [8], we created ontological descriptions of virtual environments. However, the project focused on support for reasoning about simulation fidelity as it relates to large-scale training exercises and simulations. In the current work, we are instead focused on modeling actions and objects on a detailed, individual level.

6   Conclusions and Future Work

We have developed the semantic models necessary for a semantically enhanced virtual learning environment. In a sense, these models constitute a simulation of the M4 rifle. A 3D environment is used to interact with this semantic simulation in order to perform a given procedural task. The steps taken by the student are automatically assessed and compared to the "gold standard" solution. There are several possible directions for future work.

Currently, the Exercise UI allows the user to try any action on any objects. With little to no modifications to our modeling, we could use the semantic models to show a user only the actions that are physically *possible* in a given situation, or the ones that are *allowed, required*, etc. This could help users better understand the environment as well as the task they are supposed to learn. In some contexts it may prove *too* helpful, by telling the student exactly what to do. For actions that are not possible or allowed, we could show explanations of *why* that is the case. This feature could be implemented using Sunflower's tracing and natural language capabilities, described in [2]. It would also be interesting to examine the use of semantics for *discovering* relevant ontologies or classes during the annotation phase. Finally, modeling a second domain would demonstrate the generalizability of our work.

Acknowledgements. This material is based upon work supported by the United States Government under Contract No. W911QY-14-C-0023. Any opinions, findings and conclusions or recommendations expressed in this material are those of the authors and do not necessarily reflect the views of the Government. Development of the Sunflower IDE was funded in part by the U.S. Office of the Assistant Secretary of Defense for Readiness under the Open Netcentric Interoperability for Training and Testing (ONISTT) project, and by TRMC (Test Resource Management Center) T&E/S&T (Test and Evaluation/Science and Technology) Program under the NST Test Technology Area. We are also indebted to the research community for developing and maintaining the open source language and software components on which Sunflower depends, especially Flora-2 (a.k.a. Ergo Lite), XSB Prolog, and InterProlog.

References

1. Soldier's manual of common tasks - warrior skills level 1. Technical report, Headquarters Department of the Army, September 2012
2. Ford, R., Denker, G., Elenius, D., Moore, W., Abi-Lahoud, E.: Automating financial regulatory compliance using ontology+rules and Sunflower. In: Proceedings of SEMANTICS (2016). (to appear)
3. Greuel, C., Myers, K.: Assessment and content authoring in semantically enabled virtual environments. In: Proceedings of Interservice/Industry Training, Simulation and Education Conference (2016). (submitted)
4. Kessing, J., Tutenel, T., Bidarra, R.: Designing semantic game worlds. In: Proceedings of the The Third Workshop on Procedural Content Generation in Games, PCG 2012, ACM, New York, NY, USA (2012). http://doi.acm.org/10.1145/2538528.2538530
5. Maderer, J., Gütl, C., AL-Smadi, M.: Formative assessment in immersive environments: a semantic approach to automated evaluation of user behavior in open wonderland. In: Proceedings of Immersive Education (iED) Summit, June 2013
6. Myers, K., Gervasio, M.: Solution authoring via demonstration and annotation: an empirical study. In: Proceedings of International Conference on Advanced Learning Technologies (2016). (submitted)
7. Nau, D., Ghallab, M., Traverso, P.: Automated Planning: Theory & Practice. Morgan Kaufmann Publishers Inc., San Francisco (2004)
8. Riehemann, S., Elenius, D.: Ontological analysis of terrain data. In: Liao, L. (ed.) ACM International Conference Proceeding Series on COM.Geo, p. 10. ACM (2011)
9. Vujosevic, R., Ianni, J.: A taxonomy of motion models for simulation and analysis of maintenance tasks. Technical report, United States Air Force Armstrong Laboratory, January 1997

Enriching Preferences Using DBpedia and Wordnet

Okan Bursa[(✉)], Özgü Can, Emine Sezer, and Murat Osman Ünalır

Department of Computer Engineering, Ege University,
Bornova, 35100 Izmir, Turkey
{okan.bursa,ozgu.can,emine.sezer,murat.osman.unalir}@ege.edu.tr

Abstract. In Facebook, every like and interest is a preference. A like is an act of acknowledgement, which can be valuable if it is processed rightfully. In this work, the Facebook Page Like preferences of users are captured and these preferences are enriched by matching them with DBpedia and DaKick entities. To semantify these Facebook preferences, the free text search abilities of DBpedia and Allegrograph are used. WordNet is used to find the word similarities between preferences and ontology entities and evaluate the similarity for matching. Matched preferences are stored in FOAF profiles and two new ontologies are presented. SociaLike ontology is created to describe each preference with its properties and connections. Facebook profiles are converted into FOAF profiles and Facebook page information is stored inside the new FacebookAPI ontology. FOAF, SociaLike and FacebookAPI ontologies are used together to define user profile, represent and store user preferences.

Keywords: Preference matching · Ontology mapping · Entity disambiguation · Profile creation · Big data sources

1 Introduction

Personal preferences depend on domain knowledge. Expressing domain knowledge, matching with appropriate concept and saving these preferences as a connected network is a compelling problem for knowledge representation. As the social networks store enormous user data, extracting personal preferences is becoming easy. Even social networks submit APIs to query their data, matching raw data with recommender systems content and making them usable in recommending is still remains as a tough challenge.

In this work, we match Facebook preferences with ontological structures of DBpedia[1] and DaKick[2] using WordNet[3] similarity by creating a semantifying algorithm. DBPedia is a connected conceptual network that is used in various

---

[1] http://wiki.dbpedia.org.
[2] http://www.dakick.com.
[3] https://wordnet.princeton.edu.

© Springer International Publishing AG 2016
E. Garoufallou et al. (Eds.): MTSR 2016, CCIS 672, pp. 94–103, 2016.
DOI: 10.1007/978-3-319-49157-8_8

researches for recommendation and preference representation [1,2]. WordNet is an ontological lexical dictionary for English Words [3]. We have gathered Facebook Page likes of Facebook profiles and matched these preference information with related ontological entities. Facebook profile and Facebook page information are transformed into ontologies and user profiles are created to represent personal information and matched personal preferences. We use Friend of a Friend (FOAF)[4] as a framework to store personal information. All personal information and preferences are connected into two new ontologies; SociaLike and FacebookAPI ontologies.

The paper is organized as follows. Firstly, other preference extraction and enrichment approaches are explained. Secondly, the generation of semantic data used in our research is represented. DaKick website data structure is given in addition to methods to query and use this data are briefly explained. Moreover, how Facebook data is gathered and how to make this data available for our work is given. Thirdly, proposed semantifying algorithm and the architecture to implement this algorithm are described. Additionally, two ontologies those are developed to capture the personal interests are defined. Further, the results of execution of semantifying algorithm are presented. Finally, in conclusion, advantages and disadvantages of our methodology and future works are discussed.

2   Related Work

Preference handling and linking raw data to semantic knowledge is an growing research area [4]. Inside [5], enrichment of user profiles is done by developing a meta ontology. A domain was picked for development and linked with user profiles and preferences. The main struggle for this process is matching the right domain entity for the right ontology description. In order to overcome this problem, Bayesian Networks was used. A Bayesian Network was trained with potential preferences and the similarity between inputs and users was calculated using spread activation technique. Possibility of picking an entity as a preference is decided based on Bayesian Network score. Association rule generation was used for frequently used Bayesian Network paths and user profiles was created by bringing the potential preferences of a user to surface. Although, Bayesian Networks are an excellent technique for discovering the preferences, it is purely based on the probability of the connection between the user and the preference. However, user preference is already there. In this work, the semantic meaning of domain entities was missed and because of low level representations of preferences, connections between domain knowledge and preferences were lost.

Connection between Facebook and Linked data is the subject of many researches. In [6], an intermediate between Facebook Graph API and Linked Data resources is trying to be established. Even the collaboration between Turtle and JSON is complete, there are some issues that Facebook does not support

---

[4] FOAF: www.foaf-project.org.

for external resources. Therefore, domain specific descriptions need to be defined for each specific item inside Facebook Graph.

Describing user preference is an ongoing research area for social network researchers. Most of the preferences are used in social recommender systems such as [7]. Preference extraction is important for solving cold-start problem for collaborative filtering recommenders. Even though, preference modeling and extraction is an content-based filtering method, it has been used inside collaborative filtering recommenders. [7] has similar methodology to represent preferences which are user tags and their types. Hypergeometric distribution is used to estimate a tag is a preference or not. Personal browsing history is used as data for user preference extraction. Sentiment analysis is done for tag extraction. Despite covering all browser information, this method is based on purely statistics and has no collaboration with the meaning of word nor the semantics.

There are also semantic preference extraction researches like [8,9]. In [9], each Facebook user preference is extracted for TV program and preference similarity is calculated by Levenshtein distance to match the user profile and the TV program information. Recommendation is done by this matching ratio and it gave weak results because of low ratio of matching between user profiles and TV programs. The reason of low ratio matching is ignoring the meaning of each word inside TV program and user profile and using no semantic matching for similarity ratios. However, inside [8], each preference is gathered by word singularization and stemming and matched using exact match of Wikipedia Category names and Levenshtein distance. Despite Wikipedia is a good source for concept matching, it does not support automatic mechanisms such as text indexing and matching.

3   Semantic Data

Most of the known data are stored in databases. Databases store no metadata or model to represent its semantic information. Semantic Data is the representation of data in ontology structures for make it available to store and use its overall information. In this work, ontologies are used to represent all data to connect it with other data stores. However, available sources of preferences are mostly represented with database or graph-based structures. For collaboration with personal information, preferences need to be represented as more connected and structured entities. This is achieved by enriching the preferences with the known data stores such as DBpedia and word based graphs such as WordNet. Moreover, most of the preferences were described in instance level with no domain information. By enriching semantically, these preferences are also be connected with domain-based entities. In order to match the preferences of Facebook domain, we used DaKick website data.

3.1 DAKICK Data

DaKick.com.tr is a Turkish event information web site with a recommendation engine support. DaKick was online until April 2014, however since 2012, overall web site database was transformed into a semantic knowledge store. DaKick Data is an ontological data with connection to DBpedia, Schema.org[5], Freebase[6] and BBC[7] data stores. Due to its well spread domain data and well structured ontology, in our work, we used DaKick to enrich Facebook preferences in a semantically manner. DaKick has 3.013.764 triple which is about 335.691 celebrity, 66.961 movie, 2.759 place, 16.654 music band, 47.282 TV program, 5.198 sports team, 4.075 stage artist. In our work, DaKick data is stored in the Allegrograph 4.2[8] for scalability and SPARQL[9] support.

3.2 Facebook Data

Facebook is the most used social network and the biggest online network of all time. After google.com, it is the most visiting web page annually [10]. Facebook creates, stores and uses the biggest personalized data on the Internet. Due to its daily usage and easy access mechanism such as Facebook Graph API[10] and Facebook Query Language(FQL)[11], we used Facebook to gather personal preferences of users. Between 2011 and 2013, we gathered 7129 Facebook user information, 19186 Facebook pages, 63596 Facebook page likes, 17332 personal interest and 71494 URL like.

4 Our Approach

In this chapter, we introduce the enrichment methodology to explain how we match the Facebook preferences with DBpedia and DaKick concepts. Further, we show the general architecture implementing our methodology with semantifying algorithm. At last, we clarify how we map the Facebook Graph API's Page, User and other Facebook tables with FOAF and our two new introduced ontologies. In this work, because of its meta information such as category type, only Facebook page likes are used as a preference.

---

[5] http://schema.org.

[6] https://developers.google.com/freebase/.

[7] http://www.bbc.co.uk/ontologies.

[8] http://franz.com/agraph/allegrograph/.

[9] http://www.w3.org/TR/sparql11-query/.

[10] https://developers.facebook.com/docs/graph-api.

[11] https://developers.facebook.com/docs/reference/fql/.

4.1   Enrichment Methodology

Inside preference matching, each preference is represented as a string of its category and demographic information. Therefore, a preference, p can be represented as a collection of its categories, name and information; $p \epsilon P, c \epsilon C, n \epsilon N$ | $Category\,(p,c) \wedge Name\,(p,n)$. Inside preference matching, there are three types of matching. Exact match is the one-to-one match of the preference name and DaKick name property value. All entities those have the exact DaKick name property value can be matched with the preference;

$$e \epsilon \{DaKick, DBpedia\} \mid$$
$$ExactMatch_{<p,e>} = DaKickName\,(n_p, n_e) \vee Name\,(n_p, n_e)$$

Likewise, Text Match uses the free text functions of DBpedia and Allegrograph. DBPedia supports the string match with *bfi:* prefix similar to *LIKE* operator in SQL. Besides, Allegrograph creates free text indexing for manually created indexes. In this research, a freetext index for DaKick name property values is created as a predicate index. This index was used in Text match such as;

$$e_{dakick} \epsilon DaKick, e_{dbpedia} \epsilon DBpedia$$
$$TextMatch_{<p,e>} = FTIMatch\,(n_p, e_{dakick}) \vee BIFContains\,(n_p, e_{dbpedia})$$

For more complex preferences those do not matched with either of these matching algorithms, we created the similarity match. Similarity match is the matching of preference and its categories with ontology entities by using WordNet's word distance. We used Java WordNet Library and Java WordNet Similarity to calculate the similarity between preference name and DaKick name property value using Cosine and Jaccard distance. This distance determines the actual matching between preference and ontology entity such as;

$$e \epsilon \{DaKick, DBpedia\} \mid SimilarityMatch_{<p,e>} = Cosine\,(n_p, n_e)$$

This calculation results as a similarity rating between $[0, 1]$. In this research, for DaKick dataset, similarity rating above λ is accepted as a match that means the preference is connected to the entity inside DaKick dataset. For DBpedia, γ is accepted as similarity threshold. Both γ and λ thresholds are fixed as half of the exact similarity (0.5) as general acceptance.

4.2   Architecture

Overall architecture can be seen in Fig. 1. Facebook user information is stored within Facebook Platform API[12] and distributed into user, page, friend and page_fan tables by querying FQL Tables. FQL tables are mapped into ontological

---

[12] https://developers.facebook.com/docs/reference/fql/.

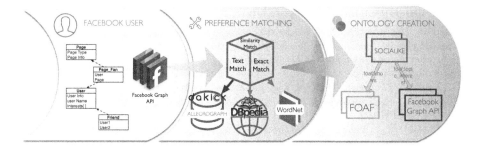

Fig. 1. Architecture

structures using RestFB API[13]. RestFB is a Java based Facebook Graph API supporting OAUTH 2.0 authentication[14]. At the middle, preference matching is done by semantifying algorithm using data DaKick, DBpedia and WordNet. This algorithm is described inside Sect. 4.3. Matched entities and user information is collaborated within the ontology creation part which is explained in Sect. 4.4.

4.3 Semantifying Algorithm

Semantifying is the process of matching personal preferences to exact or related ontological entities. Enrichment methodology is implemented inside overall architecture and user preferences are placed into ontologies. All three matching algorithms are implemented and entities of DaKick and DBpedia are matched with user preferences. Each user preference is linked with related entities from DaKick and DBpedia data stores. If both data stores have related ontological structures with the preference, the best matching entity is decided by ranking of matched entities based on WordNet similarity. If one of data sources has related entities, these entities are accepted as a preference candidate and the first entity is accepted as the matched entity of the preference. If none of data sources have related entities with the preference, we used the type of preference to match with the class definitions of DBpedia and DaKick. Likewise, the first match is accepted as the type of preferences and added as the preference of the person.

4.4 Ontology Mapping

Matched preferences are saved inside our ontology structure from our previous work [11]. In this structure, FOAF definition is extended for preference handling by defining descriptions to connect *foaf:Person* with Preference. In this work, we have created two ontologies to represent Facebook information (Fig. 2). FacebookAPI ontology is used to store Facebook Page information and its matched

---

Algorithm 1. Semantifying Algorithm

---

forall the *user* u *in Facebook Graph* **do**

    forall the *preference* n *in Facebook Graph* **do**

        $query \leftarrow$ `createIndexedQuery`(nName)

        $matchedInterestsDBpediaN \leftarrow$ `getAllRelated`$(query, DBpedia)$;

        $matchedInterestsDaKickN \leftarrow$ `getAllRelated`$(query, daKick)$;

        if $matchedInterestsDaKick_N\ AND\ matchedInterestsDBpedia_N$ **then**

           $bestMatch \leftarrow$

                 `rank`$(matchedInterestsDaKick_N, matchedInterestsDBpedia_N)$

           `addPreference`$(bestMatch)$;

        end

        else if $matchedInterestsDaKick_N$ **then**

           `addPreference`$(matchedInterestsDaKick_N)$

        end

        else if $matchedInterestsDBpedia_N$ **then**

           `addPreference`$(matchedInterestsDBpedia_N)$

        end

        else

           forall the *classes* class *in DaKick* **do**

               if `getWordNetSimilarity`(`getCategory`(n),class) $> \lambda$ **then**

                  `addPreference`(class)

               end

           end

           forall the *classes* class *in DBpedia* **do**

               if `getWordNetSimilarity`(`getCategory`(n),class) $> \gamma$ **then**

                  `addPreference`(class)

               end

           end

        end

    end

end

---

categories. We have mapped the Facebook Graph API's Page table to FacebookAPI ontology. Each Facebook Page is a possible candidate for user preference. Thus, we have created SociaLike ontology to save the overall preference creation, matching and storing. Each preference is defined as a Facebook Page in FacebookAPI ontology and used inside SociaLike ontology by connecting the preference with Person. We have mapped the Facebook User table to FOAF *Person*, Facebook Friend table to FOAF *knows* and Facebook Page_fan table to FOAF *topic_interest*. Each preference is defined as a Facebook Page inside FacebookAPI ontology, connected by using SociaLike ontology and connected with *topic_interest* to FOAF *Person*.

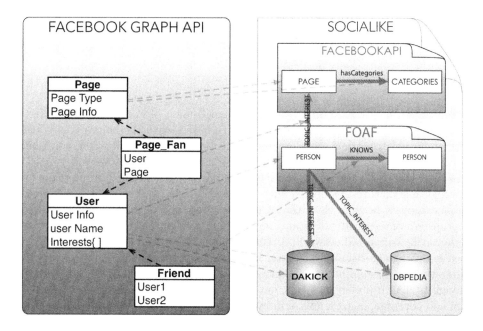

Fig. 2. Ontology creation

5 Implementation

Implementation is done inside a Windows 10 machine with Ubuntu 10.1 Virtual Machine running AllegroGraph version 4.1. DaKick data store is queried from virtual machine and DBpedia is queried by online DBpedia SPARQL endpoint[15]. Due to DBpedia Endpoint's query limit, we experimented with 581 Facebook Page Likes. Facebook Demographic information is gathered but discarded in matching algorithm as it stores no meta information to be matched with.

From all 581 preferences captured from Facebook Graph API, 433 preferences are matched with their suitable equivalent inside data stores. The enriching ratios of each matching algorithm can be seen in Table 1. Each matching factor is making the similarity ratio better. However, a small set of user preference was

Table 1. Matching success of methods

Matching algorithm	Matching success
Exact match	27 %
Exact and text match	52 %
Exact, text and similarity match	74.5 %

---

[15] DBpedia Endpoint: http://dbpedia.org/sparql.

semantically matched due to DBpedia's online query limit. This problem will be fixed with querying DBpedia set offline in further research.

6   Conclusion

In this paper, a new method for preference enrichment of Facebook user preferences using DBpedia and DaKick ontology resources with the help of WordNet is introduced. To enrich user preferences, a new algorithm with collaboration of Exact, Text and Similarity Match methods is developed. Moreover, to capture each matched preference, the Facebook Graph API tables are mapped to ontological structures connected with FOAF. By doing this, overall enrichment methodology is become clear that Facebook preferences can be matched with appropriate ontological entity and represented with ontological structures in a semantically connected network. As the developed algorithm is examined, adding each matching algorithm gives better results with an incremental success rate. It shows that in a social network, representation of preferences as raw data is not enough to match preferences with ontological structures. For better representation and coverage, using a lexical network for word matching and a semantically connected network with domain knowledge is needed. As a future work, for better similarity success rate, similarity thresholds will be determined based on the dataset structure. Moreover, DBpedia dataset will be used offline for full similarity calculation of Facebook User Data. Later, enriched user preferences will be used to calculate the similarity between Facebook users. Further, this similarity rate and enriched preferences will be jointly used to increase the recommendation systems' success. Likewise, user preferences inside other social networks such as Twitter[16] and Foursquare[17] will be enriched with addition of other data sources like Freebase[18], Knowledge Vault [12] and YAGO [13]. By doing this, all user preferences of different social networks will be represented in a single user profile.

References

1. Mirizzi, R., Di Noia, T., Ragone, A., Claudio Ostuni, V., Di Sciascio, E.: Movie recommendation with DBpedia. In: CEUR Workshop Proceedings, vol. 835, pp. 101–112 (2012)
2. Paulo Leal, J., Rodrigues, V., Queirós, R.: Computing semantic relatedness using DBPedia. In: 1st Symposium on Languages, Applications and Technologies (SLATE 2012), pp. 133–147 (2012)
3. Miller, G.A.: WordNet: a lexical database for English. Commun. ACM **38**(11), 39–41 (1995)
4. Bizer, C., Heath, T., Berners-Lee, T.: Linked data - the story so far. Int. J. Semant. Web Inf. Syst. **5**(3), 1–22 (2009)

---

[16] http://www.twitter.com.
[17] http://www.foursquare.com.
[18] https://developers.google.com/freebase/.

5. Eyharabide, V., Amandi, A.: Ontology-based user profile learning. Appl. Intell. **36**(4), 857–869 (2012)
6. Weaver, J., Tarjan, P.: Facebook linked data via the graph API. Semant. Web **4**(3), 245–250 (2013)
7. Sato, T., Fujita, M., Kobayashi, M., Ito, K.: Recommender system by grasping individual preference and influence from other users. In: Proceedings of the IEEE/ACM International Conference on Advances in Social Networks Analysis and Mining - ASONAM 2013, pp. 1345–1351 (2013)
8. Cantador, I., Castells, P., Bellog'ın, A.: An enhanced semantic layer for hybrid recommender systems: application to news recommendation. Int. J. Semant. Web Inf. Syst. **7**(1), 44–78 (2011)
9. Krauss, C., Braun, S., Arbanowski, S.: Preference ontologies based on social media for compensating the cold start problem. In: Proceedings of the 8th Workshop on Social Network Mining, Analysis, SNAKDD 2014, New York, NY, USA, pp. 12:1–12:4. ACM (2014)
10. Wittkower, D.E.: Facebook and Philosophy: What's on Your Mind?. Popular Culture and Philosophy. Open Court, Chicago (2010)
11. Bursa, O., Sezer, E., Can, O., Unalir, M.O.: Using FOAF for interoperable and privacy protected healthcare information systems. In: Closs, S., Studer, R., Garoufallou, E., Sicilia, M.-A. (eds.) MTSR 2014. CCIS, vol. 478, pp. 154–161. Springer, Heidelberg (2014). doi:10.1007/978-3-319-13674-5_15
12. Dong, X., Gabrilovich, E., Heitz, G., Horn, W., Lao, N., Murphy, K., Strohmann, T., Sun, S., Zhang, W.: Knowledge vault: a web-scale approach to probabilistic knowledge fusion. In: Proceedings of the 20th ACM SIGKDD International Conference on Knowledge Discovery and Data Mining, KDD 2014, New York, NY, USA, pp. 601–610. ACM (2014)
13. Suchanek, F.M., Kasneci, G., Weikum, G.: Yago: a core of semantic knowledge. In: Proceedings of the 16th International Conference on World Wide Web, WWW 2007, New York, NY, USA, pp. 697–706. ACM (2007)

RDF Data in Property Graph Model

Dominik Tomaszuk[✉]

Institute of Informatics, University of Bialystok, Białystok, Poland
d.tomaszuk@uwb.edu.pl

Abstract. This paper proposes a formalization of the Property Graphs (PG) model, which now does not have a commonly agreed-upon formal definition. The paper shows how to store Resource Description Framework (RDF) triples in the form that can be easily processable in PG databases. We propose methods for mapping from one model to another. This is important because of existing many graph databases, in which we enable to load RDF data. Moreover, we propose a new serialization, called YARS, for RDF that is compatible with PG solutions.

1 Introduction and Motivations

Graphs are useful in understanding a wide variety of datasets in areas such as government, science, social network, life sciences, media and geographic. The real world is interlinked. In some parts it is uniform, in others it is irregular. Such specificity can be easily represented precisely by graphs. There are two main models that allow it: Property Graphs (PG) model and Resource Description Framework (RDF) model.

This paper show how to how to store the RDF triples in the form that can be easily processable in PG databases. In this paper we propose methods for interoperability between PG and RDF data stores. Our proposals contribute to enable a user who is familiar with PG databases to load and access RDF data. To accomplish this, we proposed a new serialization for PG databases, which is compatible with RDF.

The paper is constructed according to sections. Section 2 is devoted to related work. Section 3 presents RDF concepts. In the Sect. 4 we formalize PG data model. Section 5 proposes a new RDF serialization, which complies PG model. In Sect. 5 we introduce an algorithm for transforming RDF graphs to property graphs and show an example of our serialization. Section 6 gives detailed results of our implementation and experiments. The paper ends with conclusions.

2 Related Work

In this section we present serializations and data stores from the Property Graphs area and RDF area.

© Springer International Publishing AG 2016
E. Garoufallou et al. (Eds.): MTSR 2016, CCIS 672, pp. 104–115, 2016.
DOI: 10.1007/978-3-319-49157-8_9

2.1   Serializations

In the Property Graphs area there are a few solutions for serializing graphs. It may be divided into two groups: ones that uses XML and ones that is text-based. The first group can be distinguished to GraphML [2] and DotML[1]. Unfortunately, XML does not allow certain characters in attributes, which can be used in RDF. The second group includes GraphSON[2] that uses JSON syntax and GML [11] that uses a hierarchical textual file format. Both formats have some limitations. GraphSON holds vertices and edges in different places, which is difficult to read for humans. GML supports only a 7-bit ASCII characters.

On the other hand there are a few solutions for RDF serialization. It may be divided into three groups: Turtle-family languages, XML-based serializations and JSON-based serializations. The first group includes Turtle [6], N-Triples [20], TriG [21] and N-Quads [5]. RDF/XML [8] and RDFa [1] are the most importent ones in the second group. The third group includes JSON-LD [14] and RDF/JSON [22,23]. Unfortunately, none of these syntaxes does not match the property graph databases. Turtle* [10] extends the Turtle grammar and can support property graphs, but this proposal extends beyond RDF standard and does not have many implementations.

There are some papers [9,12,19] that formalize some parts of PG model. In [9] Hartig proposes a formalization of the PG model and introduces transformations between PGs and RDF* [10]. In [12] Jouili et al. propose another definition of PG based on Blueprints[3]. In [19] Schätzle et al. present a formalization of PG in the RDF context.

2.2   Data Stores

There are a few data stores in Property Graph world [12,13]. Neo4j [13] is native graph database purpose-built to leverage not only data but also its relationships. It uses Cypher and Gremlin as well. Titan [12] is another graph data store that is distributed and transactional. It supports Gremlin query language. Dex/Sparksee[4] is yet another data store that uses Gremlin.

On the other hand there a lot of RDF data stores [4,16,17]. Jena [16] is a framework that supports SPARQL. It allows store RDF in a memory and in a relational database. Sameas [4] is yet another framework for querying and analyzing RDF data. RDF-3X is a implementation of SPARQL that uses a relational database to store RDF triples. Other RDF store proposals are discussed in [15].

There are also data stores that support Property Graph and RDF: Oracle database [7] and Bigdata/Blazegraph [9].

---

[1] http://martin-loetzsch.de/DOTML/.

[2] https://github.com/tinkerpop/blueprints/wiki/GraphSON-Reader-and-Writer-Library.

[3] https://github.com/tinkerpop/blueprints/wiki.

[4] http://sparsity-technologies.com/.

3 RDF Basics

The RDF data model rests on the concept of creating web-resource statements in the form of subject-predicate-object expressions, which in the RDF terminology, are referred to as *triples* (or *statements*). Following [24], we provide definitions of RDF triples below. The elemental constituents of the RDF data model are RDF terms that can be used in reference to resources: anything with identity. The set of RDF terms is divided into three disjoint subsets: IRIs, literals, and blank nodes.

Definition 1 (IRIs). IRIs *serve as global identifiers that can be used to identify any resource.*   □

Definition 2 (Literals). Literals *are a set of lexical values.*   □

Definition 3 (Blank nodes). Blank nodes *are defined as existential variables used to denote the existence of some resource for which an IRI or literal is not given.*   □

Definition 4 (RDF triple). *An* RDF triple t *is defined as a triple* $t = \langle s, p, o \rangle$ *where* $s \in \mathcal{I} \cup \mathcal{B}$ *is called the* subject, $p \in \mathcal{I}$ *is called the* predicate *and* $o \in \mathcal{I} \cup \mathcal{B} \cup \mathcal{L}$ *is called the* object. $\mathcal{I}$ *is the set of all Internationalized Resource Identifier (IRI) references,* $\mathcal{B}$ *an infinite set of blank nodes,* $\mathcal{L}$ *the set of RDF literals.*   □

Example 1. The example presents an RDF triple consisting of subject, predicate and object.
$\langle$`http://example.net/me#j`,`foaf:name`,`John Smith`$\rangle$

A collection of RDF triples intrinsically represents a labeled directed multi-graph. The nodes are the subjects and objects of their triples. RDF is often referred to as being *graph structured data* where each $\langle s, p, o \rangle$ triple can be interpreted as an edge $s \xrightarrow{p} o$.

Definition 5 (RDF graph). *Let* $\mathcal{O} = \mathcal{I} \cup \mathcal{B} \cup \mathcal{L}$ *and* $\mathcal{S} = \mathcal{I} \cup \mathcal{B}$, *then* $G \subset \mathcal{S} \times \mathcal{I} \times \mathcal{O}$ *is a finite subset of* RDF triples, *which is called* RDF graph.   □

Example 2. The example in Fig. 1 presents an RDF graph of a FOAF[5] profile in Turtle syntax. This graph includes the following elements:

```
1  @prefix rdf: <http://www.w3.org/1999/02/22-rdf-syntax-ns#>.
2  @prefix foaf: <http://xmlns.com/foaf/0.1/>.
3  <http://example.org/p#j> rdf:type     foaf:Person.
4  <http://example.org/p#j> foaf:name  "John Smith".
```

Definition 6 (RDF data store). *An* RDF data store *is any storage system that uses RDF graphs to represent data.*   □

---

[5] http://xmlns.com/foaf/spec/.

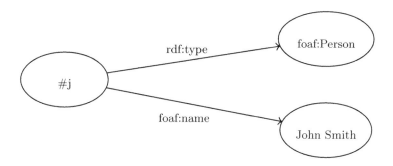

Fig. 1. An RDF graph with two triples

4   Formalization of the PG Model

The PG data model rests on the concept of creating directed and key/value-based graphs. It means that there is a tail and head to each edge and both vertices and edges can have properties associated with them. Following [18], a property graph has the following characteristics:

– A property graph contains vertices[6] and edges[7].
– Vertices can be labeled with one or more labels.
– Vertices contain key-value pairs called properties.
– Edges are named and directed.
– Edges have a start and end vertices.
– Edges can also contain properties.
– Properties are in the form of arbitrary key-value pairs.
– The keys are strings and the values are arbitrary datatypes.

Following above characteristics we provide formal definition below.

Definition 7 (Property Graph). *A* Property Graph *is a tuple*
$PG = \langle V, E, S, P, h_e, t_e, l_v, l_e, p_v, p_e \rangle$, *where:*

1. *V is a non-empty set of vertices,*
2. *E is a set of edges,*
3. *S is a set of strings,*
4. *P contains each properties that has a form $p = \langle k, v \rangle$, where $k \in S$ and $v \in S$,*
5. *$h_e : E \to V$ is a function which yields the source of each edge (head),*
6. *$t_e : E \to V$ is a function which yields the target of each edge (tail),*
7. *$l_v : V \to S$ is a function mapping each vertex to label,*
8. *$l_e : E \to S$ is a function mapping each edge to label,*
9. *$p_v : V \to 2^P$ is a function used to assign vertices to their multiple properties.*
10. *$p_e : E \to 2^P$ is a function used to assign edges to their multiple properties.*   □

---

[6] Another name for a vertex is a node.
[7] Another name for an edge is an arc.

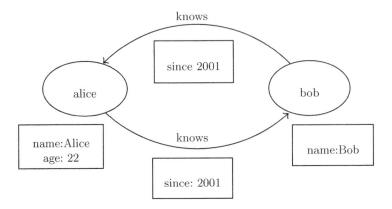

Fig. 2. A property graph with two vertices and two edges

Note that $\langle V, E, h_e, t_e, l_e \rangle$ is an edge-labeled directed multigraph. Properties can be implemented as an associative array that is an unordered list of attributes with associated values.

Example 3. The example in Fig. 2 presents a Property Graph. This graph includes the following elements:

- $S = \{name, Alice, Bob, age, 22, since, 2001, knows, alice, bob\}$,
- $V = \{v_1, v_2\}$,
- $p_v(v_1) = \{\langle name, Alice \rangle, \langle age, 22 \rangle\}$,
- $p_v(v_2) = \{\langle name, Bob \rangle\}$,
- $l_v(v_1) = alice$,
- $l_v(v_1) = bob$,
- $E = \{e_1, e_2\}$,
- $h_e(e_1) = bob$,
- $t_e(e_1) = alice$,
- $l_e(e_1) = knows$,
- $p_e(e_1) = \{\langle since, 2001 \rangle\}$
- $h_e(e_2) = alice$,
- $t_e(e_2) = bob$,
- $l_e(e_2) = knows$,
- $p_e(e_2) = \{\langle since, 2001 \rangle\}$.

Definition 8 (Property graph data store). *A property graph data store is any storage system that uses property graph structures with vertices, edges, and properties to represent data.* □

5   Serializing RDF in Property Graphs Style

In section we present RDF serialization in PG style and propose algorithm that transform RDF to our serialization.

We propose Yet Another RDF Serialization (YARS), which allow prepare RDF data to exchange on the property graph data stores. Our serialization is textual. It has three different parts:

1. prefix directives – a part where prefixes are defined,
2. vertex declarations – parts where vertices are created,
3. relationship declarations – parts where edges and properties are created.

Prefix directives should be written in the starting lines. Vertex and relationship can be defined in different places. Values of subjects and objects are stored in vertex properties. The same vertices have the same names. Predicates are edge labels.

Example 4. The example presents YARS serialization that represents the same triples as in Example 2. This property graph includes the following triples:

```
1   :rdf: <http://www.w3.org/1999/02/22-rdf-syntax-ns#>
2   :foaf: <http://xmlns.com/foaf/0.1/>
3   (a {value:<http://example.org/p#j>})
4   (b {value:<http://xmlns.com/foaf/0.1/Person>})
5   (a)-[:rdf:type]->(b)
6   (c {value:"John Smith"})
7   (a)-[:foaf:name]->(c)
```

We also provide a method for transforming an RDF graph to our serialization. At the input our poposal requires RDF graph in the abstract syntax so there

---

input : RDF Graph G
output: YARS Y
1 $P \leftarrow \varnothing$;
2 **foreach** $g \in G$ **do**
3       $s \leftarrow \mathrm{subj}(g)$;
4       $p \leftarrow \mathrm{pred}(g)$;
5       $o \leftarrow \mathrm{obj}(g)$;
6       $(p_{id}, p_{name}) \leftarrow \mathrm{generatePrefix}(p)$;
7   **if** $p_{id} \notin P$ **then**
8           $\lfloor$ addPrefix(p_{id}, P);
9       $s_{id} \leftarrow \mathrm{hash}(s)$ ◁   md5(), sha512(), ...;
10      $o_{id} \leftarrow \mathrm{hash}(o)$ ◁   md5(), sha512(), ...;
11      $s_{rel} \leftarrow \mathrm{createVertex}(s_{id}, s)$;
12      $o_{rel} \leftarrow \mathrm{createVertex}(o_{id}, o)$;
13      $Y \leftarrow \mathrm{createRel}(s_{rel}, p_{name}, o_{rel})$;
14 moveBackToBeginning(Y);
15 $Y \leftarrow \mathrm{addPrefixes}(P)$;
16 **return** Y;

Algorithm 1: YARS generation

is no need to provide specific RDF serialization. Algorithm 1 presents creation of YARS serialization. The algorithm takes subject (`subj()` function), predicate (`pred()` function) and object (`obj()` function) from RDF graph and divides a predicate into two parts. The first part is used to shorten IRI. The second part is an edge label. Hash strings of subject and object are vertex names with values in properties. The next step is vertices (`createVertex()` function) and relationships (`createRel()` function) creation.

YARS can have more than one possible representation in the syntax level. For example vertex declarations and relationship declarations can be mixed with each other. This feature is desirable for humans, because of readability. To easier processing by property graph data stores, we also propose a canonical form of our serialization. A canonical YARS (YARSC) has the following additional constraints:

- prefix directives do not exist, all IRI are stored in the absolute form,
- edges have a key called `iriref` and a value, which is vocabulary IRI or ontology IRI namespace,
- edge labels have predicate name without prefix,
- vertex declarations should be at the top of the file,
- relationship declarations should be at the bottom of the file.

The grammar (see Appendix A) for the language is the same. The new feature is a `iriref` key, which define vocabulary IRI or ontology IRI namespace for edge label.

Example 5. The example presents a canonical YARS serialization that is equivalent to YARS in Example 4.

```
1  (a {value:<http://example.org/p#j>})
2  (b {value:<http://xmlns.com/foaf/0.1/Person>})
```

> **input** : YARS Y
> **output:** Canonical YARS Y_c
> 1 $P \leftarrow$ getPrefixes(Y)   ◁   prefixes structure;
> 2 $Y_c \leftarrow$ removePrefixDirectives(Y) ;
> 3 **foreach** $y \in Y_c$ **do**
> 4   **if** y *is vertex declaration* **then**
> 5 $D_v \leftarrow$ addVertexDeclaration(y) ;
> 6   **else**
> 7 $D_r \leftarrow$ addRelationshipDeclaration(y) ;
> 8   **foreach** $p \in P$ **do**
> 9 $Y_c \leftarrow$ ContextEnrich(y, p) ;
> 10 $Y_c \leftarrow$ removePrefix(y, p) ;
> 11 $Y_c \leftarrow$ createCanonical(D_v, D_r) ;
> 12 **return** Y_c;

Algorithm 2: YARS canonicalization

```
3   (c {value:"John Smith"})
4   (a)-[type {iriref:<http://www.w3.../22-rdf-syntax-ns#>}]->(b)
5   (a)-[name {iriref:<http://xmlns.com/foaf/0.1/>}]->(c)
```

Algorithm 2 presents canonicalization of YARS serialization. At the input the algorithm requires YARS. In the first step prefix directives are removed. In the second step the algorithm divide content into vertex declarations (`addVertexDe-claration()` function) and relationship declarations (`addRelationshipDeclar-ation()` function). In the next step edges are enriched with a property consists of a vocabulary. In this step label is devoided of prefix. The last is vertex declarations and relationship declarations merging (`createCanonical()` function). YARS canonicalization has a worst-case space complexity $O(|Y| \cdot |P|)$.

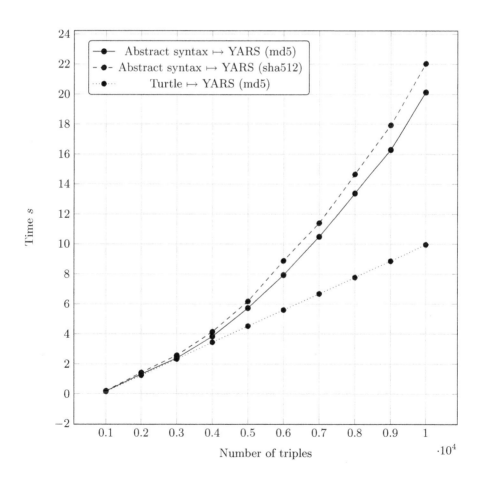

Fig. 3. The YARS generation time result

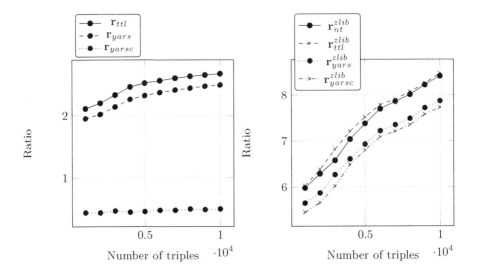

Fig. 4. Format size ratios

6   Implementation and Evaluation

In this Section we evaluate the creating YARS based on our implementation including Algorithms 1 and 2. All experiments have been executed on a Intel Core i7-4770K CPU @ 3.50 GHz (4 cores, 8 thread), 8 GB of RAM (clock speed: 1600 MHz), and a HDD with reading speed rated at ∼160 MB/sec[8]. We have been used Linux Mint 17.3 Rosa (kernel version 3.13.0) and Python 3.4.3 with RDFLib 4.2.1[9].

To test our serialization we implemented N-Triples generator and transformation tool into YARS and Turtle. We prepare 10 datasets in YARS, N-Triples and Turtle. The YARS generation times are presented in Fig. 3. We consider two version of the implementation with MD5 and SHA512 algorithms. The plot shows the arithmetic mean encoding time from 10 runs. It presents that times are nearly quadratic to the number of RDF triples in both cases. In this case we assume that we do not know how many prefixes should be shorten so we use RDF graph abstract syntax. The results can be improved while we consider specific RDF serialization with prefixes at the file beginning i.e. Turtle[10].

In the next step we tested serialization file size. We define size ratio r_y^x as $\frac{nt\_size}{y\_size}$, where nt_size is the size of an N-Triples file and y_size is the size of Turtle and YARS files. We also test ZLib[11] compressed serializations. Figure 4 analyzes size ratio of YARS and YARSC. It shows that plain YARS serialization

---

[8] We test it in `hdparm -t`.

[9] https://github.com/RDFLib/rdflib.

[10] If we assume that this serialization has all prefixes are shorten.

[11] http://www.zlib.net/.

(r_{yars}) has similar ratios to Turtle r_{ttl} and better ratios compared to N-Triples. In compressed serialization YARS results (r_{yars}^{zlib}) are are similar to N-Triples (r_{nt}^{zlib}) and Turtle (r_{ttl}^{zlib}). Both YARSC (r_{yarsc}) and YARSC with ZLib compression (r_{yarsc}^{zlib}) have the worst ratios, because they have additional data included to make processing easier and faster.

7   Conclusions and Future Work

Graphs are used in many areas of our lives. There are two main graph models: RDF model and PG model. The first one is well studied and formalized. This paper proposes a formalization of the PG model. Moreover, we present how to store RDF triples in the form that can be easily processable. We propose a new serialization for RDF that is compatible with PG databases and based on Cypher syntax.

Future work will focus on preparing algorithms for mapping SPARQL into PG query languages i.e. Cypher and Gremlin. Another challenges is reducing repeated nodes from our serialization to reduce the size of a document and speed up processing. It should also be considered supporting RDF named graphs.

Acknowledgements. The author gratefully acknowledges the members of the Neo4j team. We thank Olaf Hartig for comments that greatly improved the paper.

A   Appendix: YARS Grammar

In this appendix we present the grammar of YARS in EBNF [3].

```
1   doc          ::= elem*
2   elem         ::= directive | declaration
3   directive    ::=":" alnum ":" S "<" alnum ">"
4   declaration  ::= vertex | rel
5   prop         ::="{" S alnum ":" S alnum S"}"
6   vertex       ::= alnum prop
7   node         ::="(" (vertex | alnum) ")"
8   rel          ::= node "-[" alnum ( prop )+ "]->" node
9   alnum        ::= (ALPHA | DIGIT | "_")+
10  S            ::= (#x20 | #x9 | #xD | #xA)+
```

References

1. Adida, B., Birbeck, M., McCarron, S., Herman, I.: RDFa Core 1.1 - Third Edition. W3C recommendation, World Wide Web Consortium, March 2015. http://www.w3.org/TR/2015/REC-rdfa-core-20150317/
2. Brandes, U., Eiglsperger, M., Herman, I., Himsolt, M., Scott Marshall, M.: GraphML progress report structural layer proposal. In: Mutzel, P., Jünger, M., Leipert, S. (eds.) Graph Drawing. LNCS, vol. 2265, pp. 501–512. Springer, Heidelberg (2002). doi:10.1007/3-540-45848-4_59

3. Bray, T., Paoli, J., Sperberg-McQueen, C.M., Maler, E., Yergeau, F.: EBNF Notation. W3C recommendation, World Wide Web Consortium, November 2008. https://www.w3.org/TR/2008/REC-xml-20081126/#sec-notation

4. Broekstra, J., Kampman, A., Harmelen, F.: Sesame: a generic architecture for storing and querying RDF and RDF schema. In: Horrocks, I., Hendler, J. (eds.) ISWC 2002. LNCS, vol. 2342, pp. 54–68. Springer, Heidelberg (2002). doi:10.1007/3-540-48005-6_7

5. Carothers, G.: RDF 1.1 N-Quads. W3C recommendation, World Wide Web Consortium, February 2014. http://www.w3.org/TR/2014/REC-n-quads-20140225/

6. Carothers, G., Prud'hommeaux, E.: RDF 1.1 turtle. W3C recommendation, World Wide Web Consortium, February 2014. http://www.w3.org/TR/2014/REC-turtle-20140225/

7. Das, S., Srinivasan, J., Perry, M., Inseok Chong, E., Banerjee, J.: A tale of two graphs: property graphs as RDF in oracle. In: EDBT, pp. 762–773 (2014)

8. Gandon, F., Schreiber, G.: RDF 1.1 XML syntax. W3C recommendation, World Wide Web Consortium, February 2014. http://www.w3.org/TR/2014/REC-rdf-syntax-grammar-20140225/

9. Hartig, O.: Reconciliation of RDF*, property graphs. arXiv preprint arXiv:1409.3288 (2014)

10. Hartig, O., Thompson, B.: Foundations of an alternative approach to reification in RDF. arXiv preprint arXiv:1406.3399 (2014)

11. Himsolt, M.: GML: a portable graph file format, Universität Passau (1997). http://www.fmi.uni-passau.de/graphlet/gml/gml-tr.html

12. Jouili, S., Vansteenberghe, V.: An empirical comparison of graph databases. In: Proceedings of Social Computing (SocialCom), pp. 708–715. IEEE (2013)

13. Lal, M.: Neo4j Graph Data Modeling. Packt Publishing, Birmingham (2015)

14. Lanthaler, M., Sporny, M., Kellogg, G.: JSON-LD 1.0. W3C recommendation, World Wide Web Consortium, January 2014. http://www.w3.org/TR/2014/REC-json-ld-20140116/

15. Martínez-Bazan, N., Muntés-Mulero, V., Gómez-Villamor, S., Nin, J., Sánchez-Martínez, M.-A., Larriba-Pey, J.-L.: Dex: high-performance exploration on large graphs for information retrieval. In: Proceedings of the Sixteenth ACM Conference on Information and Knowledge Management, pp. 573–582. ACM (2007)

16. McBride, B.: Jena: implementing the RDF model and syntax specification. In: Proceedings of SemWeb (2001)

17. Neumann, T., Weikum, G.: RDF-3X: a RISC-style engine for RDF. Proc. VLDB Endow. **1**(1), 647–659 (2008)

18. Robinson, I., Webber, J., Eifrem, E.: Graph Databases. O'Reilly Media Inc., California (2013)

19. Schätzle, A., Przyjaciel-Zablocki, M., Berberich, T., Lausen, G.: S2X: graph-parallel querying of RDF with graphX. In: Wang, F., Luo, G., Weng, C., Khan, A., Mitra, P., Yu, C. (eds.) Big-O(Q)/DMAH -2015. LNCS, vol. 9579, pp. 155–168. Springer, Heidelberg (2016). doi:10.1007/978-3-319-41576-5_12

20. Seaborne, A., Carothers, G.: RDF 1.1 N-triples. W3C recommendation, World Wide Web Consortium, February 2014. http://www.w3.org/TR/2014/REC-n-triples-20140225/

21. Seaborne, A., Carothers, G.: RDF 1.1 triG. W3C recommendation, World Wide Web Consortium, February 2014. http://www.w3.org/TR/2014/REC-trig-20140225/

22. Tomaszuk, D.: Flat triples approach to RDF graphs in JSON. In: Proceedings of W3C Workshop - RDF Next Steps. World Wide Web Consortium (2010)
23. Tomaszuk, D.: Named graphs in RDF/JSON serialization. Zeszyty Naukowe Politechniki Gdańskiej 273–278 (2011)
24. Wood, D., Lanthaler, M., Cyganiak, R.: RDF 1.1 concepts and abstract syntax. W3C recommendation, World Wide Web Consortium, February 2014. http://www.w3.org/TR/2014/REC-rdf11-concepts-20140225/

Track on Agriculture, Food and Environment

Arguing About End-of-Life of Packagings: Preferences to the Rescue

Bruno Yun[1], Pierre Bisquert[2]([✉]), Patrice Buche[2], and Madalina Croitoru[1]

[1] LIRMM, University of Montpellier, Montpellier, France
[2] IATE, INRA, Montpellier, France
bisquert@supagro.inra.fr

Abstract. Argumentation methods and associated tools permit to analyze arguments against or in favor of a set of alternatives under discussion. The outputs of the argument methods are sets of conflict-free arguments collectively defending each other, called extensions. In case of multiple extensions, it is often difficult to select one out of many alternatives. We present in this paper the implementation of an complementary approach which permits to filter or rank extensions according to the expression of preferences. Methods and tools are illustrated on a real use case in food packagings. The aim is to help the industry choose among different end-of-life possibilities by linking together consumer behavior insights, socio-economic developments and technical properties of packagings. The tool has been used on a real use-case concerning end-of-life possibilities for packagings.

1 Introduction

Communication is a pillar of our society, humans have always been concerned with debating and arguing as it constitutes a great part of our daily interactions. Argumentation dialogues are of important effect on our lives as it is implied in debates and decision-making. It is within such argumentation dialogues that opinions of different stakeholders are confronted against each other and arguments are advanced to support them [8]. One can then extract the several coherent viewpoints from an argumentation framework called extensions (sets of conflict-free arguments collectively defending each others). For a more detailed formalization of arguments other than the one of abstract argumentation, one can take the road of structured argumentation where the construction of arguments is based on a formal language. In this approach, arguments have a specific structure and attacks are defined with respect to this structure.

Within the framework of the European project EcoBioCap [11–13], a Decision Support System (DSS) based on the *ASPIC+ argumentation framework* has been implemented as a java GXT/GWT web application[1]. This DSS takes as input a collection of textual arguments in favor or against a set of alternatives under debate. It implements the entire process from argument elicitation

---

[1] Accessible online at http://pfl.grignon.inra.fr/EcoBioCapProduction/ (although the access is restricted).

© Springer International Publishing AG 2016
E. Garoufallou et al. (Eds.): MTSR 2016, CCIS 672, pp. 119–131, 2016.
DOI: 10.1007/978-3-319-49157-8_10

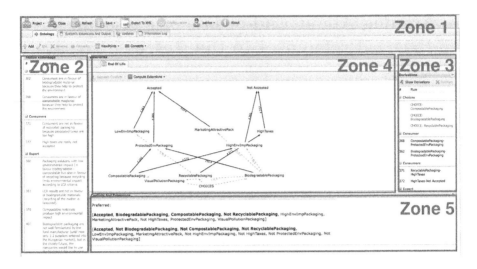

Fig. 1. Main interface of the argumentation system.

to extension computation and it also provides several GUIs for visualization purposes. The process is composed of four steps: formalizing text arguments, processing arguments, computing extensions. Hereinafter, some user interfaces are displayed showing the obtained result in the case of the viewpoint "end of life" within EcoBioCap. The main interface of the system is illustrated in Fig. 1. It is divided into five zones. Zone 1 corresponds to the task bar implementing user accounts management and general functions applied on projects (create, load, close, refresh, export, etc.). Zone 2 lists the text arguments by stakeholders. Zone 3 displays the extracted concepts and rules from the text arguments; they are also listed by stakeholders. Zone 4 displays the graphical representation of the formalized concepts and arguments. Zone 5 is a notification area displaying the computed conflicts and extensions.

It was decided that this decision support system based on argumentation could be used to select the best end-of-life according to possibly conflicting requirements provided by multiples stakeholders. For instance, one can discuss the pros and cons of incineration, anaerobic digestion or landfill for the end-of-life of a packaging. Incineration may produce energy but may hurt the human health by producing dioxin. Packagings that are processed by anaerobic digestion will also be used to produce gazes but these packagings may disturb the sorting of recyclable packagings. Likewise, landfill is a good alternative because it is low-cost but it also have long-term effects on grounds.

In this paper, we address a crucial problem for decision-making tools that are using argumentation frameworks, that is the existence of *multiple extensions*. For instance, in Zone 5 of Fig. 1, there are two extensions with justifications for each of them: one promoting the use of biodegradable and compostable packagings because they protect the environment but can induce visual pollution and high environment impact, and the other promoting to not use them. Indeed, argumentation frameworks are able to extract several coherent viewpoints from the arguments but in the event that the argumentation system returns more than one extension, it is often difficult to select one out of many alternatives. Many researchers have studied this problem and came up with various ideas. In [1], the authors suggested to vote on extensions. Another idea introduced in [6,9] was to use preferences on pieces of information that are used to generate the arguments. These preferences can represent either the importance or the confidence of the information and are usually gathered from experts. We chose to focus on preferences as they are widely studied in the field of argumentation and constitute a simple and comprehensive way to explain decisions to users.

The next section recall the notions needed to comprehend the fundamental components of the web application, i.e. the *ASPIC+ framework* and the *propositional language*.

2   Preliminaries

In this section, we present useful notions: Dung's semantics (Sect. 2.1), the logical language used in this application (Sect. 2.2) and the *ASPIC+* argumentation framework (Sect. 2.3).

2.1   Dung's Semantics

Here, we briefly recall the acceptability semantics introduced by Dung [8] and used in the rest of this paper.

Definition 1. *Given an argumentation framework* $\mathcal{AS} = (\mathcal{A}, Att)$, *where* $\mathcal{A}$ *is a set of arguments and Att is a binary attack relation between arguments of* $\mathcal{A}$. *We say that an argument* $a \in \mathcal{A}$ *is* acceptable *w.r.t a set of arguments* $\varepsilon \subseteq \mathcal{A}$ *iff* $\forall b \in \mathcal{A}$ *such that* $(b, a) \in Att, \exists c \in \varepsilon$ *such that* $(c, b) \in Att$. *Moreover, an extension can follow different semantics:*

- ε *is* conflict-free *iff* $\nexists a, b \in \varepsilon$ *such that* $(a, b) \in Att$.
- ε *is* admissible *iff* ε *is conflict-free and all arguments of* ε *are acceptable w.r.t* ε.
- ε *is* preferred *iff it is maximal (for set inclusion) and admissible.*
- ε *is* stable *iff it is conflict-free and* $\forall a \in \mathcal{A} \backslash \varepsilon, \exists b \in \varepsilon$ *such that* $(b, a) \in Att$.

2.2   The Language

Formally, we consider a propositional language and we denote by $\mathcal{L}$ the set of well formed formulas of this language given the usual connectives $\wedge, \vee, \rightarrow, \neg$, the constants $\bot, \top$ and extended with the defeasible inference $\Rightarrow$. The set of symbols in the language is denoted by $\mathcal{V}$. A *strict rule* (or strict implication) is a propositional sentence of the form $P \rightarrow Q$ where P and Q are propositions. Strict rules are important because they enable us to infer certain information from a knowledge base. Likewise, a *defeasible rule* (or defeasible implication) is a propositional sentence of the form $P \Rightarrow Q$ where P and Q are propositions. Defeasible rules represent reasonings that are not always true. A *negative constraint* (or simply a constraint) is a strict rule (resp. defeasible rule) of the form $P \wedge Q \rightarrow \bot$ (resp. $P \wedge Q \Rightarrow \bot$) where P and Q are propositions. In order to simplify the notation, we will introduce the function *Incompatible* that takes as input a set of propositions $\{P_1, \ldots, P_n\}$ and returns the set of corresponding negative constraints $(P_i \wedge P_j \rightarrow \bot)$ for all pairs of propositions $(P_i, P_j), i \neq j$.

2.3   ASPIC+ Argumentation System

The *ASPIC+* argumentation framework was proposed as a simple tool for structured argumentation. It is based on a logical language, a set of strict and defeasible rules, a contrariness function and a preference ordering over the defeasible rules.

Definition 2. *As expressed in [11–13], an ASPIC+ argumentation system is a tuple* $AS = (\mathcal{L}, cf, \mathcal{R}, \geq)$ *where:*

- $\mathcal{L}$ *is the logical language of the system.*
- *cf is a contrariness function which associates to each formula f of $\mathcal{L}$ a set of its incompatible formulas (in $2^{\mathcal{L}}$): in our case, cf corresponds to classical negation $\neg$.*
- $\mathcal{R} = \mathcal{R}_s \cup \mathcal{R}_d$ *is the set of strict ($\mathcal{R}_s$) and defeasible ($\mathcal{R}_d$) inference rules where $\mathcal{R}_s \cap \mathcal{R}_d = \emptyset$. Please note that for each strict rule $P \rightarrow Q$, the transposed rule $\neg Q \rightarrow \neg P$ is generated to ensure the completeness and the consistency of reasoning.*
- $\geq$ *is a preference ordering over defeasible rules, not used in this work.*

A knowledge base in an $AS = (\mathcal{L}, \mathcal{R}, cf, \geq)$ is $\mathcal{K} \subseteq \mathcal{L}$, which contains the concepts defined in the domain and the alternative choices under discussion.

Argument Structure. An argument in *ASPIC+* can be in two forms. Form 1 represents basic arguments that are deduced from the knowledge base. Arguments in Form 2 are more complex arguments that are constructed from other arguments using strict and defeasible rules.

Definition 3. *An* ASPIC+ *argument A can be of the following forms:*

1. $\emptyset \Rightarrow C$ *with $C \in \mathcal{K}$, such that $Prem(A) = \{C\}, Sub(A) = \{A\}$ and $Conc(A) = C$, with Prem returns premises of A, Sub returns its sub-arguments and Conc returns its conclusion,*
2. $A_1, ..., A_m \rightarrow C$ *(resp. $A_1, ..., A_m \Rightarrow C$), such that there exists a strict (resp. defeasible) rule in $\mathcal{R}_s$ (resp. $\mathcal{R}_d$) of the form $Conc(A_1), ..., Conc(A_m) \rightarrow C$ (resp. $Conc(A_1), ..., Conc(A_m) \Rightarrow c$), with $Prem(A) = Prem(A_1) \cup \cdots \cup Prem(A_m)$, $Conc(A) = C$, $Sub(A) = Sub(A_1) \cup \cdots \cup Sub(A_m) \cup \{A\}$.*

The Attack Relation. The engine only considers the rebutting attack as defined in [10]. This attack relation represents the incompatibility between two arguments with conflicting conclusions.

Definition 4. *Argument A rebuts argument B on B' if and only if $Conc(A) \in cf(\phi)$ (where ϕ is an atom in the language) for some $B' \in Sub(B)$ of the form $B'_1, \ldots, B'_m \Rightarrow \phi$. Finally, A defeat B if A rebuts B.*

Example 1. Let $\mathcal{AS}$ be an ASPIC+ argumentation framework defining the set of rules $\mathcal{R} = \mathcal{R}_s \cup \mathcal{R}_d$.

- $\mathcal{R}_s = \{BP \rightarrow HIP, \neg HIP \rightarrow \neg BP, HIP \rightarrow \neg ACC, ACC \rightarrow \neg HIP\}$
- $\mathcal{R}_d = \{BP \Rightarrow PEV, PEV \Rightarrow ACC\}$

The following structured arguments can be built on the knowledge base $\mathcal{K} = \{BP\}$:

- $A_0 : \emptyset \Rightarrow BP$
- $A_1 : A_0 \rightarrow HIP$
- $A_2 : A_1 \rightarrow \neg ACC$
- $B_1 : A_0 \Rightarrow PEV$
- $B_2 : B_1 \Rightarrow ACC$
- $B_3 : B_2 \rightarrow \neg HIP$
- $B_4 : B_3 \rightarrow \neg BP$

Following the definition of the attack, we have that argument B_4 rebuts argument A_1 on A_0.

3 Use-Case

In this section we will describe the use-case we obtained from several meetings with experts concerning the end-of-life of packagings. The use-case presents the text arguments (see Fig. 2) given by several stakeholders (consumers, restaurateurs, etc.) regarding the end-of-life possibilities for packagings (Anaerobic digestion, Incineration, etc.). From these text arguments, we first formalized a set $\mathcal{C}$ of *propositional constants* (also called *concepts*), corresponding to the several important notions of the text arguments (see Fig. 3a) and identified a set $\mathcal{A}$ of specific concepts that correspond to the alternative choices under discussion (AE, I, LF, C, R). Then, we formalized the inferences contained in the text arguments as strict rules and sorted concepts between positive and negative by linking them to either *Accepted* or *Not Accepted* using defeasible rules to represent that such a concept is a justification for accepting or rejecting the associated alternative. Moreover, we added negative constraints to represent that each of the end-of-life possibilities are mutually exclusive. This use-case can be represented by a knowledge base $\mathcal{K} = \{AE, I, LF, C, R\}$ in an argumentation framework $\mathcal{AS} = (\mathcal{L}, \mathcal{R}, cf, \geq)$ with $\mathcal{R} = Incompatible(\{AE, I, LF, C, R\}) \cup \mathcal{R}'$, where $\mathcal{R}'$ is the set of rules displayed in Fig. 3b.

Stakeholder	Argument
Consumer	Consumers are in favor of biodegradable materials because they help to protect the environment .
Consumer	Consumers are in favor of compostable materials because they help to protect the environment.
Consumer	Concerning incineration, consumers express concerns because of dioxin production which has an impact on human health.
Consumer	Consumers are not ready to pay higher prices for biodegradable packagings.
Restaurateur	Restaurants are not in favors of compostable materials because they need heavy procedures to function (designated bin, trained employees and consumers).
Restaurateur	Restaurants have to contact their local composting facility to arrange a pick-up or drop-off procedure.
Expert	LCA results are in favor of recycling.
Expert	LCA results are not in favor of biodegradable materials.
Expert	In France, recyclable materials benefit from eco-tax bonus (Eco-emballage).
Expert	A European directive forbids new landfill centers in the horizon of 2020.
Researcher	Biodegradable materials could encourage people to throw their packagings in nature, causing visual pollution.
Researcher	Plastic materials cause pollution of oceans.
Researcher	The bio-polyesters (compostable) materials as PLA are disturbing PET recycling (non-organic polyesters).
Waste Management	Biodegradable materials may disturb the sorting of recyclable packagings.
Waste Management	In France, landfill is encouraged because it is low-cost. (around 80 euros per ton).
Waste Management	In France, composting is not encouraged because of high treatment cost (around 130 euros per ton).
Waste Management	Incineration (other pack) permits to produce energy.
Waste Management	Anaerobic digestion permits to produce gazes.
Waste Management	Compostable materials permit to produce fertilizers.
Waste Management	Landfill have long-term negative effects on grounds (residues heavy metals).

Fig. 2. Set of arguments obtained during meetings with experts.

We inputed this model in the web application presented in [11–13] and after calculation, the argumentation framework used the preferred semantics and produced five extensions. The preferred semantics was introduced in *Dung*'s seminal paper [8] alongside three other argumentation semantics: *grounded*, *stable* and *complete* semantics. We chose this semantics because it captures the intuition of the *stable semantics* and avoids its drawbacks (non-existence of extensions, etc.). Each of those preferred extensions[2] corresponds to one alternative:

- $\varepsilon_1 = \{\mathbf{AE}, PGS, PEV, HIP, VPL, DPR, LCD\}$
- $\varepsilon_2 = \{\mathbf{I}, DXP, PEN\}$
- $\varepsilon_3 = \{\mathbf{LF}, FNL, LCT, LTE\}$
- $\varepsilon_4 = \{\mathbf{C}, PFZ, PEV, HIP, VPL, DPR, HEP, APP, LCD\}$
- $\varepsilon_5 = \{\mathbf{R}, LCA, ETX\}$.

In [6,7], the authors introduced the notion of base of an extension to denote the elements of the knowledge base representing the arguments of the extension. We reused this term here to represent the concepts corresponding to the

---

[2] Please note that for simplicity purposes, we write that a concept belongs to an extension instead of writing that the argument with this concept as a conclusion is contained in the extension.

Name	Concepts
AE	Uses Anaerobic digestion
I	Uses Incineration
LF	Uses Landfill
C	Uses Compostable
R	Recycling
PEV	Protects the environment
DXP	Produces dioxin
HIP	Has higher prices
HEP	Needs heavy procedures
APP	Needs to arrange pick-up procedures
LCA	LCA results in favor
LCD	LCA results in disfavor.
ETX	Has Eco-tax
FNL	Forbids new landfills
VPL	Induces Visual pollution
LCT	Is Low-cost
DPR	Disturbs plastic recycling
PEN	Produces energy
PGS	Produces gazes
PFZ	Produces fertilizers
LTE	Causes long term effect on grounds.

(a) Concepts and their initials.

Strict rules	Defeasible rules
$AE \to PEV$	$PEV \Rightarrow Accepted$
$AE \to HIP$	$LCT \Rightarrow Accepted$
$AE \to VPL$	$LCA \Rightarrow Accepted$
$AE \to DPR$	$ETX \Rightarrow Accepted$
$AE \to LCD$	$HIP \Rightarrow NotAccepted$
$I \to DXP$	$VPL \Rightarrow NotAccepted$
$LF \to FNL$	$DPR \Rightarrow NotAccepted$
$LF \to LCT$	$LCD \Rightarrow NotAccepted$
$LF \to LTE$	$DXP \Rightarrow NotAccepted$
$C \to PEV$	$FNL \Rightarrow NotAccepted$
$C \to HIP$	$LTE \Rightarrow NotAccepted$
$C \to VPL$	$HEP \Rightarrow NotAccepted$
$C \to DPR$	$APP \Rightarrow NotAccepted$
$C \to HEP$	
$C \to APP$	
$C \to LCD$	
$C \to DPR$	
$R \to LCA$	
$R \to ETX$	

(b) Rules of the knowledge base.

Fig. 3. Rules and concepts extracted from the text arguments.

alternatives and appearing in an extension. Please note that arguments of an extension ε that appear in bold are said to belong to the base of that extension (denoted by $Base(\varepsilon)$). At this point, we introduce two new methods for decision-making using preferences: refining the set of extensions using the *globally optimal extension semantics* or using *scores* to rank extensions and extract a ranking.

4   Preferences Module in Argumentation Software

After instantiating an argumentation framework, one can choose to add preferences to refine the output of the framework. Preferences can either occur in the computation of extensions or in the refining of the solutions as described in [2]. In the latter, we do not change the computation of extensions and only extract different subsets of extensions (locally optimal, Pareto optimal and globally optimal extensions) from the extensions produced. It was shown that this preference-based argumentation system satisfies rationality postulates [7]. Please note that these methods do not always produce a strict order on extensions.

In [6], preferences are viewed as a relation $\geq$ on facts (not necessarily total) to represent the confidence we have in the pieces of information. However, it appeared that in the area of decision-making, a preference relation on the facts that are induced may be more useful because preferences are often stated on the effects of decisions rather than on the decisions themselves. A preference is a statement of the form:"I am ready to pay higher prices in order to protect the environment" and is formalized as a binary relation on concepts ($LCT < PEV$).

4.1 Refining Extensions Using Semantics

In this section, we introduce a new method for refining a set of extensions E using semantics (locally, Pareto and globally optimal) inspired from [6]. These semantics return subsets of the original set of extensions. We introduce here the three notions.

An extension ε is said to not be locally optimal if we can find another extension ε' such that the concepts of ε are either included in ε' or dominated by elements of ε'(there is at most one concept dominated).

Definition 5. *We say that an extension ε of E is locally optimal if and only if $\nexists x \in \varepsilon\backslash Base(\varepsilon)$ and $y \in C$ such that $\exists \varepsilon' \in E\backslash\{\varepsilon\}, (((\varepsilon\backslash Base(\varepsilon))\backslash\{x\})\cup\{y\}) \subseteq \varepsilon'$ and $x < y$.*

An extension ε is said to not be Pareto optimal if we can find another extension ε' such that the concepts of ε are either included in ε' or dominated by elements of ε'(they are dominated by a single concept).

Definition 6. *We say that an extension ε of E is Pareto optimal if and only if $\nexists X \subseteq \varepsilon\backslash Base(\varepsilon)$ and $y \in C$ and $X \neq \emptyset$ such that $\exists \varepsilon' \in E\backslash\{\varepsilon\}, (((\varepsilon\backslash Base(\varepsilon))\backslash X)\cup \{y\}) \subseteq \varepsilon'$ and $\forall x \in X, x < y$.*

An extension ε is said to not be globally optimal if we can find another extension ε' such that the concepts of ε are either included in ε' or dominated by elements of ε' (no restrictions).

Definition 7. *We say that an extension ε of E is globally optimal if and only if $\nexists X \subseteq \varepsilon\backslash Base(\varepsilon)$ and $Y \subseteq C$ and $X \neq \emptyset$ such that $\exists \varepsilon' \in E\backslash\{\varepsilon\}, (((\varepsilon\backslash Base(\varepsilon))\backslash X) \cup Y) \subseteq \varepsilon'$ and $\forall x \in X, \exists y \in Y$ such that $x < y$.*

These semantics enable us to obtain a simple refining, i.e. we obtain four subsets of the initial set of extensions. The following example shows the approach.

Example 2. Suppose that $E = \{\varepsilon_1, \varepsilon_2, \varepsilon_3, \varepsilon_4, \varepsilon_5\}$ is the set of extensions returned by the argumentation system as described in Sect. 3; we add the following preferences:

– Having good LCA results is better than producing dioxin: $DXP < LCA$,
– Beneficing of the EcoTax is preferred to producing energy: $PEN < ETX$.

The module removes the extension ε_2 corresponding to the alternative "Incineration" from the set of globally optimal extensions because the set of concepts of ε_2 (DXP, PEN) is dominated by the concepts of ε_5. If we further add the preference:

– Producing fertilizers is more important than producing gazes: $PGS < PFZ$.

Not locally optimal	Locally optimal	Pareto optimal	Globally optimal
	ε_5	ε_5	ε_5
	ε_4	ε_4	ε_4
	ε_3	ε_3	ε_3
	ε_1	ε_1	
ε_2			

Fig. 4. Overview of the results after application of the preferences.

The module no longer considers the extension ε_1 corresponding to the alternative "anaerobic digestion" as being important because its concepts are included in ε_4 and PGS is dominated by PFZ. The preferences module removes it from the set of locally optimal extensions (and Pareto/globally optimal extensions). Please find the results in Fig. 4.

Following this result, we can say that according to the preferences stated, the three more preferred end-of-life possibilities for packagings are "Recycling", "Landfill" and "Compostable". Moreover, an ordering can be deduced from these semantics:

$$\{\varepsilon_5, \varepsilon_4, \varepsilon_3\} > \{\varepsilon_1\} > \{\varepsilon_2\}$$

Note that while those semantics allow to refine the extensions, they may be unable to output only one extension as it is the case in the previous example. This is of course dependent of the preferences the user has used: the more preferences are used, the more refining is going to happen. Note as well that it is possible to use the preferences differently, namely in a more "quantitative" fashion. We study this new approach in the next section.

4.2 Ranking Methods Using Scores

This new approach using scores is interesting in many ways. First, it is obviously easier and faster to compute that the approach introduced in [6]. Furthermore, an extension can be accurately scored (using the preferences) even if we do not have the entire set of extensions. This can be useful in the event that we do not have enough time to compute all the extensions. In this section, we introduce two scores for ranking extensions.

First Scoring: High Score Means Less Dominated. The first method gives the highest points to the extension that is the least dominated. Namely, the score of an extension ε is $Score_1(\varepsilon) = \sum_{a \in (\varepsilon \setminus Base(\varepsilon))} |\{c \in \mathcal{C} | c < a\}|$. It is obvious that with this score, the best extension is the one with the highest score. If we reuse the previous preferences, we get the following scores (Fig. 5):

Extension	$Score_1$
ε_1	0
ε_2	0
ε_3	0
ε_4	1
ε_5	2

$\{\varepsilon_5\} > \{\varepsilon_4\} > \{\varepsilon_1, \varepsilon_2, \varepsilon_3\}$

Fig. 5. Scores obtained with $Score_1$ and the associated rank.

Second Scoring: High Score Means More Dominated. The second method gives the highest points to the extension that is the most dominated. Namely, the score of an extension ε is $Score_2(\varepsilon) = \sum_{a \in (\varepsilon \setminus Base(\varepsilon))} |\{c \in \mathcal{C} | a < c\}|$. With this score, the best extension is the one with the lowest score. If we reuse the previous preferences, we get the following scores (Fig. 6):

Extension	$Score_2$
ε_1	1
ε_2	2
ε_3	0
ε_4	0
ε_5	0

$\{\varepsilon_5, \varepsilon_4, \varepsilon_3\} > \{\varepsilon_1\} > \{\varepsilon_2\}$

Fig. 6. Scores obtained with $Score_2$ and the associated rank.

We noticed that with the first score, the ranking obtained for the most preferred extensions is more detailed ($\varepsilon_5 > \varepsilon_4$) than the globally optimal semantics (ε_5 and ε_4 are ranked equally). However, it is less accurate for the least preferred extensions ($\varepsilon_1, \varepsilon_2$ and ε_3 have the same score). On the contrary, with the second score, the ranking obtained for the least preferred extensions is as detailed ($\varepsilon_1 > \varepsilon_2$) as the one obtained with globally optimal semantics. However, it is less accurate for the most preferred extensions ($\varepsilon_3, \varepsilon_4$ and ε_5 have the same score).

A research issue is to find a way to combine the two scores in order to produce a more efficient ranking. This can be achieved by using multi-criteria methods. We provide a naive way to combine the two scores, namely $Score_3(\varepsilon) = Score_1(\varepsilon) - Score_2(\varepsilon)$. Using this new score, we get the following results (Fig. 7):

Extension	$Score_3$
ε_1	-1
ε_2	-2
ε_3	0
ε_4	1
ε_5	2

$\{\varepsilon_5\} > \{\varepsilon_4\} > \{\varepsilon_3\} > \{\varepsilon_1\} > \{\varepsilon_2\}$

Fig. 7. Scores obtained with the combination of $Score_1$ and $Score_2$ and the associated rank.

In our example, this new score leads to a strict total order, which is arguably more useful in terms of decision-making.

4.3 Implementation

We integrated a simple and intuitive interface in the web application for inputting preferences which enables users to clearly visualize the preferences implied and the possible incoherences (see Fig. 8c). The preferences are saved in a database and are specific to a particular argumentation. We also implemented all the preferences methods discussed in this paper. The processing of the argumentation framework is hidden to the user and only the different extensions produced are displayed (see Fig. 8a). The user can then add preferences and use the refining method introduced in Sect. 4.1 (see Fig. 8b). Although the process has been simplified, more work is required to make it easier to understand for end users that are not expert in argumentation theory.

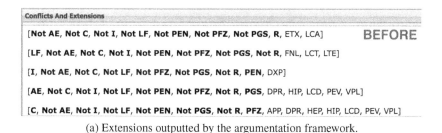

(a) Extensions outputted by the argumentation framework.

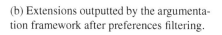

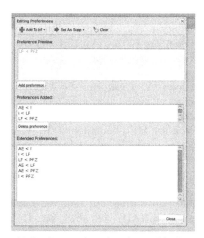

(b) Extensions outputted by the argumentation framework after preferences filtering.

(c) Preference interface in EcobioCap.

Fig. 8. Different interfaces of the web application.

5   Conclusion

In this paper, we described a real life use-case we obtained from several meetings with experts concerning the end-of-life of packagings. We applied an argumentation approach and showed how preferences can have a real impact on the selection of alternatives in a decision-making problem from a real agronomy inspired use case. This new approach is implemented as a web application and a demonstration of the tool can also be provided upon request.

Future work includes the investigation of a natural language processing module that will be able to semi-automatically extract arguments from text files. Another current research avenue includes the investigation of explanatory dialogues with the users that will help better understand the output of our system [3–5].

Acknowledgements. The authors would like to thank the partners of the Pack4Fresh project, for all their help during the argument elicitation phase as well as for their constant feedback.

References

1. Amgoud, L., Prade, H.: Explaining qualitative decision under uncertainty by argumentation. In: Proceedings, The Twenty-First National Conference on Artificial Intelligence and the Eighteenth Innovative Applications of Artificial Intelligence Conference, July 16–20, Boston, Massachusetts, USA, pp. 219–224 (2006)
2. Amgoud, L., Vesic, S.: Two roles of preferences in argumentation frameworks. In: Liu, W. (ed.) ECSQARU 2011. LNCS (LNAI), vol. 6717, pp. 86–97. Springer, Heidelberg (2011). doi:10.1007/978-3-642-22152-1_8
3. Arioua, A., Croitoru, M.: Formalizing explanatory dialogues. In: Beierle, C., Dekhtyar, A. (eds.) SUM 2015. LNCS (LNAI), vol. 9310, pp. 282–297. Springer, Heidelberg (2015). doi:10.1007/978-3-319-23540-0_19
4. Arioua, A., Croitoru, M.: Dialectical characterization of consistent query explanation with existential rules. In: Proceedings of FLAIRS 2016, pp. 621–625 (2016)
5. Arioua, A., Croitoru, M.: A dialectical proof theory for universal acceptance in coherent logic-based argumentation frameworks. In: Proceedings of ECAI (2016, to appear)
6. Croitoru, M., Thomopoulos, R., Vesic, S.: Introducing preference-based argumentation to inconsistent ontological knowledge bases. In: Chen, Q., Torroni, P., Villata, S., Hsu, J., Omicini, A. (eds.) PRIMA 2015. LNCS (LNAI), vol. 9387, pp. 594–602. Springer, Heidelberg (2015). doi:10.1007/978-3-319-25524-8_42
7. Croitoru, M., Vesic, S.: What can argumentation do for inconsistent ontology query answering? In: Liu, W., Subrahmanian, V.S., Wijsen, J. (eds.) SUM 2013. LNCS (LNAI), vol. 8078, pp. 15–29. Springer, Heidelberg (2013). doi:10.1007/978-3-642-40381-1_2
8. Dung, P.M.: On the acceptability of arguments and its fundamental role in non-monotonic reasoning, logic programming and n-person games. Artif. Intell. **77**(2), 321–358 (1995)
9. Kaci, S.: Working with Preferences: Less Is More. Cognitive Technologies. Springer, New York (2011)

10. Modgil, S., Prakken, H.: The ASPIC$^+$ framework for structured argumentation: a tutorial. Argument Comput. **5**(1), 31–62 (2014)
11. Tamani, N., Mosse, P., Croitoru, M., Buche, P., Guillard, V.: A food packaging use case for argumentation. In: Closs, S., Studer, R., Garoufallou, E., Sicilia, M.-A. (eds.) MTSR 2014. CCIS, vol. 478, pp. 344–358. Springer, Heidelberg (2014). doi:10.1007/ 978-3-319-13674-5_31
12. Tamani, N., Mosse, P., Croitoru, M., Buche, P., Guillard, V., Guillaume, C., Gontard, N.: Eco-efficient packaging material selection for fresh produce: industrial session. In: Hernandez, N., Jäschke, R., Croitoru, M. (eds.) ICCS 2014. LNCS (LNAI), vol. 8577, pp. 305–310. Springer, Heidelberg (2014). doi:10.1007/ 978-3-319-08389-6_27
13. Tamani, N., Mosse, P., Croitoru, M., Buche, P., Guillard, V., Guillaume, C., Gontard, N.: An argumentation system for eco-efficient packaging material selection. Comput. Electron. Agric. **113**, 174–192 (2015)

A *Datalog*± Domain-Specific Durum Wheat Knowledge Base

Abdallah Arioua[1,2(✉)], Patrice Buche[2], and Madalina Croitoru[1]

[1] LIRMM, University of Montpellier, Montpellier, France
abdallaharioua@gmail.com
[2] IATE, INRA, Montpellier, France

Abstract. We consider the application setting where a domain-specific knowledge base about Durum Wheat has been constructed by knowledge engineers who are not experts in the domain. This knowledge base is prone to inconsistencies and incompleteness. The goal of this work is to show how the state of the art knowledge representation formalism called *Datalog*± can be used to cope with such problems by (1) providing inconsistency-tolerant techniques to cope with inconsistency, and (2) providing an expressive logical language that allows representing incomplete knowledge.

1 Introduction

The Dur-Dur research project[1] aims at restructuring the Durum Wheat agrifood chain in France by reducing pesticide and fertilizer usage while providing a protein-rich Durum Wheat. The project relies on constructing a multidisciplinary knowledge base (involving all actors in the agrofood chain) which will be used as a reference for decision making. This knowledge base is collectively built by several knowledge engineers from different sites of the project. Due to various causes (errors in the factual information due to typos, erroneous databases/Excel files, incomplete facts, unspoken obvious information "everybody knows" etc.) the collectively built knowledge base (KB) is prone to *incompleteness and inconsistencies*. Incompleteness has many forms, in our case it reflects itself as a lack of precision and explicitness. For instance, an expert may say that the Durum Wheat is contaminated by a mycotoxin but he/she may, for some reasons, do not specify which mycotoxin. Inconsistency appears as logical contradictions due to the causes stated above. The problem is that in presence of inconsistencies the knowledge base becomes unreliable and not trustworthy, let alone the fact that reasoning under inconsistency is challenging for logical formalisms.

To solve the above mentioned problems, we propose in this paper a methodology of representing Durum Wheat knowledge in the logical framework of *Datalog*± [5,10]. *Datalog*± is expressive enough to allow the representation of unknown individual in the knowledge base and cope with heterogeneous data as it allows for n-ary predicates. Moreover, *Datalog*± has an interesting equivalent

---

[1] http://www.agence-nationale-recherche.fr/?Projet=ANR-13-ALID-0002.

© Springer International Publishing AG 2016
E. Garoufallou et al. (Eds.): MTSR 2016, CCIS 672, pp. 132–143, 2016.
DOI: 10.1007/978-3-319-49157-8_11

relation with conceptual graphs [13], in fact any logical formula in *Datalog*± can be translated to a graphical representation, which significantly helps experts in other domains in the process of knowledge acquisition. We present with detailed examples how this methodology is used to construct the Durum Wheat knowledge base in the French project Dur-Dur. The knowledge base is available online at http://www.lirmm.fr/~arioua/dkb/ where the reader can find downloadable materials.

2 The Logical Language *Datalog*±

There are two major approaches in the knowledge representation community: Description Logics (DL) (such as $\mathcal{EL}$ [2] and DL-Lite [6] families) and rule-based languages (such as *Datalog*± language [5,10], a generalization of Datalog that allows for existentially quantified variables in rule's head). Despite its undecidability when answering conjunctive queries, different decidable fragments of *Datalog*± are studied in the literature [4]. These fragments generalize the above mentioned DL families and overcome their limitations by allowing any predicate arity as well as cyclic structures.

The *Datalog*± corresponds to *the positive existential* conjunctive fragment of first-order logic, which is composed of formulas built with the connectors $(\wedge, \rightarrow)$ and the quantifiers $(\exists, \forall)$, with constants but no function symbol.

An *atom* is of the form $p(t_1, \ldots, t_k)$ where p is a predicate of arity k and the t_i are terms, i.e., variables or constants (we use vectors, e.g. $\vec{x}$, to denote a sequence of variables). A finite set of atoms F is called an *atomset* (a *fact*), we denote by $terms(F)$ (resp. $vars(F)$) the set of terms (resp. variables) that occur in F. A *homomorphism* π from two atomsets A_1 to A_2 is a substitution of $vars(A_1)$ by $terms(A_2)$ such that $\pi(A_1) \subseteq A_2$. An *existential rule* (or a rule) is of the form $R = \forall \vec{x} \forall \vec{y} (B \rightarrow \exists \vec{z} H)$, where B and H are conjunctions of atoms, with $vars(B) = \vec{x} \cup \vec{y}$, and $vars(H) = \vec{x} \cup \vec{z}$. B and H are respectively called the *body* and the *head* of R. Chase is the mechanism by which one deduce new facts by rule application on the initial set of facts $\mathcal{F}$. We denote by $\mathtt{Cl}_{\mathcal{R}}(\mathcal{F})$ the set of all facts that can be deduced from $\mathcal{F}$ by a set of rules $\mathcal{R}$. A *knowledge base* $\mathcal{K} = (\mathcal{F}, \mathcal{R}, \mathcal{N})$ is composed of a finite set of facts $\mathcal{F}$, rules $\mathcal{R}$ and negative constraints $\mathcal{N}$ (i.e. a rule whose head is set to $\perp$). A *Boolean conjunctive query* (BCQ or query in the following) has the form of a fact. We say a query Q is entailed from $\mathcal{K}$ iff $\mathtt{Cl}_{\mathcal{R}}(\mathcal{F}) \models Q$. We say $\mathcal{K}$ is inconsistent iff $\mathtt{Cl}_{\mathcal{R}}(\mathcal{F}) \models \perp$.

3 The Durum Wheat Knowledge Base

The Durum Wheat knowledge base has been constructed within the French National Project DUR-DUR. The goal of this knowledge base is to integrate scientific knowledge acquired from different tasks during the project to redesign the durum wheat chain. The Dur-Dur project suggests developing a systematic approach to investigate issues related to the management of the nitrogen,

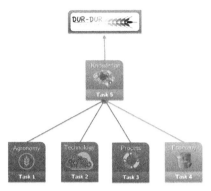

Fig. 1. The different tasks of the Dur-Dur project. The knowledge tasks aims at integrating multidisciplinary knowledge from other tasks.

energy and contaminants, to guarantee a global quality of products throughout the production and the processing chain. Started in 2014 and planned over 4 years, the project aims at integrating the 3 dimensions of the sustainability (environmental, economic, and social), at 4 levels of investigation (4 tasks) with a complementary task (task 5). Figure 1 depicts the different tasks of the project where the fifth task's central role is to integrate knowledge from different tasks. The Durum Wheat knowledge base is the product of the fifth Task. It will be used in many computational tasks, notably analyzing and comparing the alternative innovative technical itineraries proposed in the project to reduce the use of chemical inputs (nitrogen fertilizers and pesticides). The knowledge base represents domain-specific knowledge about Agronomy. It is composed of four main parts:

- **Vocabulary:** it contains knowledge about concepts and relations.
- **Rules:** they represent rules that encode generic knowledge.
- **Negative constraints:** this part contains constraints about crops and Agronomy-related constraints.
- **Facts:** this part contains factual knowledge about Agronomy-related subjects (fertilizers, pesticides, diseases, etc.).

In the next section we start by highlighting the guidelines which were followed to author the knowledge base (Subsect. 3.1) then we turn to the internal structure or the architecture of the knowledge base including the *vocabulary*, the *rule-base* alongside with the *constraints* and the *factual knowledge* (Subsect. 3.2).

3.1   The Authoring

A multidisciplinary process of knowledge acquisition and representation was deployed to author the knowledge base. We used technical reports to define the scope of the knowledge base and the relevant concepts of our vocabulary. Taking into account the recommendation of [15], we followed three steps *specification, conceptualization and formalization* to build the knowledge base.

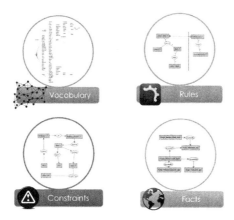

Fig. 2. An overview of the Durum Wheat knowledge base. The circles contains knowledge examples represented in the conceptual graph framework.

Specification. The scope of the Durum Wheat knowledge base has been defined by exclusively focusing on *Durum Wheat Sustainability* management. The goal is improving Durum Wheat sustainability in France and reduce the use of nitrogen fertilizers and pesticides and optimize energy consumption using a systematic approach that makes use of innovative technical itineraries. The contribution of the knowledge base lays in offering an expressive way of representing domain-knowledge.

Conceptualization. The concepts and the relations among them alongside to rules, facts and constraints have been defined and collected from technical reports (see Fig. 3) and online materials (see [1]). It is worth mentioning that in the vocabulary part we have built on the vocabulary of Agropedia indica [12] with an increase (and modification) in content that approximates 60 %.[2]

Formalization. Since understanding logical formulas is quite difficult for experts who are not familiar with KRR formalism we have chosen a graphical framework (conceptual graphs; [13,14]) to author the knowledge base. Moreover, the conceptual graphs (CGs) made it easy for the Agronomy experts to understand the content of the knowledge base. Furthermore, CGs enjoy the same expressive power as *Datalog±*. In fact, it is an equivalent formalism of *Datalog±* as shown in [7]. Therefore, our choice was to choose CGs for knowledge acquisition and *Datalog±* as a framework for reasoning. For CGs, we used *CoGui 1.6b* which is an IDE for representing and reasoning with CGs.[3] We shall explain in-depth in Sect. 3.2 the graphical and logical representation for each part of the knowledge base. The facts within the knowledge base are exported to an RDF/XML format whereas the vocabulary, rules and constraints are exported as DLGP format

---

[2] http://www.agropedia.net.
[3] http://www.lirmm.fr/cogui/.

1- taux varietal d'après leur bonne qualité technologique a priori (pour leur transformation en semoule) puis définition des pratiques culturales les plus adaptées pour ces variétés

Les variétés sont classées en fonction de leur aptitude à produire des grains de bonne qualité technologique (bon taux de protéines et faible mitadinage) et ceci de façon régulière.

Variété	Obtenteur	Type de gluténines*	Rdt (% du témoin)	Maladies feuilles	Fusa.	Besoin N (kg/t/q)
Pescadou	FD	G3	103%		+	3.5
Remarque : Absorbe bien l'N en fin de cycle						
Dakter	LG	G3	90%	+	-	3.5
Remarque : Potentiel plus faible mais suffisant pour atteindre une production de 50 à 60 q/ha						
Miradoux	FD	G3	100%		-- (DON)	3.7
Remarque : Variété peu connue. A noter que Miradoux est la principale variété cultivée dans les différentes régions et pourra donc servir pour notre témoin des ITKs de référence régionaux et du N0 et du Phyto+o et qui permettra de fournir des lots à l'UMR IATE avec différents niveaux de teneur en protéines						
Isildur	R2N	G1	95%	+		?
Fabulis	LG	G4	98%	+	+	?
Anvergur	RAGT	?	113%	++	+	3.7
Remarque : Potentiellement la prochaine référence car combine différentes caractéristiques.						
Joyau	SYN	G2	90%	-	++	3.7
Auris	LG	G3	?	+	+	?
Remarque : Peu de surfaces cultivées						

* Les variétés sont classées selon la composition en protéines de leur grain. Sans rentrer dans les détails, les groupes G2 et G3 sont considérés comme ayant les compositions les plus intéressantes pour la qualité technologique (information transmise par MF Samson, UMR IATE Montpellier). les

ITK proposées par le groupe à partir du choix différencié des variétés :

1.1. ITK « Référence »
• Variété Miradoux
• Précédent tournesol
• Déchaumage
• Semis à 280 grains /m² autour du 20-25 octobre
• Fertilisation : 40U au tallage, 50U fin tallage (avant E1cm), 50U à 1 nœud et 60U DFL ou gonflement
• Désherbage : A l'automne si risque fort ou si problème de résistance type raygrass en privilégiant dans ce cas des herbicides racinaires aux herbicides foliaires. Complément de désherbage en sortie hiver si besoin avec un herbicide foliaire en fonction des types de résistance rencontrés
• Fongicides : 3 traitements (sous règles de décision) à savoir un premier à 2 nœuds, 1 second à Dernière Feuille Etalée contre la rouille et la septoriose) et un à la Floraison contre la fusariose
• Insecticide si besoin sous règle de décision • molluscicide si besoin à 3 feuilles/tallage

1.2. ITK « Réduction d'intrants (objectif DurDur) »
• Variété Miradoux
• Précédent adapté (pas de maïs, ni de céréale paille, ni de sorgho) avec enfouissement des résidus par double déchaumage (pouvant faire office de faux semis).
• Culture intermédiaire mixte avec légumineuse et crucifère ou graminées couvrant rapidement couvrant pour réduire l'enherbement avec une destruction mécanique fin octobre-début novembre

SI   le seuil d'infestation est atteint   ALORS on utilisera tel traitement à telle dose

SI   la structure du sol ne convient pas   ALORS telle technique de travail du sol sera mise en
œuvre

SI   les conditions sont réunies (humidité et portance du sol et date de développement du blé ad
hoc et adventices pas trop développées)   ALORS on utilisera la herse étrille   SINON   la

Fig. 3. Some snapshots of the technical reports.

(**DataLoG Plus**; [8]). The vocabulary of the knowledge base contains 279 concepts and 116 relations, the rule-base contains 23 rules and the constraints part contains 25 constraints. The factual part has around 900 atoms. The knowledge base is available online at http://www.lirmm.fr/~arioua/dkb/.

3.2   The Structure

As depicted in Fig. 2 the knowledge base is composed of four parts. It is worth mentioning that on the logical level the vocabulary and the rule-base are the same. However, we adapt here the Semantic web notation and we differentiate between them. Therefore, we distinguish between those rules that express logical consequences (in the rule-base) and those that encode generalizations and classes inclusions (in the vocabulary).

The Vocabulary. The vocabulary represents an explicit specification of the terms and concepts used in Agronomy. The vocabulary is composed of two parts: (1) concept types hierarchy and (2) relation types hierarchy.

1. **Concept types hierarchy:** concepts are organized within a hierarchy as super-concepts and sub-concepts. For instance, the concept disease and its sub-concepts (e.g. viral disease, fungal disease, etc.), types of pesticides (e.g. herbicide, insecticide, fungicide) are all of organized in a hierarchy.
2. **Relation types hierarchy:** in CGs the concepts are related by relationships. Since concepts are divided into super-concepts and sub-concepts, relationships are divided in the same way. In the relation types hierarchy we find super-relations and sub-relations. For instance, the relation "useSowingProcess" which relates the *seeding and sowing* production step with the process of *sowing* (which is a super-concept of *broadcasting, behind plough* and a sub-concept of *process*). This relation is a sub-relation of the super-relation "useProcess" that relates any production step with any process.

In CGs the hierarchy of concept types is represented as in the upper graph of Fig. 4. Rectangles represent concepts and the arrow represents the generalization between them where the source of the arrow is the sub-concept and the target of the arrow is super-concept. In the relation types hierarchy (the lower graph), the circles are the relations and the arrows are generalizations.

To better illustrate the relation between existential rules and CGs let us take an example that shows the transformation of some part of the graphs of Fig. 4 to their logical form.

Example 1. The left-most part of the concept types hierarchy that indicates that "Viral disease is a disease" is represented logically by a rule as follows:

- $\forall x (Viral\_disease(x) \rightarrow Disease(x))$.

The part of the relation types hierarchy that indicates that "Using Herbicide is using Pesticide" is represented logically by a rule as follows:

- $\forall x, y (useHerbicide(x, y) \rightarrow usePesticide(x, y))$.

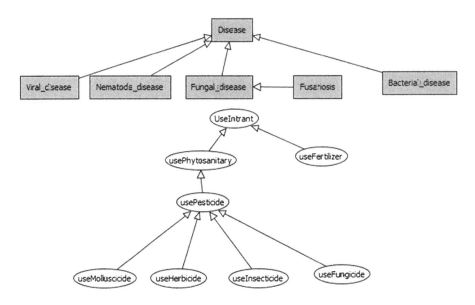

Fig. 4. Concept and relation types hierarchy.

The Rule-Base. Rules in the rule-base encode general-purpose domain-specific knowledge. For instance, consider the following rules:

(a) If a Durum Wheat x has a fusariosis disease y then there exists a mycotoxin z that has contaminated the Durum Wheat x.
$\forall x, y \exists z (Durum\_wheat(x) \land hasDisease(x,y) \land Fusariosis(y) \to$
$isContaminatedBy(x,z) \land Mycotoxin(z))$
(b) If the soil is rich of organic matters and it contains seeds of weed then these seeds will develop in this soil.
$\forall x, y, z, w (Soil(x) \land Organic\_matter(y) \land richOf(x,y) \land contains(x,z) \land$
$Seed(z) \land seedOf(z,w) \land Weed(w) \to developIn(w,x))$

The mycotoxin z is unknown (it could be Aflatoxins, Deoxynivalenol, etc.) but still the information that "there is necessarily a mycotoxin" is present, which is an important information when it comes to risk management where a possible contamination by any mycotoxin is taken to be critical. Moreover, the importance of such representation manifests also in helping knowledge elicitation where the knowledge base can make use of incomplete information and then be updated incrementally by identifying the existential variables.

In conceptual graphs the rule (b) is depicted in Fig. 5. In a rule, the rectangles are called *concept nodes* and the circles are called *relation nodes*. A concept node has a *concept type* and a *marker* which can be either an *individual marker* (constant) or a *generic marker* (a variable denoted as *). For instance, the concept *richOf* has a generic marker (*) which represents a variable. If the marker were an individual marker we should have found a constant name like *Nitrogen*. The relation nodes are predicates that relate different concepts. A rule in conceptual

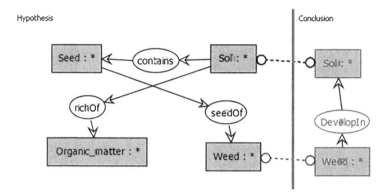

Fig. 5. The rule (b) in the CGs framework.

graphs is composed of two parts, a hypothesis (left) and a conclusion (right). The dashed lines link those concepts that share the same variables (called frontier variables). That means, variables that appear in the hypothesis and in the conclusion. In the rule (b), the concept *weed* in the hypothesis part shares the same variable with the concept *weed* in the conclusion.

As we said earlier, certain classes of *Datalog±* render the inference undecidable. However, there are some classes that ensure decidability. Most notably, FUS (Finite Unification Set) and FES (Finite Expansion Set) classes. The online tool *Kiabora* deploys syntactical and semantic analysis on any set of rules written in the DLGP format.[4] The tool, for a given rule-base, classifies all the rules with respect to the known classes. From the analysis we found that our rule-base lays within the decidable classes. Specifically, FUS and FES.

The Negative Constraints. Representing what cannot be allowed within certain domain of interest is called *negative constraint* (or constraint). Consider the following negative constraint:

(c) $\forall x, y, z(Soil(x) \land Maize(y) \land Durum\_wheat(z) \land hasPrecedent(x, y)$
    $\land isCultivatedOn(z, x) \rightarrow \bot)$.

This negative constraint forbids using Maize as a precedent on a soil if we want to cultivate Durum Wheat on this soil. Figure 6 represents the CGs representation of this negative constraint. Besides this type of constraints we have the banned types constraints. These are particular forms of constraints that express concept disjointness. For instance, a soil x cannot be a disease, $\forall x(Soil(x) \land disease(x) \rightarrow \bot)$. In the Durum Wheat knowledge base all concepts are disjoint except those concepts which have a generalization/specialization relations among them.

---

[4] Kiabora 0.1 website: http://www.lirmm.fr/~mugnier/graphik/kiabora/, see [8] for a detailed explanation.

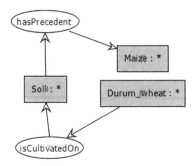

Fig. 6. The negative constraint (c) in the CGs framework.

The Factual Knowledge. In the Durum Wheat knowledge base the factual part represents domain-specific knowledge. This knowledge is divided into two parts: (1) general factual knowledge and (2) knowledge about different technical itineraries. According to [11] a *technical itinerary* is a *"logical organized course of technical actions* applied to a cropped species".

General factual knowledge is the part of the knowledge base that represents general facts about the domain, for instance, *Miradoux* is a variety of Durum Wheat or the fungal disease *Fusarium Flag smut* is cause by, among other causes, the fungi *Urocyctis agropyri* of the family Fusarium. The following is an example of a set of facts. Recall that commas are interpreted as conjunctions.

(d) {$Fungal\_disease(Flag\_smut), isCausedBy(Fusarium\_ear\_blight,$
    $Urocyctis\_agropyri), fungi(Urocyctis\_agropyri)$}.

Here we have the relation *isCausedBy* instantiated on the individuals *Flag_smut* and *Urocyctis_agropyri*. The former is a fungal disease as stated by the concept *Fungal_disease* and the latter is a fungi. Figure 7 depicts the set of facts in the CGs framework.

Fig. 7. The set of facts (d) about fungal diseases and fungus.

The second part of the factual knowledge part are those facts about the technical itineraries. In what follows we give a real-world example of a well-known technical itinerary in France.

Example 2. This example represents the reference technical itinerary in France which is followed by farmers to cultivate their fields.

"The variety to be seeded in the soil is Miradoux, the culture precedent is sunflower. The soil is prepared by means of harrowing. The seeding is done with density of 280 grains/m^2. Fertilization is to be performed at the growing stage when the tiller begins with dose 40u and 50u at the end of the tiller."

This technical itinerary is a set of facts, e.g. "variety is Miradoux", "Fertilization is to be performed at the growing stage", etc. However, not any set of facts. Particularly, it is a precise set of describing facts. Actually, any ITK (according to the studied reports) should precisely account for the following steps:

1. Variety to bee seeded.
2. Date of seeding alongside the density.
3. Cultural precedent.
4. Inter-cropping techniques.
5. Soil preparation method.
6. Disease management method.
7. Weed management method.
8. Insect control method.

Thus a technical itinerary should be mainly composed of these describing facts. The fowling is a snippet of the technical itinerary described in Example 2.

$$
\mathcal{F}_{ITK} = \begin{cases}
Soil(Soil_1) & Durum\_wheat(D1) \\
isOfVariety(D1, Miradoux) & Variety(Miradoux) \\
isCultivatedOn(D1, Soil_1) & Seeding\_and\_sowing(Seeding_1) \\
Seed(Seed_1) & useSeed(Seeding1, Seed_1) \\
seedOf(Seed_1, Durum_w1) & isAppliedOn(Seeding_1, Soil_1) \\
withDensity(Seeding_1, Density_1) & Density(Density_1) \\
Unit(grain\_mm) & hasValue(280) \\
Value(280)
\end{cases}
$$

What has been presented so far is the Durum Wheat knowledge base we have constructed manually within the project, which can be seen as a contribution on itself. since what is mainly proposed by researchers in this field are ontologies. Besides that, our knowledge base provides querying facilities not only on the ontological layer but also on the factual layer where real knowledge about the domain is represented in form of facts.

In the Durum Wheat knowledge base each technical itinerary is stored separately from the other technical itineraries (we have three in total). On can query them all together or separately.

3.3   Reasoning

The first and foremost reason to acquire knowledge and store it in knowledge bases is to provide querying facilities for the end-user. Like in classical database systems, in *Datalog±* the main reasoning task is *query answering*. The main and important difference is that in our case the querying is enriched by a rule-base layer. Thus the reasoner takes into account the domain-knowledge represented within the rules while querying.

Formally, a conjunctive query has the form of a fact but with possibly free variables. For instance $Q(x) = Fungal\_disease(x) \wedge isCausedBy(x, culmorum)$ is a conjunctive query that looks for "the fungal disease that is caused by *culmorum*".

In order to perform reasoning in forward chaining in presence of rules, the reasoner applies all the rules in the rule-base on the set of facts in the factual part then query the knowledge in a classical manner. Given a set of facts $\mathcal{F}$ and a set of rules $\mathcal{R}$ This means that the chase computes all deducible knowledge of $\mathcal{F}$ by the application of all the rules of $\mathcal{R}$ on all the facts on $\mathcal{F}$ until no rule will be applicable. This process is also called *saturation*. Note that if the closure of a set of facts $\mathcal{F}$ is the same as $\mathcal{F}$, i.e. $\mathtt{Cl}_\mathcal{R}(\mathcal{F}) = \mathcal{F}$, then we say that $\mathcal{F}$ is closed under the application of rules (or deductively closed). A query Q has an answer within a knowledge base $\mathcal{K}$ iff $\mathtt{Cl}_\mathcal{R}(\mathcal{F}) \models Q$ where $\models$ refers to the usual first-order entailment.

Example 3. Consider the following knowledge base $\mathcal{K}$:
   $\mathcal{F} = \{D(a), S(b)\}$, $\mathcal{R} = \{\forall x(D(x) \rightarrow C(x)), \forall x, y(S(x) \wedge C(y) \rightarrow M(x, y))\}$, $\mathcal{N} = \emptyset$. The closure is $\mathtt{Cl}_\mathcal{R}(\mathcal{F}) = \{D(a), S(b), C(a), M(a, b)\}$.

It may happens that the set facts $\mathcal{F}$ contains contradictory knowledge (i.e. inconsistencies). We say that a set of facts is inconsistent iff $\mathtt{Cl}_\mathcal{R}(\mathcal{F})$ triggers a negative constraint. The solution [9] is to construct maximal (with respect to set inclusion) consistent subsets of $\mathcal{F}$. Such subsets are called *repairs* and denoted by $\mathcal{R}epair(\mathcal{K})$. They actually represent possible distribution of facts to restore consistency. Once the repairs are computed, different semantics can be used for query answering over the knowledge base.

Example 4. Consider: $\mathcal{F} = \{D(a), S(b), P(c)\}$, $\mathcal{R} = \{\forall x(D(x) \rightarrow C(x))\}$, $\mathcal{N} = \{\forall x, y(S(x) \wedge C(x)) \rightarrow \bot\}$. Then the negative constraint will be triggered after the application of the rule which infers $C(a)$. Therefore our repairs would be $\mathcal{A}_1 = \{D(a), P(c)\}$ and $\mathcal{A}_2 = \{S(b), P(c)\}$ and $\mathcal{R}epair(\mathcal{K}) = \{\mathcal{A}_1, \mathcal{A}_2\}$. While $\mathtt{Cl}_\mathcal{R}(\mathcal{A}_1) = \{D(a), C(a), P(c)\}$ and $\mathtt{Cl}_\mathcal{R}(\mathcal{A}_2) = \mathcal{A}_2$.

After repairing the knowledge base we can query it using different semantics. The most common semantics is to query the intersection of all repairs. This is a cautious strategy because the intersection is practically those facts which are not involved in any inconsistency.

4   Conclusion

In this paper we have presented a general methodology to build Durum Wheat knowledge bases within the logical language *Datalog*±. We presented detailed examples and a real-world Durum Wheat knowledge base which has been built within the French national project Dur-Dur. The expressiveness of *Datalog*± lays in its ability to deal with incompleteness and inconsistency. Moreover, it has an interesting relation with Conceptual Graphs which makes it easy to non-experts to manipulate and understand logical formulae. In addition, DLGP format (**DataLoG P**lus; [8]) can be translated to semantic web languages as OWL/RDFS using COGui or GRAAL framework [3].

References

1. Arvalis. Les fiches arvalis: Varits produits accidents en grandes cultures. Online materials published by ARVALIS Plant Institue (2015). http://www.fiches. arvalis-infos.fr/. Accessed 19 Mar 2015
2. Baader, F., Brandt, S., Lutz, C.: Pushing the EL envelope. In: Proceedings of IJCAI 2005 (2005)
3. Baget, J.-F., Leclère, M., Mugnier, M.-L., Rocher, S., Sipieter, C.: Graal: a toolkit for query answering with existential rules. In: Bassiliades, N., Gottlob, G., Sadri, F., Paschke, A., Roman, D. (eds.) RuleML 2015. LNCS, vol. 9202, pp. 328–344. Springer, Heidelberg (2015). doi:10.1007/978-3-319-21542-6_21
4. Baget, J.-F., Mugnier, M.-L., Rudolph, S., Thomazo, M.: Walking the complexity lines for generalized guarded existential rules. In: Proceedings of IJCAI 2011, pp. 712–717 (2011)
5. Calì, A., Gottlob, G., Lukasiewicz, T.: A general datalog-based framework for tractable query answering over ontologies. J. Web Sem. **14**, 57–83 (2012)
6. Calvanese, D., De Giacomo, G., Lembo, D., Lenzerini, M., Rosati, R.: Tractable reasoning and efficient query answering in description logics: the DL-Lite family. J. Autom. Reason. **39**(3), 385–429 (2007)
7. Chein, M., Mugnier, M.: Graph-Based Knowledge Representation - Computational Foundations of Conceptual Graphs. Advanced Information and Knowledge Processing. Springer, London (2009)
8. Leclère, M., Mugnier, M.-L., Rocher, S.: Kiabora: an analyzer of existential rule bases. In: Faber, W., Lembo, D. (eds.) RR 2013. LNCS, vol. 7994, pp. 241–246. Springer, Heidelberg (2013). doi:10.1007/978-3-642-39666-3_22
9. Lembo, D., Lenzerini, M., Rosati, R., Ruzzi, M., Savo, D.F.: Inconsistency-tolerant semantics for description logics. In: Hitzler, P., Lukasiewicz, T. (eds.) RR 2010. LNCS, vol. 6333, pp. 103–117. Springer, Heidelberg (2010). doi:10.1007/ 978-3-642-15918-3_9
10. Mugnier, M.-L., Thomazo, M.: An introduction to ontology-based query answering with existential rules. In: Koubarakis, M., Stamou, G., Stoilos, G., Horrocks, I., Kolaitis, P., Lausen, G., Weikum, G. (eds.) Reasoning Web 2014. LNCS, vol. 8714, pp. 245–278. Springer, Heidelberg (2014). doi:10.1007/978-3-319-10587-1_6
11. Sebillotte, M.: Itineraires techniques et evolution de la pensee agronomique. Comptes Rendus des Séances de l'Académie d'Agriculture de France (1978)
12. Sini, M., Yadav, V.: Building knowledge models for Agropedia Indica v 1.0 requirements, guidelines, suggestions (2009). http://agropedia.iitk.ac.in/km_guidlines. pdf. Accessed 19 Mar 2015
13. Sowa, J.F.: Conceptual graphs for a data base interface. IBM J. Res. Dev. **20**(4), 336–357 (1976)
14. Sowa, J.F.: Conceptual structures: information processing in mind and machine (1983)
15. Thunkijjanukij, A., Kawtrakul, A., Panichsakpatana, S., Veesommai, U.: Lesson learned for ontology construction with Thai rice case study. In: World Conference on Agricultural Information and IT, IAALD AFITA WCCA 2008, pp. 495–502 (2008)

Machine-Understandable and Processable Representation of UNECE Standards for Meat. Bovine Meat - Carcases and Cuts Case Study

Robert Trypuz[1]([✉]), Piotr Kulicki[1], Przemysław Grądzki[1], Rafał Trójczak[1], and Jerzy Wierzbicki[2]

[1] Faculty of Philosophy, The John Paul II Catholic University of Lublin, al. Racławickie 14, 20-950 Lublin, Poland
trypuz@kul.pl
[2] Polish Beef Association, ul. Smulikowskiego 4, 00-389 Warszawa, Poland

Abstract. The paper offers a semantic representation of one of the sixteen UNECE standards for meat. The content of the standard is briefly presented in a natural language. Then, the methodology is discussed. Simple Knowledge Organisation System (SKOS) is chosen as the basis of the representation. A limited number of ontological object properties extending SKOS are used to represent relations among beef cuts such as the relation of *being prepared from*. The advantages of the representation are discussed, especially the possibility of the use of reasoning mechanisms, as well as its flexibility and robustness. This paper also sketches a vision of a larger project aimed at semantic representation of all UNECE meat standards and their joining into a network of Web vocabularies such as Schema.org and GS1. A web application to browse the UNCECE standard for beef and SPARQL endpoint to query it are also provided.

Keywords: Ontology · UNECE meat standards · E-commerce trade standards · Bovine meat · Beef cut

1 Introduction

It is generally agreed that standards, such as GS1[1], facilitate trade by providing a common vocabulary that can be used by both buyers and sellers during transactions. GS1 DataBars (e.g. GS1-128 Bar Codes) take advantage of GS1 standards to code information about products that can then be easily exchange between the trade partners within the GS1 Global Data Synchronization Network.

The United Nations Economic Commission for Europe (UNECE) provided a coding system enabling the identification of a broad range of meat and poultry products that has been incorporated in to the GS1 system (the details of this implementation are not relevant for this paper).

---

[1] http://www.gs1.org.

© Springer International Publishing AG 2016
E. Garoufallou et al. (Eds.): MTSR 2016, CCIS 672, pp. 144–154, 2016.
DOI: 10.1007/978-3-319-49157-8_12

Each standard for meat, established and promoted by UNECE, is an example of a good practice. However, it is currently a static document (printed or in either PDF or DOC format). Such documents are themselves difficult to process, search for particular fragments, and compare. For instance, it is problematic to compare versions of catalogs, to track changes and updates of the files.

Thus, to take a full advantage of the UNECE standards their computer processable versions are needed. In this paper we present a representation of the UNECE standards for beef - one of sixteen UNECE standards for meat[2]. We treat it as a case study that shows benefits of such a representation. It is the first stage of a larger project that is targeted towards translating all of the UNECE meat standards into a machine-understandable representation.

The success of this program requires that every standard within the UNECE beef standards finds is counterpart inside such a representation and that the final product has advantages (thereby justifying its creation). In the following sections we will show and discuss both of these aspects.

We also intend to place our representation of the UNECE standards for meat within the wider context of Web vocabulary. It is a fact that GS1 Web vocabulary[3] is being aligned with schema.org vocabulary and their future integration is expected. Both are designed to support richer product description in the Internet enabling a more precise Web search.

That is why our representation follows the Linked Data good practice. All the elements from the UNECE standards have their own dereferenceable IRI. Their content of the representation is accessed by a SPARQL Endpoint. We would also like to propose our representation to be used by Schema.org. IRIs of cuts of beef or muscles can serve as the external enumerations for Schema.org types. We are going to submit to Schema.org a proposal of hosted extension consisting of a few types and properties that are related to meat[4].

The proof of concept of an ontology-based computer processable representation of UNECE standards for meat was already presented in [5]. In the present paper this idea is developed and a different representation is used. We find it at the same time both simpler to apply and more flexible.

2   UNECE Standard for Beef

The UNECE beef standards [1] recommend an international language for beef carcases and cuts marketed as being fit for human consumption. It provides purchasers with a variety of options for meat handling, packing, and conformity assessment for meat intended to be sold on the international market. It has been established in 2004 and has been regularly reviewed since then. The most recent version comes from 2014 (althought it is still named "2012 Edition").

The UNECE standards for meat consist of several elements. First of all they stipulate the minimal requirements for meat to be approved for trade such as

---

[2] All from are here: http://www.unece.org/trade/agr/standard/meat/meat_e.html.

[3] http://gs1.org/voc/.

[4] See: http://sdo-meat.appspot.com and https://www.w3.org/community/meat/.

FOREQUARTER 1063

Forequarter is prepared from a side (1000) by the separation of the forequarter and hindquarter (1010) by a cut along the specified rib and at right angles to the vertebral column through to the ventral portion of the flank.

To be specified:

- Rib number required (5 to 13 ribs).
- Diaphragm retained or removed.

ITEM NO.

1060 (10-rib)	1065 (5-rib)
1061 (11-rib)	1066 (7-rib)
1062 (12-rib)	1067 (9-rib)
1063 (13-rib)	1068 (8-rib)
	1064 (6-rib)

Fig. 1. Forequarter specified in [1, p. 23]

being free from visible blood clots, bone dust, visible foreign matter, offensive odors, freezer-burn etc. Furthermore they stipulate a system of codes allowing to describe different features of meat such as the way it is refrigerated (chilled, frozen, deep-frozen or other), the properties of animals it comes from (bovine category understood as a combination of sex and age of slaughter), production system (intensive, extensive, organic, other specified) feeding system (grain fed, forage fed or mixed, other specified) slaughter system (conventional, kosher, halal, other specified). Post-slaughter system is also discussed but, instead of codes, several suggestions on specification are provided. Moreover, codes representing maximum fat thickness and structure of carcases, sides and cuts, different options of packing, and weight ranging of carcases and cuts are provided. A separate coding system is devoted to the different levels of bovine quality systems (official, company or industry). Standard values of meat color, fat color and meat marbling are provided using photographs.

Parts of the carcase are defined along with a list of bovine muscles. For identification purposes each part and each muscle receives a numerical symbol. Cuts and parts are specified in different language versions: English, French, Russian, Spanish and Chinese. The way a beef cut is obtained is provided for most of the meat products (see Fig. 1).

Muscle names are presented in Latin, as it is the international language for anatomical practice. Muscle content of particular cuts is provided. The standard is illustrated with photographs and figures showing cuts and relations between muscles and cuts (see Fig. 2).

In this paper we limit ourselves only to the representation of cuts of beef and parts of a carcase but our approach is flexible enough to include other properties of meat and its production, environment and other factors as described by UNECE standards.

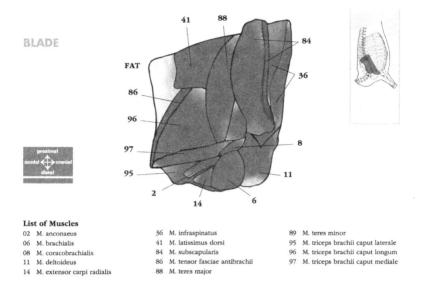

Fig. 2. Blade muscles, [1, p. 53]

List of Muscles

02	M. anconaeus	36	M. infraspinatus	89	M. teres minor
06	M. brachialis	41	M. latissimus dorsi	95	M. triceps brachii caput laterale
08	M. coracobrachialis	84	M. subscapularis	96	M. triceps brachii caput longum
11	M. deltoideus	86	M. tensor fasciae antibrachii	97	M. triceps brachii caput mediale
14	M. extensor carpi radialis	88	M. teres major		

3   Machine-Understandable and Processable Representation of UNECE Standard for Bovine Meat

3.1   Methodological Assumptions

We decided to represent the UNECE standards for beef as it is, allowing in some places for ambiguity that can be eliminated later when the standard changes itself as a result of discussions within the community. So we did not introduce any "improvements" that would change the meaning of information contained in these standards.

A minimal ontological engagement was expected. That allowed us to postpone some ontological choices for the moment until we will have an opportunity to discuss some controversial aspects of the UNECE standards for meat with the Working Party on Agricultural Quality standards. That was also a driving reason to choose SKOS (Simple Knowledge Organisation System) Core Vocabulary[5] as a basis of our representation. The representation is expressed in RDF Turtle serialization of SKOS and below we shall also use this serialization for its explanation.

SKOS expressive power was enough to represent many important aspects of the UNECE standard. We used skos:member to link collections with concepts and lexical labels to provide names of cuts, muscles and other structures in the various languages. We have also used SKOS definition to explain the meaning of the concepts (see Sect. 3.2 for more details). Using SKOS language has also this advantage that any representation expressed in it is easily understandable and queried.

---

[5] http://www.w3.org/TR/skos-reference/.

It was intended to express explicitly all important relation between objects described in the UNECE standard. That requirement moved us beyond SKOS. We added three ontological object properties together with their inverses and three data properties. We also added one OWL2 ObjectPropertyChain axiom to improve reasoning (see Sect. 3.3 for more details).

In this paper we present only the semantic representation of carcases, cuts, and muscles. So the following information from the standard will not be discussed here: refrigeration methods, bovine category, production system, feeding system, slaughter system, post-slaughter system, fat thickness, meat quality, weight ranging, meat and fat color and pH and packing system.

3.2 SKOS Part of the Representation

In the UNECE standards for bovine meat we have identified four collections of concepts. We list them below taking into account that the canonical URI of our representation is http://onto.kul.pl/unece/beef[6].

```
1 :BeefBoneIn a owl:NamedIndividual , skos:Collection ;
2    skos:prefLabel"Beef bone-in"@en .
3 :BoneLessBeef a owl:NamedIndividual , skos:Collection ;
4    skos:prefLabel "Boneless beef"@en .
5 :Muscle a owl:NamedIndividual , skos:Collection ;
6    skos:prefLabel "Muscle"@en .
7 :OtherStructure a owl:NamedIndividual , skos:Collection ;
8    skos:prefLabel "Other structure"@en .
```

Listing 1.1. Instances of skos:Collection

By skos:member each of the collections are linked to its members. Thus, for instance :BeefBoneIn by skos:member is related to all concepts and ordered collections representing beef cuts containing bones and :OtherStructure collects parts of carcase other than beef cuts and muscles.

skos:Concept contains (as instances): 188 beef cuts (i.e. bone-in and boneless beef cuts), 100 muscles and 8 other structures. Each instance of skos:Concept is linked with only one skos:Collection. In addition some beef cuts are also members of skos:OrderedCollection.

The UNECE standards for beef has 21 groups or collections of cuts. Each such a group collects different ways of preparing a cut. For instance, Forequarter can have 5, 6,... or 13 ribs (see Fig. 1). Each of the specific Forequarter cuts has its own item code (e.g. Forequarter (5-rib) has code 1065). All of the Forequarter group members are grouped and are identified by under code 1063. Such collections of beef cuts are instances of skos:OrderedCollection.

One can notice that there is an ambiguity here. The group and Forequarter (13-rib) have the same code (see Fig. 1). In general it is often the case that the number of a set of cuts is the same as one of its member. This ambiguity has further consequences. For almost every cut it is stated from which other cut is

---

[6] Our representation is available here: http://onto.kul.pl/unece/beef/unece-beef.ttl.

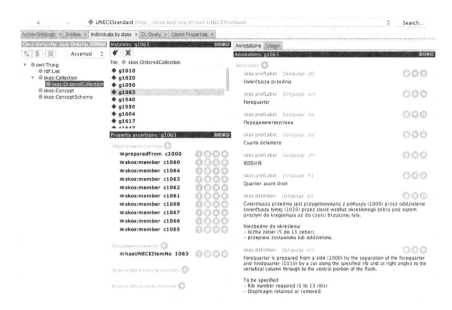

Fig. 3. Description of Forequarter from UNECE Standard for Bovine Meat - Carcases and Cuts, 2012 Edition, in our representation in Protege.

it prepared. For instance, having in the UNECE standards a description "Blade is prepared from a forequarter (1063) by following the natural seam between the ribs and [...]" We cannot be sure what it is meant here. That the Blade can be obtain from:

- all members of Forequarter group or
- some members of Forequarter group or
- Forequarter (13-rib)

In such cases we chose to connect a cut with a group of cuts which can be interpreted as one of the two first options listed above. Representation of Forequarter ordered collection in Protege editor is depicted in Fig. 3. It is also worth noting that each skos:Concept and skos:Collection is described by the following annotations:

1. *skos:prefLabel* that provides names of cuts and other structures in many languages. Muscles have both Polish and Latin names, and the rest of structures have at least Polish and English names. Some of them have also Spanish, Chinese, Russian or French names.
2. *skos:definition* that describes an element. Members of ordered groups do not have definitions.

3.3   OWL Part of the Representation

In our representation of the standard we have three data properties. :hasUNE-CEItemNo provides codes for each beef cut, muscle and other structures. :hasUNECEItemNoFrom and :hasUNECEItemNoTo should appear together and provide an interval of UNECE codes for groups of beef cuts.

Important aspect of the representation is from which element a certain cut is prepared. :preparedFrom object property is used for that purpose.

```
1 :preparedFrom a owl:ObjectProperty ;
2    rdfs:subPropertyOf :partOf.
3 :partOf a owl:ObjectProperty , owl:TransitiveProperty.
```

Listing 1.2. Prepared from

We have generalized the :preparedFrom relation by parthood (:partOf) relation. It is reasonable assumption. If a cut of beef X is prepared from other cut Y then X has to be a part of Y. Of course, this relations is true before the real cutting will take place. :partOf is assumed to be transitive. This assumption gives interesting new information when reasoners are used. For instance for Blade Bolar (:c2302) described as below and :preparedFrom relations established between other beef cuts Pellet gives us that Blade Bolar is a part of (:partOf) Blade (clod) (:c2300), Bone-In Shoulder (:c1621), Forequarter (:g1063), Side (:c1000), Whole Carcase (:c1001) (see also Fig. 4).

```
1 :c2302 a owl:NamedIndividual , skos:Concept ;
2    skos:prefLabel"Blade Bolar"@en
3    :memberOf :BoneLessBeef ;
4    :preparedFrom :c2300 .
```

Listing 1.3. Blade Bolar (code 2302)

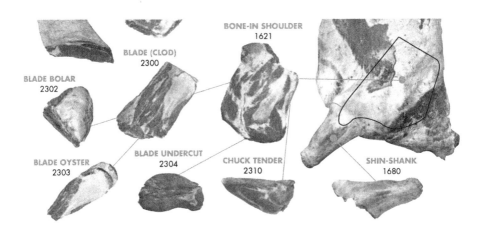

Fig. 4. Some standard bovine primal cuts, [1, p. 19]

The mereological relation between muscles and primal cuts is expressed by object property :overlaps. In Fig. 2 we see how the relation is expressed in UNECE standard. We represent information from the figure in Turtle as follows:

```
1  :c2300 a owl:NamedIndividual , skos:Concept ;
2     :memberOf :BoneLessBeef ;
3     :overlaps :m0002 , :m0006 , :m0008 , :m0011 , :m0014 , :m0036 , :m0041 ,
          :m0084 , :m0086 , :m0088 , :m0089 , :m0095 , :m0096 , :m0097 ;
4     :partOf :c1621 ;
5     :preparedFrom :g1063 ;
6     skos:prefLabel "Blade (Clod)"@en .
```

Listing 1.4. Parthood and Overlap

For reasoning purposes we assumed that :overlaps is symmetric and that property Chain Axiom stating that the composition of properties :overlaps and :partOf is a subproperty of :overlaps.

```
1  :overlaps a owl:ObjectProperty , owl:SymmetricProperty;
2     owl:propertyChainAxiom (:overlaps :partOf).
```

Listing 1.5. Overlap

That gives us that all muscles that overlap a cut :c2300 overlap also all cuts that :c2300 is a part of (e.g. :c1621 or :g1063).

Shin-shank (:c1680) is interesting in this context. We have that Forequarter Shin - Shank (c:1682) and Hindquarter Shin - Shank (c:1683) are parts of Shin-shank (:c1680). We also have axioms stating that some particular muscles overlap Forequarter Shin - Shank and some other overlap Hindquarter Shin - Shank. From that we get that all of them (24 muscles in total) also overlap Shin-shank.

To finalize and complete this section it should be mentioned that we have introduced also three inverses for object properties:

```
1  :hasPart a owl:ObjectProperty ;
2     owl:inverseOf :partOf ;
3     a owl:TransitiveProperty .
4  :contains a owl:ObjectProperty ;
5     rdfs:subPropertyOf :hasPart ;
6     owl:inverseOf :preparedFrom .
7  :memberOf a owl:ObjectProperty ;
8     owl:inverseOf skos:member .
```

Listing 1.6. Inverses

3.4   Problems

We have already mentioned in Sect. 3.2 that there are situations within the UNECE standards in which an item code of a collection of cuts is the same as a code of one of the members of this collection. Such a situations does not cause serious problems in static version of the standard, except for the risk of misunderstandings. However, in a computer processable representation the consequences of such ambiguities may be more destructive. Thus, we had to propose a solution: we have provided different IRI for cuts and their collections having the same UNECE item codes.

We also address a few mistakes or typos within the UNECE standards. For instance in description of cut named "Inside cap off side muscles removed" we could find that "Topside Cap Off is prepared from the Topside (item 2000)". Name "Topside" was used here instead "Inside" and the item number "2000" was also misleadingly provided instead of 2010. We used corrected version following Polish translation of the English version of the UNECE standards.

Another problem concerns the fact that not all cuts "prepared from" property was provided (see e.g. cuts with UNECE code 2180 and 2133). In such cases we did not try to figure out ourselves what the proper cut for the "prepared from" relations are. We have just omitted this information.

Some definitions of cuts are ambiguous. For instance in the following description "Pectoral Meat is remaining portion of the M. pectoralis profundus muscle located in the chuck [...]". It is unclear to which cut the name "chuck" refers to. We have e.g. Chuck - Square Cut, Chuck Roll, Chuck Eye, Chuck Tender and a few more. In such cases we did not decide ourselves which cut is meant by the standard's authors.

Some further ambiguities concern item numbers. For instance in three places of UNECE catalog for beef we find a reference "butt (1500-1503)". It is unclear whether "1500-1503" is meant to be and interval or not. In our opinion it is not an interval because there is no item no. 1501 and a cut possessing item number 1502 has a different name, i.e. "butt & rumb". We interpreted "1500-1503" as "1500 and 1503".

There were also some problems in the Polish translation of the UNECE catalog for beef. For instance a cut number 2483 had confusing translation (See http://onto.kul.pl/unece/beef/c2483 for comparison.).

The above remarks show a need for another revision of these standards, taking into account problems discovered during the preparation of its computer processable representations. It is clear that such a representation of the UNECE standards makes it easier to carry out any data correctness and coherence check.

3.5   Advantages of Semantic Representation of UNECE Standards

First of all semantic representation requires the precision and consistency of the content. Preparation of it has already proven its usefulness by pointing out the internal problems of the UNECE standards discussed in the previous section.

We have already shown how reasoning expands the knowledge about connections between parts of the carcase. That new knowledge is a direct result of applying semantic representation and the standard automatic reasoning tools. Even if the new information is not of a great importance it enriches our system. The reasoning mechanisms allows also to prove the consistency of the standard.

Moreover, the representation is flexible enough to be extended by new elements, e.g. new names of cuts and their descriptions in chosen languages; new parthood and overlaps links or any connections between items established by new relations extending the representation.

The description of meat cuts and parts and the information about they relations to animal muscles taken from the UNECE catalog and represented in the semantic way can be incorporated into information system designed for meat research such as the Animal Trait Ontology for Livestock (ATOL) [2] or the OntoBeef thesaurus and knowledge base [3,4]. In the later, notions taken from the UNECE catalog are already present but the OntoBeef ontology lacks an adequate representation of their relations. The relation between muscles and cuts of meat are especially important for the purpose of connecting research with industry since researchers usually investigate the properties of muscles and industry uses parts and cuts.

Our semantic representation can be used as a component of different information systems. A simple application allowing for user friendly browsing of the content of the standard can be found here: http://onto.kul.pl/unece/beef. One can easily check that the IRIs of the standard resources are dereferenceable – by using http://onto.kul.pl/unece/beef/c2030 an agent gets a website with information about the resource and by adding .rdf, .ttl or .json to the IRI he/she/it gets information about it in a desired serialization.

The content of our representation can be also accessed by SPARQL Endpoint[7]. It can be used by external agents to obtain information about parts of carcase to their own purposes. We can imagine many possible applications based on our representation including the ones dealing with cooking recipes, cut optimization, etc. Below we propose an example of a query that looks for all beef cuts (and their codes) overlapped by M. serratus ventralis thoracis (:m0079):

```
1 PREFIX : <http://onto.kul.pl/unece/beef/>
2 PREFIX skos: <http://www.w3.org/2004/02/skos/core#>
3 SELECT DISTINCT ?beefCutLabel ?UNECENumber
4 WHERE {
5     :m0079 ^:overlaps/:preparedFrom* ?beefCutIRI.
6     ?beefCutIRI skos:prefLabel ?beefCutLabel;
7         :hasUNECEItemNo ?UNECENumber
8     FILTER(lang(?beefCutLabel) ="en") }
9 ORDER BY DESC(?UNECENumber)
```

Listing 1.7. Find all beef cuts overlapped by M. serratus ventralis thoracis (:m0079)

As mentioned in the introduction, we would also like to propose our representation to be used by Schema.org. IRIs of beef cuts or muscles can serve as the external enumerations for Schema.org types. We are going to submit to Schema.org as an external extension consisting of a few types and properties that are related to meat.

4   Conclusions

In this paper we presented a semantic representation of beef standard – one of the sixteen standards for meat promoted by UNECE. We focused on the representation of codes for all parts of the carcase and relations that hold among beef cuts (e.g. "prepared from" property) and between them and muscles. Importance and benefits of such representation were discussed.

The main advantage of the representation is its usefulness for potential applications. To show how they can work we provided a web application presenting the cuts and muscles from the UNECE catalog. We also established a SPARQL endpoint to enable external access to it.

It is a first stage of a larger project that targets towards translating of all remaining UNECE meat standards into machine-understandable and open data representation. We also plan to extend current representation of UNECE standard for beef on the remaining information encoded in the UNECE standard, i.e.: refrigeration methods, bovine category, production system, feeding system, slaughter system, post-slaughter system, fat thickness, meat quality, weight ranging, meat and fat color and pH and packing system. The integration of the presented representation within schema.org and GS1 web vocabularies is also part of prospective future work.

Acknowledgments. We would like to thank Dagmara Siek for her help in the initial phase of building the semantic representation of UNECE standard for beef.

References

1. UNECE Standard, Bovine Meat - Carcases and Cuts. Ece/trade/326/rev.1, United Nations Economic Commission for Europe (2013). http://www.unece.org/trade/agr/standard/meat/meat_e.html
2. Golik, W., et al.: ATOL: the multi-species livestock trait ontology. In: Dodero, J.M., Palomo-Duarte, M., Karampiperis, P. (eds.) MTSR 2012. CCIS, vol. 343, pp. 289–300. Springer, Heidelberg (2012). doi:10.1007/978-3-642-35233-1_28
3. Kulicki, P., Trypuz, R., Trójczak, R., Wierzbicki, J., Woźniak, A.: Ontology-based representation of scientific laws on beef production and consumption. In: Garoufallou, E., Greenberg, J. (eds.) MTSR 2013. CCIS, vol. 390, pp. 430–439. Springer, Heidelberg (2013). doi:10.1007/978-3-319-03437-9_42
4. Trójczak, R., Trypuz, R., Gradzki, P., Wierzbicki, J., Wozniak, A.: Evaluation of beef production and consumption ontology and presentation of its actual and potential applications. In: Proceedings of FedCSIS (2013)
5. Trypuz, R., Kulicki, P., Wierzbicki, J., Woźniak, A., Carlhian, B.: Vers des standards automatisés dans la production de viande? Viandes & Produits Carnés, 30 (2014). VPC-30-4-1 Date de publication: 08 juillet 2014

PO^2 - A Process and Observation Ontology in Food Science. Application to Dairy Gels

Liliana Ibanescu[1]([✉]), Juliette Dibie[1], Stéphane Dervaux[1], Elisabeth Guichard[2], and Joe Raad[1]

[1] UMR MIA-Paris, AgroParisTech, INRA, Université Paris-Saclay,
75005 Paris, France
{liliana.ibanescu,juliette.dibie,stephane.dervaux,
joe.raad}@agroparistech.fr
[2] UMR 1324 INRA, UMR 6265 CNRS, Université de Bourgogne,
21000 Dijon, France

Abstract. This paper focuses on the knowledge representation task for an interdisciplinary project called Delicious concerning the production and transformation processes in food science. The originality of this project is to combine data from different disciplines like food composition, food structure, sensorial perception and nutrition. Available data sets are described using different vocabularies and are stored in different formats. Therefore there is a need to define an ontology, called PO^2 (Process and Observation Ontology), as a common and standardized vocabulary for this project. The scenario 6 of the NeON methodology was used for building PO^2 and the core component is implemented in OWL. By making use of PO^2, data from the project were structured and an use case is presented here. PO^2 aims to play a key role as the representation layer of the querying and simulation systems of Delicious project.

Keywords: Process and Observation Ontology · Domain ontology

1 Introduction

Recently, Europe faces two societal challenges: the increasing of overweight and obesity and the population aging. These problems, while having a tremendous impact on population life quality (e.g. poor health, social exclusion, increase in the need of assistance), are challenging the food industry to develop new strategies to produce well-balanced products in terms of nutritional requirements (e.g. less fat, sugar and salt) while using sustainable transformation processes. It is therefore crucial to better understand the food production system and a very interesting issue is to combine data and knowledge from different disciplines, like food composition in terms of nutrition, food digestion as a physiological process and sensorial perception of food.

Delicious project addresses the problem of analyzing the production and transformation processes of dairy gels using information available from different

© Springer International Publishing AG 2016
E. Garoufallou et al. (Eds.): MTSR 2016, CCIS 672, pp. 155–165, 2016.
DOI: 10.1007/978-3-319-49157-8_13

collaborative projects concerning the food composition, food structure, mobility/bioavailability of flavor compounds and nutrients, sensory perception and digestibility. It involves domain experts and computer scientists researchers from INRA, the French National Institute for Agricultural Research. The expected result of Delicious project is to collect and structure the available data and knowledge into a data warehouse in order to enhance the analysis of the production process according to different cross-domain criteria. However, it is very difficult to take advantage of all the available data and knowledge from Delicious project. The main difficulty comes from the heterogeneity of their sources, the different inter-domain or cross-domain vocabularies, the different formalisms used according to the involved domain. A second challenge concerning the data integration task is the uncertainty quantification such as randomness, incompleteness, imprecision, vagueness, resulting from the natural variability of the domain and the lack of information. In order to address the question of the integration of knowledge and data, a relevant solution is the use of an ontology [4]. An ontology can be defined as a formal common vocabulary of a given domain, shared by the domain experts [7].

This paper present the Process and Observation Ontology, called PO^2, designed for Delicious project. The scenario 6 of the NeON methodology [2], i.e. reusing, merging and re-engineering ontological ressources, was used for building PO^2. The core component is implemented in OWL[1] and the domain component is under development.

By making use of the PO^2 vocabulary, the data sets available for the project were well-structured for the integration task. An use case is presented in order to show the complexity of this task.

This first step of building the PO^2 ontology allows to structure and organize the knowledge into a meaningful model at the knowledge level. This will lead to the possibility of designing more complex decision support systems allowing to compare different production scenarios and therefore suggesting improvements concerning the product quality while reducing the environmental impact. It may also help the field by giving hints about what data should be collected in order to perform an analysis concerning a target population (e.g. children or old people) or an cause and effect analysis. It may also provide the French food industry with the necessary tools to anticipate and develop future food products.

The paper is organized as follows. In Sect. 2, we present the ontology specification. In Sect. 3, the conceptualisation of PO^2 is detailed. In Sect. 4, we illustrate PO^2 through a use case. Finally, we conclude in Sect. 5 and present our further work.

2   Ontology Specification

Ontology specification was done during an iterative process. The ontology developers and the domain experts had a lot of meetings in order to identify (1) why

---

[1] https://www.w3.org/2001/sw/wiki/OWL.

the domain experts want to build an ontology (i.e. for what purpose), (2) what its intended users will be and (3) what are the main entities.

First, the purpose of building an ontology is to provide a consensual model of the production and transformation of dairy gels and to solve the lack of communication between domain experts. Available data were gathered for many different purposes by different experts with their own experimental itineraries, vocabularies and technical materiel and methods. There is an obvious need to build a common and shared structured vocabulary.

Second, the intended users are researchers in several distinct domains: nutrition, microbiology, biochemistry, physico-chemistry, chemistry, process engineering, food science and sensory analysis. Reaching a consensus about a common vocabulary was therefore a hard task. The ontology developers and the 15 domain experts involved in Delicious project spent about 20 h using CMap Tool[2] to identify a vocabulary common to all the involved experts. The resulting vocabulary was unstructured and composed of approximately 500 entities dealing with composition, structure, technical and physiological transformation processes, mobility and bioavailability of small molecules in relation with sensory perception and nutritional value. It proposes a first representation of the explicite and implicite knowledge of all the involved domain experts.

Third and finally, in order to investigate how to structure the vocabulary, we focused on a small representative subset of data and knowledge concerning the *In the mouth* process. Taking into account the previously identified entities, relying on available documents [1,5] and data and in close collaboration with domain experts of the target domain, entities were grouped into three main parts (see Fig. 1):

- the part concerning the production and transformation process which contains the concepts: process, itinerary and step;
- the part concerning the participant which contains the concepts: product, mixture, material and sensing device;
- the part concerning the observation which contains the concepts: observation, scale, sensor output, computed observation, method and measure.

We therefore reached a consensus about a common structured vocabulary with the following specifications. An itinerary is an execution of a production or transformation process, i.e. a set of interrelated steps. A step is characterized by its participants and its temporal duration/interval. A participant may be a mixture, a material or a sensing device. Each participant is characterized by its experimental conditions. Moreover a mixture is characterized by its composition. An observation observes a participant at a certain scale during a step. It is characterized by some participants such as a given material or a sensing device and implements a method. It has for result a sensor output and/or a computed observation, each of them can have for value a function or a simple measure. A measure is characterized by either a quantity and a unit of measure or a symbolic concept and a measurement scale.

---

[2] http://cmap.ihmc.us.

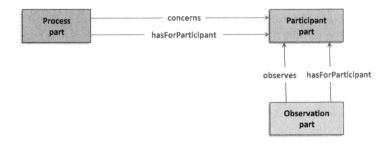

Fig. 1. The three main parts of the ontology for Delicious

3 Ontology Conceptualisation

The ontology conceptualization follows the Scenario 6 of the NeON methodology [2], i.e. reusing, merging and re-engineering ontological ressources. A number of existing ontologies have been analyzed: the supply chain ontology [6], the bussiness process ontology [9], the ontology for wine production [8], SSN^3, BFO^4, IAO^5 and $[MS]^2O$ (Multi Scales and Multi Steps Ontology)[6].

Based on our experience and after a careful analysis, it was decided that the best method to adopt for building the PO^2, Process and Observation Ontology, is to re-engineer the core component of $[MS]^2O$, an ontology designed for a project concerning the representation of the production of stabilized micro-organisms (see [3] for more details). This re-engineering task of $[MS]^2O$ was done with the two following main concerns:

- establish a clear distinction between a process and its participants which was achieved by reusing BFO;
- link all together the observations with the step where they occur, their participants, their materials and methods and their measures reusing IAO (Information Artifact Ontology) an ontology of information entities.

The PO^2 core component is given in Fig. 2. The concepts identified in Sect. 2 during the ontology specification are represented as nodes and the relations between the concepts are represented as arrows.

The PO^2 core component is implemented in OWL and it is available at http://agroportal.lirmm.fr/ontologies/PO2. The domain component is under development.

---

[3] http://www.w3.org/2005/Incubator/ssn/XGR-ssn-20110628.

[4] https://bioportal.bioontology.org/ontologies/BFO.

[5] https://bioportal.bioontology.org/ontologies/IAO.

[6] http://lovinra.inra.fr/2015/12/16/multi-scale-multi-step-ontology/.

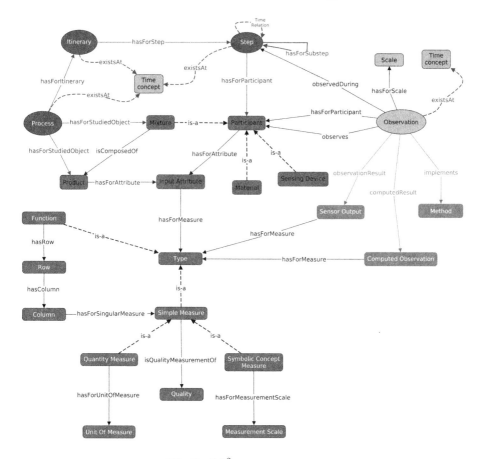

Fig. 2. PO^2 core component

4 PO^2 Use Case

This section presents an use case concerning the *In the mouth* process, in order to show the complexity of the representation task.

At the beginning of the Delicious project, data and knowledge concerning this process were available in different vocabularies and formats. By making use of the vocabulary from the PO^2 core component presented in Sect. 3, the data concerning the studied use case were structured into 20 EXCEL files:

- 2 files describe the *In the mouth* process (e.g. Fig. 3),
- 11 files describe the mixture composition (e.g. Fig. 4),
- 6 files describe experimental observations (e.g. Fig. 6), and
- 1 file describes the materials and methods with 29 methods and 16 materials (e.g. Figs. 8 and 9).

Date	2012
Project name	CARREDAS PhD BOISARD
Sample name	cheese model
Sample code number	L20P28

Step	Sub-step	Measure	Scale	Files numbre	Data type
Before putting in the mouth	NA	Composition	Mixture	1	Raw
Before putting in the mouth	NA	Itinerary	Mixture	1	Raw
Before putting in the mouth	NA	Characteristics	Mixture	10	Raw and computed
In the mouth	Chewing	Study of the sodium release in the saliva	Mixture	1	Raw
In the mouth	Chewing	Chewing activity	Step	1	Raw and computed
In the mouth	Chewing	Study of the flavour components release in the air	Mixture	1	Computed
In the mouth	Chewing	Sensory properties	Mixture	1	Raw and computed
In the mouth	Swallowing	Swallowing number	Step	1	Raw
In the mouth	Swallowing	Study of the flavour components release in the air	Mixture	1	Raw and computed

Fig. 3. The EXCEL file which describes the *In the Mouth* process

Let us notice that these EXCEL files allow the domain experts to collect and re-structure the available data using the PO^2 vocabulary. Moreover, these files can be automatically translated into instances of PO^2 (see e.g. Figs. 5 and 7).

In Fig. 3 the description of *In the mouth* process is given: it contains one itinerary which is composed of two steps: the *Before putting in the mouth* step and the *In the Mouth* step. The last step is composed of two sub-steps: *Chewing* and *Swallowing*.

This process has for studied object a sample of the mixture *cheese model* identified by the code number *L20P28*. This mixture is composed of ten products as described in Fig. 4, each product being characterized by the input attribute *Weight*.

Figure 5 gives an example of an instance extracted from Fig. 4: the mixture *L20P28* is composed of the product *Rennet casein* where its input attribute *Weight* has for simple measure the value 238.3 of unit of measure g/kg *of cheese model.*

Product description	
Sample code number	L20P28
Name of the description file	2012-CAREDAS-001-FicheDescriptif

Raw data		
Characteristics	Value	Unit
Rennet casein	238.3	g/kg of cheese model
Acide casein	59.6	g/kg of cheese model
AMF	200.2	g/kg of cheese model
Deionized water	464.4	g/kg of cheese model
Melting salt	25	g/kg of cheese model
Added Nacl	0	g/kg of cheese model
Citric acid	2.5	g/kg of cheese model
Aromatic solution + water	10	g/kg of cheese model
lipids	20	g/kg of cheese model
proteins	28	g/kg of cheese model

Fig. 4. The EXCEL file which describes the composition of the mixture *L20P28*

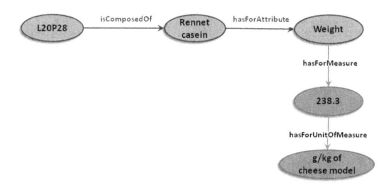

Fig. 5. An example of instance concerning a mixture and its composition

Let us now focused on an experimental observation of the *In the Mouth* process as described in the EXCEL file of Fig. 6. This instance of observation, called in the following *Observation1*, has the following properties (see Fig. 7):

- is observed during the **sub-step** *Chewing* of the **step** *In the mouth*;
- observes the **mixture** *L20P28*;
- has for **participants** the two **materials**: *Material 1* and *Material 14* as described in Fig. 8;
- has for scale the molecular **scale**;
- has for **date** 10/09/2012;
- implements the two **methods**: *Method 22* and *Method 23*, both described in Fig. 9;
- has for observation result the **sensor output** *Sodium concentration in the saliva* which is function of the sodium concentration during time;
- has for computed result the **computed observation** *yield curve of the release* which has for **measure** the value 2.75 of **unit** mM.

General informations	
Date of the measure	10/09/2012
Time	
Duration of the manipulation	
Step	In the mouth
Sub-step	Chewing
Scale	Molecular
Repetitions	1

Product description	
Sample code number	L20P28
Name of the description file	2012-CAREDAS-001-FicheDescriptif

Material used	
Identifier	Material 1
Identifier	Material 14
File	2012-carredas etapesenbouche-001-M&M

Method used	
Identifier	Method 22
Identifier	Method 23
File	2012-carredas etapesenbouche-001-M&M

Raw data

Characteristics	value	Unit	Incertainty	Description
Sodium concentration in the saliva at T0	0	mM		
Sodium concentration in the saliva at T5	10	mM		
Sodium concentration in the saliva at T15	35	mM		
Sodium concentration in the saliva at T30	55	mM		
Sodium concentration in the saliva at Td	70	mM		

Computed data

Characteristics	value	Unité	Incertainty	Description
Yield curve of the release = sodium concentration by tooth bite	2.75	mM		

Fig. 6. The EXCEL file which describes an experimental observation during the sub-step *chewing* of the step *In the mouth* for the mixture *L20P28*

What it is interesting to report about our experience with this use case is that the process of building the ontology is an iterative one. Notice that the EXCEL files of Figs. 3, 4, 6 and 8 contain well structured data and knowledge, but the EXCEL file of Fig. 9 describing the methods with many textual informations, is currently unusable for automatic querying. Domain experts were not able up to now to express their needs about the querying concerning the different methods they used in the different domains. The lessons they learned while they organized and structured their data and knowledge according to the concepts from PO^2 give them the understanding that allow to refine the specification concerning the methods. This is an ongoing process.

To conclude, we would like to stress on the fact that the complexity of the knowledge representation task of this use case allows us to identify a common and shared structured vocabulary that encompasses almost all the domains involved in the Delicious project.

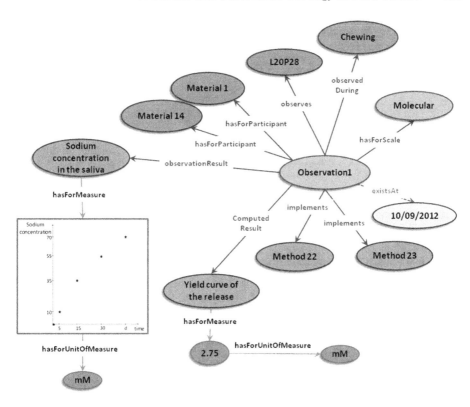

Fig. 7. An example of instance representing an experimental observation *Observation1* during the sub-step *Chewing* for the mixture *L20P28*

Material 14		
Sensing device	x	
Transformation device		
Device used	HPLC (dionex ICS 3000)	
Column	IonPac CS12A 5µm	
Column temperature	25	°C
Eluent	Sulphuric acid 11mM	
Eluent flow	0,5	mL/min
Detector	conductometer	
Suppressor	CSRS 300 in 2mm (courrent 33 mA)	
Software	UCI 100 Chromeleon	
Additional information	The sodium concentrations were obtained from a calibration curve	

Fig. 8. The EXCEL file which describes the *Material 14* used in the experimental observation *Observation1* of Fig. 7

Method 22	
Method name	Itinerary sample cut and saliva gathering
Method description at step 1	The sample of 5g of the cheese model L20P28 is proposed at 13°C
Method description at step 2	The saliva was spat during the chewing at T5 (5 tooth bites), T10, T15, T30 and Td (before swallowing) in 5 tubeS eppendorfs put after in a basin of crushed ice
Method description at step 3	The tubs were centrifuged during 15min at 15300g at 4°C
Method description at step 4	50µL were levyed from thes du supernatant et dropped again in a tube eppendorf with 950µL of water Milli-Q
Method description at step 5	2mL were levyed with a syringe in order to be filtered (filter in nylon 13mm, pore size 0.45 µL)
Method description at step 6	then put in a flat base flask of 2mL.

Method 23	
Method name	Determination of the sodium concentration in the saliva by HPLC
Method description at step 1	used of HPLC

Fig. 9. The EXCEL file which describes the *Method 22* and the *Method 23* used in the experimental observation *Observation1* of Fig. 7

5   Conclusion

In this paper we presented the building of PO^2, a Process and Observation Ontology, designed for a cross-domain project concerning the production and transformation of dairy gels. The core component of PO^2 is the result of re-engineering $[MS]^2O$, using BFO and IOA. A use case on an *In the Mouth* process was presented.

Further work is to express users requirements through competency questions and prioritizing those requirements. Then the domain component will be developed and the ontology will be validated against the competency questions.

PO^2 aims to play a key role as the representation layer of the querying and simulation system of Delicious project. This leads to the possibility of comparing different production systems and may also help to develop a decision support system taking into account the uncertainty of data.

The developed ontology could be further adapted to other types of food products, such as bakery, vegetable or meat products. This may provide to the French food industry tools in order to develop food products according to the nutritional recommendation for a healthy population while increasing efficiency and adopting an eco design approach.

References

1. Boisard, L., Andriot, I., Martin, C., Septier, C., Boissard, V., Salles, C., Guichard, E.: The salt and lipid composition of model cheeses modifies in-mouth flavour release and perception related to the free sodium ion content. Food Chem. **145**, 437–444 (2014)

2. Suárez-Figueroa, M.C., Gómez-Pérez, A., Fernández-López, M.: The NeOn methodology for ontology engineering. In: Suárez-Figueroa, M.C., Gómez-Pérez, A., Motta, E., Motta, A. (eds.) Ontology Engineering in a Networked World, pp. 9–34. Springer, Heidelberg (2012). doi:10.1007/978-3-642-24794-1_2
3. Dibie, J., Dervaux, S., Doriot, E., Ibanescu, L., Pénicaud, C.: $[MS]^2O$ – a multiscale and multi-step ontology for transformation processes: application to microorganisms. In: Haemmerlé, O., Stapleton, G., Faron Zucker, C. (eds.) ICCS 2016. LNCS (LNAI), vol. 9717, pp. 163–176. Springer, Heidelberg (2016). doi:10.1007/978-3-319-40985-6_13
4. Doan, A., Halevy, A.Y., Ives, Z.G.: Principles of Data Integration. Morgan Kaufmann (2012)
5. Feron, G., Ayed, C., Qannari, E.M., Courcoux, P., Laboure, H., Guichard, E.: Understanding aroma release from model cheeses by a statistical multiblock approach on oral processing. PLoS ONE **9**(4), 1–15 (2014)
6. Grubic, T., Fan, I.S.: Supply chain ontology: review, analysis and synthesis. Comput. Ind. **61**(8), 776–786 (2010)
7. Guarino, N., Oberle, D., Staab, S.: What is an ontology? In: Staab, S., Staab, R. (eds.) Handbook on Ontologies. International Handbooks on Information Systems, vol. 2009, pp. 1–17. Springer, Heidelberg (2009). doi:10.1007/978-3-540-92673-3_0
8. Muljarto, A.-R., Salmon, J.-M., Neveu, P., Charnomordic, B., Buche, P.: Ontology-based model for food transformation processes - application to winemaking. In: Closs, S., Studer, R., Garoufallou, E., Sicilia, M.-A. (eds.) MTSR 2014. CCIS, vol. 478, pp. 329–343. Springer, Heidelberg (2014). doi:10.1007/978-3-319-13674-5_30
9. Rospocher, M., Ghidini, C., Serafini, L.: An ontology for the business process modelling notation. In: Garbacz, P., Kutz, O. (eds.) Formal Ontology in Information Systems - Proceedings of the Eighth International Conference, FOIS , 22–25 September 2014, Rio de Janeiro. Frontiers in Artificial Intelligence and Applications, vol. 267, pp. 133–146. IOS Press (2014)

Track on Cultural Collections and Applications

Rich Semantics for Interactive 3D Models
of Cultural Artifacts

Leslie F. Sikos$^{(\boxtimes)}$

Flinders University, Adelaide, Australia
leslie.sikos@flinders.edu.au

Abstract. The automated processing of 3D models of cultural artifacts can be significantly improved with formally defined high-level structured descriptors. Despite the large number of multimedia ontologies, however, the semantic enrichment of 3D models still has open issues. Many 3D ontologies are semi-structured only, cover a very narrow knowledge domain, do not provide comprehensive coverage for geometric primitives, or do not exploit the full expressivity of the implementation language. This paper presents the first attempt to transform the entire XML Schema-based vocabulary of the latest version of the X3D ISO standard (ISO/IEC 19775, 19776, and 19777) to OWL 2, complemented by fundamental concepts and roles of the 3D modeling industry not covered by X3D. The result of this effort is the most comprehensive formally grounded 3D multimedia ontology to date with standard alignment, which can be used for the representation, annotation, and efficient indexing and retrieval of 3D models.

Keywords: Multimedia ontology · X3D · 3D annotation · 3D model retrieval · Cultural heritage

1 Introduction

Photos of artifacts do not always provide sufficient information demanded by researchers, students, and enthusiasts who, however, might benefit from interactive 3D models of the objects. Interactive 3D models are already utilized by a number of high-profile museums, such as the Smithsonian,[1] the British Museum,[2] and the Victoria and Albert Museum.[3] These 3D models have aided preservation efforts, broadened public accessibility, and are used for research and education alike [1].

The precise reconstruction of real-world objects requires shape measurements and spectrophotometric property acquisition, typically performed using 3D laser scanners, RGB-D depth cameras, Kinect depth sensors, structured light devices, photogrammetry, and photomodeling [2]. Beyond the dimension and shape of a model, additional information must be captured, such as textures, diffuse reflection, transmission spectra, transparency, reflectivity, opalescence, glazes, varnishes, enamels, etc. Many of these properties can be represented by low-level descriptors [3], which can be used for

---

[1] http://3d.si.edu/browser
[2] https://sketchfab.com/britishmuseum/models
[3] http://www.3dcoform.eu/x3domCatalogue/

© Springer International Publishing AG 2016
E. Garoufallou et al. (Eds.): MTSR 2016, CCIS 672, pp. 169–180, 2016.
DOI: 10.1007/978-3-319-49157-8_14

training machine learning systems for efficient information retrieval [4]. Low-level features, however, when stored as unstructured data, are limited in terms of machine-processability. Also, most low-level descriptors do not provide the meaning of the corresponding 3D scene. These issues can be addressed using Semantic Web standards, so that the structured representation of the corresponding concepts becomes machine-interpretable by linking it to formal definitions and related concepts. High-level descriptors can represent sophisticated concepts and the actual meaning of the visual content, however, they are typically annotated manually, because they are based on human background, knowledge, and experience. Since manual annotation is time-consuming, some software provide the option for collaborative annotation [5].

The formal structured knowledge representation of 3D models enables efficient annotation, segmentation, indexing, and retrieval. Ontology-based 3D model retrieval can be performed not only by textual descriptions, but also by 3D characteristics such as shape or material [6]. The corresponding concepts and properties leverage defini-tions from OWL ontologies and related concepts from the Linked Open Data (LOD) Cloud [7], and map them to geometric primitives [8]. However, the full exploitation of cultural heritage datasets requires ontology matching to solve hetero-geneity issues by discovering semantic links [9].

3D annotations have not been evolving head to head with 3D modeling. The most prominent specification for 3D annotations, *Extensible 3D (X3D)*, is now the industry standard for representing interactive 3D computer graphics in web browsers without proprietary plugins, and is described in ISO/IEC 19775, 19776, and 19777. The standard is supported both by industry-leading proprietary and open source 3D com-puter graphics software, such as AutoDesk 3ds Max (via a free plugin), AutoDesk Maya, AC3D, Modo, Blender, and Seamless3d. X3D is used directly on web sites as the successor of the *Virtual Reality Modeling Language* (VRML, ISO/IEC 14772–1:1997) for encoding 3D models and scenes. In 2009, Behr et al. introduced X3DOM, a scalable HTML/X3D integration model to integrate and update declarative X3D con-tent directly in the DOM tree of HTML documents [10].

Developed with XML schemas, the vocabulary of X3D is semi-structured, machine-readable, but not machine-interpretable, making the standard inefficient for automation, data sharing, and data reuse. This issue can be addressed by mapping the application-specific XML schemas of X3D to domain-specific machine-interpretable metadata terms using Semantic Web standards.

2 Related Work

The first OWL-based 3D graphics ontology, *OntologyX3D*, was created in 2004 by mapping X3D node elements into OWL classes [11]. This ontology, together with upper ontologies, served as the basis for a platform aimed at supporting the develop-ment of intelligent, interoperable 3D applications [12]. In the following years, the evolution of X3D resulted in several different approaches to map the standard to structured data. For example, a small subset of the X3D XML schemas was mapped to

RDFS by Gilson and Silva in 2006 as part of an ontology mapping aimed at translating SVG to X3D.[4]

In 2012, while attempting to map the X3D XML schemas to OWL, Petit et al. were faced with the consistency issues of the X3D standard [13]. The XSD to OWL transformation was done using XSLT. While the authors created a basic role hierarchy, the resulting ontology was not without flaws. Firstly, it defined property ranges with VRML and X3D datatypes, e.g., SFBool, MFInt32, rather than the globally deployed XML Schema datatypes, such as `xsd:boolean` and `xsd:integer`, which would have been a better choice to maximize interoperability. Secondly, the mapping was based on the assumption that X3D nodes are logically equivalent to OWL classes, which did not take ambiguous X3D objects into account. The XLink cross references have been mapped to OWL object properties, and the X3D element attributes to string literals. The ontology defined default property values as RDFS comments. Considering the large share of datatype properties in X3D, this approach did not give satisfactory results for all roles. The mapping also contained typos (e.g., `LocalyDefinedType`). At the time of writing, the ontology was not available online anymore.

Not all X3D-based research focus on the improvement of the X3D vocabulary. In 2014, for example, Yu and Hunter developed a 3D annotation software tool based on domain-specific ontologies and X3D terms, using open technologies and specifications, such as W3C's Open Annotation model, HTML5, jQuery, and WebGL [14]. While the user interface was quite intuitive for annotating cultural heritage objects, the authors mixed semi-structured data with structured annotations by using concepts from the X3D vocabulary directly (without semantic mapping to OWL). The 3D models have been generated using a 3D laser scanner as unstructured data (in VRML), then converted to the also unstructured Polygon File Format (PLY), and finally to semi-structured X3D, which was mixed with structured definitions from ontologies like Dublin Core to provide RDF output, rather than generating structured RDF data right from the start.

3   Ontology Engineering

To address the aforementioned issues, and to cover all the new features added to X3D since the previous OWL mappings of X3D, a novel OWL 2 ontology, the *3D Modeling Ontology*,[5] has been developed with standards alignment and mathematical grounding using description logic (DL) formalism. The X3D terms have been complemented by a large set of new concepts and roles used by industry-leading 3D modeling software, including 3ds Max and Maya.

---

[4] http://www.cs.swan.ac.uk/~csowen/SVGtoX3D/examples/X3D_OntologyRDFS.htm

[5] http://vidont.org/3d/3d.ttl

3.1   Modeling Challenges

During ontology engineering, a number of modeling issues have been identified and resolved, as detailed below. A new namespace structure has been introduced for the OWL 2 translation of the X3D standard with versioning support at http://vidont.org, taking into account the proposed (and not yet standardized) namespace structure, which is http://www.web3d.org/specifications/x3d-version.xsd. The file extension has been omitted from the new namespace structure, and the specifications directory replaced by 3d, i.e., http://vidont.org/3d/, which points to the latest version at any given time.[6] The web server that hosts the ontology has been set up to serve an RDF file for Semantic Web agents and HTML5 to web browsers using content negotiation, as suggested by the World Wide Web Consortium [15].

Transparent translation of the X3D XML schemas to first-order logic (FOL), description logics, or OWL was not possible, because the X3D nodes and fields do not directly correspond to the unary and binary predicates and constants of first-order predicate logic, nor to the description logic concepts, roles, and individuals, or the OWL classes, properties, and individuals. Furthermore, no relationships are defined for properties in the standard. For this reason, the XSD file of the current version of the standard proved to be inadequate for mapping the X3D vocabulary to OWL 2. This issue has been addressed by manually creating a new concept and role hierarchy. Due to the different aims and designs of its predecessors, no previously released mappings have been used in the proposed ontology. However, terms from general-purpose ontologies, such as Creative Commons, Dublin Core, DBpedia, FOAF, and Schema.org, as well as the domain-specific ontology VidOnt,[7] have been reused according to Semantic Web best practices [16].

Regarding the naming conventions, X3D properties have a greatly varying scope, as some properties apply to a large group of classes, while others to a particular class only. Also, many X3D terms represent objects that correspond to a class and a property (node/field) at the same time, leading to conceptual ambiguity issues (e.g., color, geoOrigin, TextureProperties). This was resolved by extending the role names to reflect more specific properties (which can be used even without context) and distinguish them from their concept counterparts. X3D also has consistency issues, such as multiple concepts have homonymous roles, e.g., color can be defined for BlendMode (the constant color used by the blend mode constant) and for Color-RGBA (the set of RGBA colors). This issue was resolved by extending the corresponding terms and declaring all domains with rdfs:domain. The multiple descriptions for a property defined for two different concepts with the same name is misleading in the standard, which was addressed by modifying, and sometimes extending, their descriptions, and declaring them using dc:description. In the X3D standard, there are ambiguous properties that are defined for multiple concepts

---

[6] This is abbreviated by the 3d: prefix for the concepts, roles, and individuals of the ontology.

[7] http://vidont.org

with a different meaning, are of a different type, or have a different range and domain. For example, the `bottom` object property of a cubemap defines the texture for the bottom of the cubemap. The `bottom` datatype property of a cone or cylinder specifies whether the bottom cap of the cone or cylinder is created. This has also been resolved by term extension. Regarding industrial 3D modeling terms, the differences between the terminology of X3D and 3D modeling software have also been considered. For example, bump map declarations used in 3ds Max are equivalent to the shader-specific vertex attributes of `FloatVertexAttribute` nodes in X3D, such as tangents and binormals.

Some X3D properties correspond to more than one datatype. For example, the `axisRotation` of `CylinderSensor` is either a vector or a floating point number. In the first case, `axisRotation` represents the local sensor coordinate system. In the second, it specifies whether the virtual cylinder's lateral surface or end-cap disks of the virtual geometry sensor are used for manipulation, or constrains the rotation output of the cylinder sensor. In such cases, two options were considered: either the extension of the property name to be more specific, or the implementation of the less restrictive datatype (applicable only if one of the datatypes is a superset of the other). The description fields have been amended accordingly by adding context to the value of `dc:description`. The X3D specification features its own datatypes, many of which are based on VRML datatypes (SFNode, SFColor, etc.). Wherever possible, all datatypes have been converted to standard XML datatypes (e.g., `x3d:SFBoolean` to `xsd:boolean`, `x3d:SFVec3f` to `xsd:complexType`). The standard declares URLs as strings, which have been declared in the new ontology using the more appropriate `xsd:anyURI` datatype instead. The majority of X3D properties are datatype properties, many of which have an array of permissible values (representing vectors or matrices) rather than just one value of a specific datatype.[8] Such datatype properties have been defined as `xsd:complexType` rather than the more specific but single-value datatypes, such as `xsd:float`, `xsd:integer`, `xsd:string`, or `xsd:anyURI`. The proposed ontology also features its own datatypes. In contrast to other controlled vocabularies, the X3D vocabulary does not contain individuals, which has been addressed in the 3D Modeling Ontology. Also, the TBox of the proposed ontology is not based solely on the X3D vocabulary; new concepts (e.g., `3d:3DModel`, `3d:Dodecahedron`, `3d:DesignStudio`,) and new roles (e.g., `3d:animated`, `3d:designedBy`, `3d:hasVertices`, `3d:baseForm`) have also been added to accommodate the needs of 3D graphic designers.

3.2 Modeling Techniques

The scope of the ontology is the knowledge domain of 3D models and scenes; the purpose of the ontology is to provide comprehensive coverage of 3D concepts and roles with X3D alignment. The ontology engineering was based on the in-depth analysis of

---

[8] The VRML-based X3D datatypes starting with SF correspond to one allowed value of the declared type, while MF indicates that multiple values are allowed for the corresponding property, whether they are floating point numbers, integers, string literals, or URIs.

X3D nodes, fields, and the corresponding semi-structured vocabulary, as well as the manual creation of OWL classes and properties. In contrast to previous RDF/XML mappings, the Turtle serialization was chosen for the proposed ontology, because the XML serialization would have been too verbose for representing the large number of X3D concepts and properties. Because it was infeasible to use automated ontology engineering techniques, such as XSLT transformation, natural language processing-based concept extraction, statistical or machine learning techniques, during ontology engineering, the components of the 3D Modeling Ontology have been individually assessed and manually coded, resulting in code optimality and an easy-to-read layout for future extensions.

3.3 Formal Grounding

The most common variant of the Web Ontology Language, the DL flavor, which is an implementation of a description logic, can be as expressive as $\mathcal{SROIQ}^{(\mathcal{D})}$ (OWL 2 DL ontologies) [17]. The main benefits of description logics are the higher efficiency in decision problems than first-order predicate logic and the expressivity higher than that of propositional logic. Consequently, description logics are ideal for modeling concepts, roles, individuals, and their relationships. For this reason, the formal grounding of the proposed ontology was written in the form of description logic axioms, and was then translated to OWL 2 (see Table 1).

Table 1. Examples for description logic to OWL 2 translation.

DL syntax	Turtle syntax
Torus $\sqsubseteq$ SpatialGeometry	`:Torus rdfs:subClassOf :X3DSpatialGeometryNode .`
hatched $\sqsubseteq$ filled	`:hatched rdfs:subPropertyOf :filled .`
LightNode $\equiv$ DirectionalLight $\sqcup$ PointLight $\sqcup$ SpotLight	`:X3DLightNode owl:disjointUnionOf (:DirectionalLight :PointLight :SpotLight) .`
Box $\sqcap$ Pyramid $\sqsubseteq$ $\bot$	`:Box owl:disjointWith :Pyramid .`
$\top \sqsubseteq \forall$visibilityRange.float	`:visibilityRange rdfs:range xsd:float .`
$\exists$topRadius.$\top \sqsubseteq$ Cone	`:topRadius rdfs:domain :Cone .`
createdBy $\equiv$ creatorOf$^-$	`:createdBy a owl:ObjectProperty ; owl:inverseOf :creatorOf .`
$0 \leq$ intensity. DirectionalLight ≤ 1	`:zeroone rdfs:range [ a xsd:float ; owl:onDatatype xsd:float ; owl:withRestrictions ( [ xsd:minInclusive "0" ] [ xsd:maxInclusive "1" ] ) ] :intensity rdfs:range :zeroone ; rdfs:domain :DirectionalLight .`

3.4   Creating the Concept and Role Hierarchies

The general structure of the ontology is based on the X3D node hierarchy described in the official specification,[9] however, not all X3D nodes can be transformed directly into a description logic concept or role. As mentioned earlier, some of the nodes are defined as the logical equivalent of both a concept and a role. To eliminate this ambiguity, the proposed ontology applies an extended name and different capitalization for such concepts and roles. Furthermore, the X3D node hierarchy provides a comprehensive concept hierarchy, but is less satisfactory in describing subrole relationships. Due to the number of role overlaps and the different meanings and characteristics of roles sharing the same name, however, the X3D roles cannot be categorized efficiently in a tree structure anyway, and role scopes are indicated mainly by their domain declarations instead.

The first version of the structured X3D ontology covers the entire vocabulary of X3D v3.3, complemented by a new set of concepts and roles. This novel ontology exploits all mathematical constructors of $\mathcal{SROIQ}^{(\mathcal{D})}$, the description logic underlying OWL 2 DL, which maximizes expressivity and at the same time retains decidability.

3.5   Integrity Checking

The integrity and correctness of the proposed ontology have been checked throughout the ontology engineering process with industry-leading reasoners, including FaCT++ and HermiT. Since the xsd:complexType datatype, which is used extensively in the proposed ontology, is not supported by FaCT++, the final integrity checking was performed using HermiT. String literals have also been checked using a U.S. English spellchecker.

4   Evaluation

The proposed ontology has been evaluated according to the five ontology engineering principles of Gruber, the researcher who introduced ontologies in the context of artificial intelligence [18]:

- Clarity: the intended semantics of the defined terms are provided in a human-readable form, complemented by machine-interpretable constraints. While the concept and role names characterize the 3D domain, their definition is independent from the 3D modeling context.
- Coherence: no RDF statement can be automatically inferred from the axioms of the ontology that would contradict any given definition, which has been tested using semantic reasoners.
- Minimal encoding bias: the conceptualization of the 3D domain has been specified at the knowledge level independent from any symbol-level encoding. The ontology

---

[9] http://doc.x3dom.org

engineering has been conducted using open standards rather than proprietary specifications, serializations, or file formats.

- Minimal ontological commitment: the proposed ontology has been designed to be as lightweight as possible, and open to more specific implementations.
- Extendibility: new concepts, roles, and individuals can be easily added to the ontology without changing the core concept or role hierarchy. The proposed ontology features standards alignment and can be easily interlinked with LOD datasets.

The description logic expressivity of the proposed ontology, $\mathcal{SROIQ}^{(\mathcal{D})}$, is significantly higher than that of other 3D ontologies, making the 3D Modeling Ontology the most expressive ontology to date in the 3D domain (see Table 2).

Table 2. Comparison of 3D ontologies.

Ontology	Language	DL expressivity	XSD alignment	X3D alignment	Linked Data integration
OntologyX3D	OWL	$\mathcal{ALIN}^{+(\mathcal{D})}$	−	+	−
Gilson-Silva Mapping	RDFS	$\mathcal{AL}$	−	+	−
Petit Mapping	OWL	$\mathcal{AL}^{(\mathcal{D})}$	−	+	−
Kinect Ontology	OWL 2	$\mathcal{ALCRIF}^{(\mathcal{D})}$	+	−	−
Ontology of Furniture	OWL	$\mathcal{ALEN}$	−	−	−
3D Modeling Ontology	OWL 2	$\mathcal{SROIQ}^{(\mathcal{D})}$	+	+	+

The ontology files of other 3D ontologies cited in the literature, such as that of the *Geometrical Application Ontology* [19] or the *Common Shape Ontology* [20], are not available online anymore, only the corresponding articles and documentation, thus they have been omitted from the comparison.

While geometric and spectrophotometric properties manipulated in a 3D modeling software can be directly represented in any X3D-compliant knowledge representation, only the proposed ontology provides machine-interpretable definitions for these features, as well as data about the modeling software used for creating the model and the geometric primitives that make up the model (see Fig. 1).

While the proposed ontology is not the first X3D-based ontology in its class, it is by far the most comprehensive formally grounded structured 3D ontology, and so the most suitable one for interlinking 3D models and model fragments with LOD concepts. Structured 3D model representations correspond to RDF graphs which, when interlinked with LOD resource identifiers, naturally merge to the LOD Cloud. The resulting structured data can be queried and updated using SPARQL [21], which can also combine federated search results retrieved from diverse cultural heritage data sources [22]. Those structured knowledge representations that implement the concepts and roles of the proposed ontology are suitable for not only annotating 3D models, but also reasoning over them, for example, to differentiate between ancient Attican and Apulian red-figure vases.

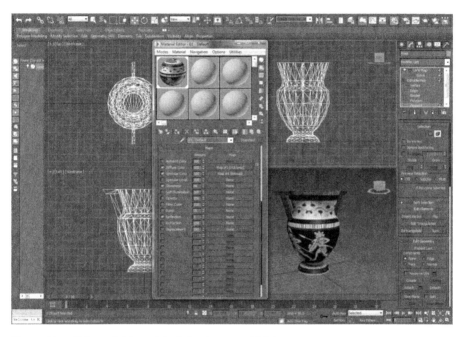

```
@prefix 3d: <http://vidont.org/3d/> .
@prefix xsd: <http://www.w3.org/2001/XMLSchema#> .
<http://vidont.org/3dmodels/ancient-greek-vase.html> a
3d:3DModel ; 3d:createdIn 3d:AutoDesk3dsMax ;
3d:baseForm 3d:prolateSpheroid ;
3d:hasCompound 3d:Sphere , 3d:Box ;
3d:hasFaces "572"^^xsd:nonNegativeInteger ;
3d:hasVertices "428"^^xsd:nonNegativeInteger ;
3d:shininess "0.145"^^xsd:decimal .
```

Fig. 1. The proposed ontology provides structured representation for 3D models not only with descriptors based on X3D, but also newly introduced descriptors for geometric primitives that constitute a model

The semantic enrichment of the above ancient Greek vase model with Linked Open Data, which would not be possible without the proposed ontology,[10] can extend the structured representation with terms from LOD datasets such as Kerameikos,[11] the

---

[10] Unless semi-structured annotations are mixed with structured annotations, as seen in the literature.

[11] http://kerameikos.org

British Museum Collection,[12] the Art & Architecture Thesaurus (AAT) dataset,[13] and the Pleiades Gazetteer of Ancient Places,[14] which yields to

```
@prefix 3d: <http://vidont.org/3d/> .
@prefix bm: <http://collection.britishmuseum.org/id/
thesauri/> .
@prefix dbpedia: <http://dbpedia.org/resource/> .
@prefix foaf: <http://xmlns.com/foaf/0.1/> .
@prefix kerameikos: <http://kerameikos.org/id/> .
@prefix pleiades: <http://pleiades.stoa.org/places/> .
@prefix rdfs: <http://www.w3.org/2000/01/rdf-schema#> .
@prefix schema: <http://schema.org/> .
@prefix xsd: <http://www.w3.org/2001/XMLSchema#> .
<http://vidont.org/3dmodels/ancient-greek-vase.html> a
3d:3DModel ; 3d:createdIn 3d:AutoDesk3dsMax ;
3d:baseForm 3d:prolateSpheroid ; 3d:hasCompound 3d:Sphere
, 3d:Box ; 3d:hasFaces "572"^^xsd:nonNegativeInteger ;
3d:hasVertices "428"^^xsd:nonNegativeInteger ;
foaf:depicts dbpedia:Vase , bm:x14796,
kerameikos:red_figure , dbpedia:Red-figure_pottery ,
3d:CompoundObject ; 3d:shininess "0.145"^^xsd:decimal ;
3d:transparency "0.0"^^xsd:decimal ;
schema:City dbpedia:Athens , pleiades:579885 ;
schema:Country dbpedia:Ancient_Greece ;
rdfs:seeAlso dbpedia:Pottery_of_ancient_Greece .
```

The structured data of 3D models in the cultural heritage domain can be used for creating, using, and sharing historical information about the ancient world—think of museum collection repositories, for example. Structured high-level descriptors are suitable for efficient indexing of declarative 3D models, especially when deployed as lightweight semantics in the web site markup.

5   Conclusions and Future Work

To use 3D models of cultural artifacts to their full potential, machine-interpretable knowledge representation of the 3D models is needed. Rich semantics make efficient indexing and retrieval possible, however, until now no controlled vocabulary or ontology was able to provide complex structured descriptors for 3D models. One of the most comprehensive 3D standards, X3D, is no exception, because it provides a semi-structured vocabulary only.

---

[12] http://collection.britishmuseum.org

[13] http://vocab.getty.edu/dataset/aat/full.zip

[14] http://pleiades.stoa.org

Early implementations of semi-structured to structured data mapping of the X3D standard came with a lack of formalism, were incomplete, and did not address the inconsistency and conceptual ambiguity issues of the standard. The 3D Modeling Ontology presented in this paper, the most expressive 3D ontology to date, overcomes the above limitations and features the latest standardized X3D nodes and fields as OWL 2 classes and properties with formal grounding and standards alignment. The ontology also extends the X3D concept list with 3D computer graphics software terminology, advancing the OWL 2 mapping of the X3D standard. The integrity of the proposed ontology has been tested with industry-leading reasoners, and the implementation potential evaluated through the semantic representation of 3D models in the cultural heritage domain, one of which has been demonstrated in this paper. The presented ontology is the very first 3D ontology with X3D alignment to provide true Linked Data integration, making it possible to interlink cultural heritage concepts with the LOD Cloud to describe the culture, provenance, and historical era related to 3D artifact models in a machine-interpretable manner, rather than just the vertices and edges of the 3D polygons. As a future work, after the standardization of X3D v4.0 (which is currently in progress), the proposed 3D ontology will be extended with the new concepts and roles of the upcoming standard.

References

1. Kim, H., Kang, Y., Cha, M., Han, S.: Cluster rendering on large high-resolution multi-displays using X3DOM and HTML. Multimed. Syst. (2015). doi:10.1007/s00530-015-0495-0
2. Callet, P.: 3D reconstruction from 3D cultural heritage models. In: Ioannides, M., Quak, E. (eds.) 3D Research Challenges in Cultural Heritage. LNCS, vol. 8355, pp. 135–142. Springer, Heidelberg (2014). doi:10.1007/978-3-662-44630-0_10
3. Sfikas, K., Pratikakis, I., Koutsoudis, A., Savelonas, M., Theoharis, T.: Partial matching of 3D cultural heritage objects using panoramic views. Multimed. Tools Appl. 75(7), 3693–3707 (2016). doi:10.1007/s11042-014-2069-0
4. Mallik, A., Chaudhury, S.: Acquisition of multimedia ontology: an application in preservation of cultural heritage. Int. J. Multimed. Inf. Retr. 1(4), 249–262 (2012). doi:10.1007/s13735-012-0021-5
5. Yu, C.-H., Groza, T., Hunter, J.: Reasoning on crowd-sourced semantic annotations to facilitate cataloguing of 3D artefacts in the cultural heritage domain. In: Alani, H. (ed.) ISWC 2013. LNCS, vol. 8219, pp. 228–243. Springer, Heidelberg (2013). doi:10.1007/978-3-642-41338-4_15
6. Wang, X., Lv, T., Wang, S., Wang, Z.: An ontology and SWRL based 3D model retrieval system. In: Li, H., Liu, T., Ma, W.-Y., Sakai, T., Wong, K.-F., Zhou, G. (eds.) AIRS 2008. LNCS, vol. 4993, pp. 335–344. Springer, Heidelberg (2008). doi:10.1007/978-3-540-68636-1_32
7. Gruber, E., Smith, T.J.: Linked Open Greek Pottery. In: 42nd Annual Conference on Computer Applications and Quantitative Methods in Archaeology, pp. 205–214. Oxuniprint, Oxford (2014)
8. Flotyński, J., Walczak, K.: Conceptual knowledge-based modeling of interactive 3D content. Vis. Comput. 31(10), 1287–1306 (2015). doi:10.1007/s00371-014-1011-9

9. Decourselle, J., Vennesland, A., Aalberg, T., Duchateau, F., Lumineau, N.: A novel vision for navigation and enrichment in cultural heritage collections. In: Morzy, T., Valduriez, P., Bellatreche, L. (eds.) ADBIS 2015. CCIS, vol. 539, pp. 488–497. Springer, Heidelberg (2015). doi:10.1007/978-3-319-23201-0_49

10. Behr, J., Eschler, P., Jung, Y., Zöllner, M.: X3DOM – a DOM-based HTML5/X3D integration model. In: 14th International Conference on 3D Web Technology, pp. 127–135. ACM Press, New York (2009). doi:10.1145/1559764.1559784

11. Brutzman, D., Blais, C., Harney, J.: X3D fundamentals. In: Geroimenko, V., Chen, C. (eds.) Visualizing Information Using SVG and X3D. Springer, London (2005). doi:10.1007/1-84628-084-2_3

12. Kalogerakis, E., Christodoulakis, S., Moumoutzis, N. Coupling ontologies with graphics content for knowledge-driven visualization. In: IEEE Virtual Reality Conference, pp. 43–50. IEEE, New York (2006). doi:10.1109/VR.2006.41

13. Petit, M., Boccon-Gibod, H., Mouton, C.: Evaluating the X3D schema with semantic web tools. In: 17th International Conference on 3D Web Technology, pp. 131–138. ACM Press, New York. doi:10.1145/2338714.2338737

14. Yu, D., Hunter, J.: X3D fragment identifiers—extending the open annotation model to support semantic annotation of 3D cultural heritage objects over the Web. Int. J. Herit. Dig. Era 3, 579–596 (2014). doi:10.1260/2047-4970.3.3.579

15. Best practice recipes for publishing RDF vocabularies. http://www.w3.org/TR/swbp-vocab-pub/. Accessed 25 July 2016

16. Simperl, E.: Reusing ontologies on the Semantic Web: a feasibility study. Data Knowl. Eng. 68(10), 905–925 (2009). doi:10.1016/j.datak.2009.02.002

17. Sikos, L.F.: A novel approach to multimedia ontology engineering for automated reasoning over audiovisual LOD datasets. In: Nguyen, N.T., Trawiński, B., Fujita, H., Hong, T.-P. (eds.) ACIIDS 2016. LNCS (LNAI), vol. 9621, pp. 3–12. Springer, Heidelberg (2016). doi:10.1007/978-3-662-49381-6_1

18. Gruber, T.R.: Towards principles for the design of ontologies used for knowledge sharing. In: Guarino, N., Poli, R. (eds.) Formal ontology in conceptual analysis and knowledge representation. Kluwer Academic Publishers, Deventer (1993)

19. Koenderink, N.J.J.P., Top, J.L., van Vliet, L.J.: Supporting knowledge-intensive inspection tasks with application ontologies. Int. J. Hum Comput Stud. 64(10), 974–983 (2006). doi:10.1016/j.ijhcs.2006.05.004

20. Vasilakis, G., Garcia-Rojas, A., Papaleo, L., Catalano, C.E., Spagnuolo, M., Robbiano, F., Vavalis, M., Pitikakis, M.: A common ontology for multi-dimensional shapes. In: 1st International Workshop on Multimedia Annotation and Retrieval Enabled by Shared Ontologies, pp. 31–43 (2007)

21. Walczak, K., Flotyński, J.: On-demand generation of 3D content based on semantic meta-scenes. In: De Paolis, L.T., Mongelli, A. (eds.) AVR 2014. LNCS, vol. 8853, pp. 313–332. Springer, Heidelberg (2014). doi:10.1007/978-3-319-13969-2_24

22. Orgel, T., Höffernig, M., Bailer, W., Russegger, S.: A metadata model and mapping approach for facilitating access to heterogeneous cultural heritage assets. Int. J. Digit. Libr. 15(2), 189–207 (2015). doi:10.1007/s00799-015-0138-2

Challenges of Mapping Digital Collections Metadata to Schema.org: Working with CONTENTdm

Patricia Lampron[1]([✉]), Jeff Mixter[2], and Myung-Ja K. Han[1]

[1] University Library, University of Illinois at Urbana-Champaign, Urbana, IL, USA
{lampron2,mhan3}@illinois.edu
[2] OCLC Membership & Research at OCLC, Dublin, OH, USA
mixterj@oclc.org

Abstract. As digitized materials stored in content management systems become more prominent as a mode of resource access, the library community is experimenting with linked data to make these collections available in new ways. Applying Schema.org semantics to curated digital collections allows for enhanced search engine discovery, as well as the dissemination of metadata in ways that can connect resources across the internet. This paper shares the challenges encountered when mapping unique digital collections metadata to Schema.org semantics, and lessons learned from experimentation on CONTENTdm collections metadata at both the University of Illinois at Champaign-Urbana Library and OCLC.

Keywords: Linked data · Digital collections · Schema.org · Metadata · Contentdm

1 Introduction

Current trends in library metadata lean toward discovery and accessibility, not just within Online Public Access Catalogs and content management systems, but as shared information that relates across systems and is searchable through highly used search engines. According to a 2015 Library edition of the Horizon Report, "Popular search engines can only touch about 10 % of the Internet; the remaining 90 % are websites that are not indexed currently because most of this data is located in library catalogs in formats that cannot be searched or is guarded in secure areas that cannot be accessed by bots," [1]. This lack of visibility in popular search engines has led to efforts by libraries to make their metadata available by incorporating Linked Open Data (LOD) technologies into their metadata management, e.g., creation, sharing and dissemination. This new direction can be seen clearly in recent efforts to overhaul current cataloging standards and practices, and the development of BIBFRAME [2], a vocabulary designed for bibliographic data that can be expressed using Resource Description Framework (RDF), led by the Library of Congress; as well, in the use of Schema.org and the Bib Extend Community Group's recommendations for describing library materials as linked data [3], in order to enable users to discover these resources on the web. However, most of

© Springer International Publishing AG 2016
E. Garoufallou et al. (Eds.): MTSR 2016, CCIS 672, pp. 181–186, 2016.
DOI: 10.1007/978-3-319-49157-8_15

the efforts are focused on the library's traditional collections whose metadata is in MARC format, and less attention has been given to the carefully curated special collections being digitized and housed in separate digital asset management systems.

The University of Illinois at Urbana-Champaign (UIUC) Library has been exploring ways to publish its bibliographic data and associated holdings and item data as linked data using Schema.org semantics [4, 5]. As a next step, the library is developing a metadata application profile consisting of a common set of properties to describe the wide variety of items across the 21 digital collections housed in their content management system, CONTENTdm [6], in attempt to make its rich and unique digital collections more discoverable on the web. OCLC has also been exploring how to map CONTENTdm data into RDF and simultaneously reconcile string values against the Virtual International Authority File (VIAF) [7] and Faceted Application of Subject Terminology (FAST) [8], and ultimately providing an N-Triple data dump of Schema.org data. These investigations both provided invaluable lessons in applying linked data principles and Schema.org semantics to digital collections described with customized, non-traditional metadata.

2 Exploiting Linked Data to Promote Digital Collections

As changes in user needs turn toward accessing information through sources outside the library (i.e. search engines), the library must find ways to make connections between the outside world and their resources. Linked data can help make these connections, but in order for libraries to benefit from the Web they must take into account W3C specifications and recommendations [9].

Schema.org is a linked data vocabulary designed and published by the major search engines and promoted as a structured vocabulary that they can all consume and understand [10]. When applied correctly, linked data using Schema.org semantics can provide search engines with well-structured data that can be harvested and that links to other resources on the Web. Notable search engines already support Schema.org semantics structured as microdata, RDFa, or more recently JSON-LD, and embedded within HTML pages, and are using this markup for indexing and display purposes, as well as building connections between information and resources, for example in Google Knowledge Graphs. These are the types of connections and exposure that libraries endeavor to create, and so embedding Schema.org enhanced metadata within CONTENTdm HTML pages could be a step toward better discoverability of digital collections.

3 CONTENTdm Collections and Their Metadata

CONTENTdm is a popular content management system used in libraries and archives for storage and access of curated digital collections. In CONTENTdm, each collection contains its own set of metadata fields, which are referred to as a "metadata profile". Each field in the collection has its own label, and can be mapped to a Simple or Qualified Dublin Core element, or the collection manager can choose to leave it unmapped. The system is organized much like a file cabinet, with the cabinet being the entire

CONTENTdm system, and the folders inside being collections that contain each metadata profile. This model provides detailed descriptive metadata specific to the individual collection. CONTENTdm also allows for the use of both local and established controlled vocabularies within a particular field or shared with multiple fields. Using controlled vocabularies ensures consistent metadata both within a collection as well as across multiple collections when the vocabulary is shared.

Many CONTENTdm fields can be mapped to the same Dublin Core element, however, they still often hold distinct descriptive information that is specified by their local field name. For example, although a field will be mapped to the Simple Dublin Core <dc:description> element, a local field name for this field could include contextual information, such as , <Inscription>, or <Translation>. The ability to refine metadata through field names while mapping to Dublin Core works well in allowing detailed descriptive metadata for discovery and access, and display in CONTENTdm, while still providing interoperable metadata for service providers and harvesters through the OAI repository. Within the CONTENTdm website, searching across collections is facilitated through both local field names as well as their mapped Dublin Core elements. The customized fields are also indexed for advanced search and discovery within an individual collection.

4   Mapping to Schema.Org

Although there are many benefits to the customizable nature of CONTENTdm collection fields, converting the metadata to linked data also presents challenges. The UIUC Library analyzed their collections' metadata fields in preparation for mapping to Schema.org semantics in order to develop a linked data based metadata profile, and discovered that while the customized fields, across collections, might share commonalities in their Dublin Core mappings, many of the Dublin Core elements are so broad in scope that overlapping fields could have a wide range of meanings. The most prominent example of this is the mapping of <dc:description>, which was mapped 116 times across 21 collections, and while <dc:description> is an extreme case, the majority of the Dublin Core terms in the library's CONTENTdm collections have multiple mappings, some even within the same collection. Another such example can be seen in <dc:contributor>. There are 13 unique field names across the 27 fields mapped to <dc:contributor>, including "Printer", "Speaker", "Architect", "Composer", "Lyricist", "Artist", and so on, all used for defining specific roles.

This analysis illustrates that a straight system wide mapping of Dublin Core elements to Schema.org semantics is not the most effective way to disseminate curated metadata to the web. As noted in the <dc:contributor> example, many of these field names can be represented by using either Schema.org properties or by employing the structured nature of Schema.org in combination with <schema:Role> to define specific terms that are common in the CONTENTdm collection metadata profiles, but this work must be performed by staff who have an understanding of the collections and how the fields are being used. It should be noted, however, that while Schema.org types and properties are designed to describe a wide variety of "things", it was originally created with commercial interests in mind, and so there are still areas in which information can be potentially lost or not represented. A number of extensions to the schema have been proposed and are in use, such as the Schema Bib Extend which has been adopted by the Schema.org vocabulary as an official extension in the bib.schema.org namespace [11]. Extensions like this can help fill the voids, but it is unclear whether extensions to Schema.org will be recognized in the future by search engines. Nonetheless, these extensions provide a set of semantics for exposing collections that contain more specific metadata through RDF.

Another difficulty in mapping CONTENTdm metadata to RDF is pulling apart conflated descriptions. It is very common for CONTENTdm records to contain statements about both a physical thing and its digital surrogate. For example, a single CONTENTdm record might contain both a <dcterms:dateCreated> value of '1904' and a <dc:format> value of 'JPEG'. It is clear that what is being described in this record is actually two items, the first being the original photograph taken in 1904 and the second being the digitized JPEG. This is problematic when mapping because in RDF both the physical item and the digital surrogate would be separately described and connected together with a property that shows the relationship between the two. In Schema.org this is done by describing a <schema:CreativeWork> (for the physical item) and a <schema:MediaObject> (for the digital item) and then connecting the *CreativeWork* to the *MediaObject* with a <schema:encoding> property. To achieve this type of granularity in mapping CONTENTdm metadata to RDF, it will be necessary to build templates that have a contextual understanding of metadata fields and can route field values to the appropriate RDF entity. This is again where local metadata fields can be useful. If the metadata fields are already mapped to Dublin Core elements it would be very difficult to distinguish between date created and date digitized using the <dc:date> element, but if local field values are retained through customized field names, there is a chance that the individual field values could carry with them enough semantics to inform a conversion template (i.e. <dcterms:dateCreated> and <dcterms:dateDigitized>).

Entity reconciliation and data inferencing during the conversion of non-RDF data into RDF is another challenge. When mapping a subject field to Schema.org, all of the various subject values become entities connected back to the item using a <schema:about> property. The idea of reconciling entities allows the subject strings to be mapped to existing linked data datasets. For example, one could take the subject string 'Ohio' and map it to the FAST URI <http://id.worldcat.org/fast/1205075>. Doing this helps connect the converted data to the wider web of linked data and alleviates the burden of having to create a new persistent URI for every subject value. Data inferencing is a result of mapping flat metadata to RDF. While it has been previously noted that the flat nature of CONTENTdm can be a benefit in the conversion process due to the simplicity involved in direct mapping, it does require that the mapper infer statements and sometimes entities that are not directly relatable to the original CONTENTdm record. For example, if a CONTENTdm record describes a recorded play there might be a customized <datePerformed> metadata field. When converting this record to Schema.org, the <datePerformed> field and value will have to spawn a new entity <schema:Event> which will be used to connect the value in the <datePerformed> field back to the play being described in the record.

5 Discussion

One of the frequent questions that comes up when discussing linked data is "Why?" There are two predominant perspectives on this question and both have valid arguments to support them. The first is an outward looking perspective that focuses on linked data syndication. The argument is that linked data can help improve the discovery of digital collections by improving the search engine optimization through metadata. As search engines put more emphasis on harvesting structured data, applying structured data to digital collections access systems using a vocabulary designed, published, and promoted by search engines seems like a logical and worthwhile effort. The second perspective is more inward looking and focuses on using linked data to help support and bolster internally maintained and curated data. As the name implies, linked data links resources on the wider web and it is believed that these connections can be leveraged to create a better end-user experience. Connecting to outside resources like FAST, VIAF, WikiData and GeoNames provides access to data such as maps, biographies, alternate names and foreign language data that can all be used to help provide a richer end-user experience.

While the work of implementing linked data in CONTENTdm collections is beneficial in many ways, it also presents its own unique challenges. Both the UIUC Library and OCLC experimentation on CONTENTdm collections metadata provided three invaluable lessons. First, it is nearly impossible to create one linked data profile that meets the needs of all special collections. Because each collection differs from the others in its descriptive metadata, the implementation of linked data transformation should be done at the collection level, rather than at the institution or system level, in order to ensure preserving and presenting the uniqueness of each collection to users. Second, unique collections require metadata reconciliation work, including the incorporation of links from various authority data, not just vocabularies that are standard to the library community, but also outside sources, for example the Internet Movie Database [12] or the Union List of Artist Names [13]. This work should be conducted with metadata creators and collection specialists to insure that the proper authority data is being chosen. Third, more communication among special collections curators, metadata specialists, and system administrators are required to make these unique digital collections available on the web. Because these collections are described through non-traditional library metadata standards, and are stored in and accessed through non-traditional library systems, sharing each other's needs and experiences would greatly benefit all stakeholders working with special collections, and ultimately users who are discovering these unique resources, both within the CONTENTdm environment and on the web.

References

1. Johnson, L., Adams Becker, S., Estrada, V., Freeman, A.: NMC Horizon Report: 2015 Library Edition. The New Media Consortium, Austin, Texas (2015). http://cdn.nmc.org/media/2015-nmc-horizon-report-library-EN.pdf
2. BIBFRAME. https://www.loc.gov/bibframe/docs/index.html
3. SCHEMA BIB EXTEND Community Group: Recipes and Guidelines. https://www.w3.org/community/schemabibex/wiki/Recipes_and_Guidelines

4. Cole, T.W., Han, M.-J., Weathers, W.F., Joyner, E.: Library MARC records into Linked open data: challenges and opportunities. J. Libr. Metadata **13**(2–3), 163–196 (2013)
5. Han, M-J., Cole, T.W., Lampron, P., Sarol, M.J.: Exposing library holdings metadata in RDF using Schema.org semantics. In: Proceedings of the International Conference on Dublin Core and Metadata Applications 2015, pp. 41–49 (2015)
6. CONTENTdm. http://www.oclc.org/en-US/contentdm.html
7. Virtual International Authority File. http://viaf.org/
8. Faceted Application of Subject Terminology. http://fast.oclc.org/searchfast/
9. Berners-Lee, T.: Linked Data (2009). https://www.w3.org/DesignIssues/LinkedData.html
10. Official Google Blog; Introducing schema.org: Search engines come together for a richer web. https://googleblog.blogspot.com/2011/06/introducing-schemaorg-search-engines.html
11. Schema.org Hosted Extension: bib. http://bib.schema.org/
12. Internet Movie Database. http://www.imdb.com/
13. Union List of Artist Names. http://www.getty.edu/research/tools/vocabularies/ulan/

Finding User Need Patterns in the World of Complex Semantic Cultural Heritage Data

Maliheh Farrokhnia[1(✉)] and Trond Aalberg[2]

[1] Oslo and Akershus University College of Applied Sciences, Oslo, Norway
Maliheh.Farrokhnia@hioa.no
[2] Norwegian University of Science and Technology, Trondheim, Norway
Trond.Aalberg@idi.ntnu.no

Abstract. When developing systems for the semantic web of cultural heritage information, it is critical to also know the real needs of users. This study discusses the notion of users' information needs and presents some patterns identified in a set of use cases. The aim of the study is to contribute in bridging the gap between experts' and users' conceptualization in order to have more efficient information retrieval system.

1 Introduction

The semantic web has created an environment where more and more information is made available and the information needs and expectations of users are getting more sophisticated. Within cultural heritage documentation, reference models such as the CIDOC Conceptual Reference Model (CRM) [3] provides an extensible semantic framework for the concepts and properties of interest. By building upon this model, museums, libraries and archives can describe their data with interoperable semantics, which potentially can provide meaningful integrated access to cultural heritage information. However, in the context of vast amounts of data that is brought together we need to have a strong focus on users' information need and expectation when developing solutions for accessing the information. Current development in the world of semantic data tends to be with the domain expert or system developer's perspective and there is a need to complement with research and development from the user's perspective.

To support end user's need for information retrieval from repositories of cultural heritage data with rich semantics, we need to understand user's conceptualization and how this maps to the formal models used to represent the information. Understanding the user's intent or information need that underlies a query has long been recognized as a crucial part of information retrieval. Literature about user's information needs in various user population and different domains has identified the importance of some elements such as name, place, genre, date, subject, and event and activity [1,2,4,6,7]. The point is that most of the identified elements are either based on the users' queries (and do not show the user's main information needs) or elements are extracted by domain experts with a well-developed understanding. Research with a more detailed analysis

© Springer International Publishing AG 2016
E. Garoufallou et al. (Eds.): MTSR 2016, CCIS 672, pp. 187–192, 2016.
DOI: 10.1007/978-3-319-49157-8_16

of users' needs includes Hennicke's study of reference questions in the scope of CRM [5]. From the preliminary findings, a common pattern was identified in which the main elements are documentation activity and the documents resulting from the documentation activity. This is a kind of research that will give a better understanding of information needs of users. However, it needs to be complemented with the actual user's query and interaction with an information system for further analysis.

In this paper, we present a qualitative user study conducted to explore and identify patterns of user's conceptualization. The aim of the study is to contribute in bridging the gap between experts' and users' perspective in order to have more efficient information retrieval system.

2   Research Design

The methodological approach of this research is to go beyond the queries and try to discover the real information needs behind a query, based on the user's task or even the user's interest and curiosity. Having a qualitative approach, this research conducted semi-structured in-depth interviews with some PhD students who are examples of active researchers with information needs within cultural heritage documentation. They are users of different memory institutions and from different domains. Following the naturalistic approach, during the interview sessions the participants have been asked to do some real searches in any preferred information systems based on their current or recent information needs related to their tasks in their PhD projects. This method has been selected because the best way to understand users' real information needs, is to hear the story behind their queries from themselves in a quite real situation.

In order to understand and formalize what the users are looking for and to conceptualize their cultural information needs, these needs have been interpreted based on the users' words and modeled in CRM and its extensions such as FRBRoo[1] and CRMinf[2].

3   Users' Information Needs

Based on the data analysis that have been done so far, no matter what the information need is or how complex this need is, most users conceptualize their information need using familiar and well-known entities such as person, place, and subject (keyword). It means that what they really need is not exactly what they query on. The important thing is to understand their main needs and how the information systems can meet their needs.

The following examples show some of the cases of the different users' information needs that have been searched in some information systems. To facilitate a better understanding of the graphical models, the query terms are identified

---

[1] http://www.cidoc-crm.org/frbr_drafts.html.
[2] http://www.ics.forth.gr/isl/index_main.php?l=e&c=713.

as the entities with solid grey colour, the suggested central entities of the search
have been shown as the entities with bold boarders, and the actual information
needs have been identified by drawing the free form lines.

3.1   Example 1

In this case, the user is a second-year PhD student in Literature working on a
project about travel writing. She has 5–6 years of research experience and good
background knowledge on her domain. As part of her project, she is writing a
book in which she is going to reproduce a particular photo taken by a travel
writer during his travel to the North where he took many photos of indigenous
people at different places. In this example, the user has a copy of the photo
at hand but she needs to know who is the rights holder of this photo. The
user also knows the photographer and the place. So, for this information need,
which is kind of fact-finding search, the user inputs a query on the name of the
photographer plus the place and the name of the specific group of people.

The interpretation of the story behind this user's query modeled in CRM
is shown in Fig. 1. There is a main event (E5) that is "traveling to the North"
carried out by a photographer (E21). This event has been documented in a
governmental document (E31) and is the subject of a book (E73). During this
event, there is an activity (E7) in which the photographer takes photos of a
group of people (E21) at a specific place (E52). This activity resulted in a photo
(E73) that has some descriptive information and right hold by (P106) a legal
body (E40) that is the Library and Archive of Canada and has a phone number
as the contact point (E51). Related to this information need, another search is
performed by the user using the same query terms but in another database (an
archive database). This time, the user is looking for all the information including
governmental documents related to the same travel.

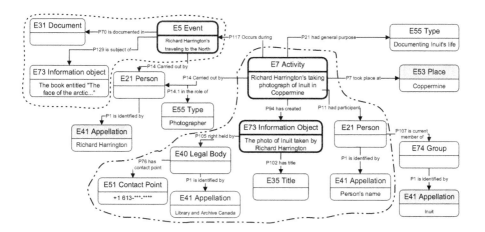

Fig. 1. Modelling information need in CIDOC CRM: example 1.

In this case, the activity of taking the photo or the photo as an information object is the central notions for the first information need and the notion of event of traveling is central for the second information need.

3.2 Example 2

The user in the second example is a last-year PhD student in Library and information science (LIS) and her project is about children's libraries. She is working on a specific concept in her project in which she is building upon a specific sociologist's theory. What this user needs, is to see how this particular theory has been used in other studies by other researchers in either LIS or Sociology. For this information need, the user performs a search task using a query containing the name of the sociologist and the concept on which she is working in her project. In this example, the user's intention for the query is to find neither the information about the sociologist nor the theories proposed by this sociologist. It is not easy to predict the user's intention only through her query.

When it comes to CRM modelling, as it is shown in Fig. 2, there are two quite separate events. On one hand, a theory has been formed as the result of an inference making activity carried out by a person (E21) who has a specific field of activity. This theory has been represented in a work (F1) created through a creation activity (F27) carried out by the same person (E21). This work (F1) is realized in an expression (F2) and is about the theory. On the other hand, there is another person (E21) working in another field of activity who performed (R14) a work conception (F27) being influenced by the mentioned theory and resulting in a work in the form of an article as an expression (F2).

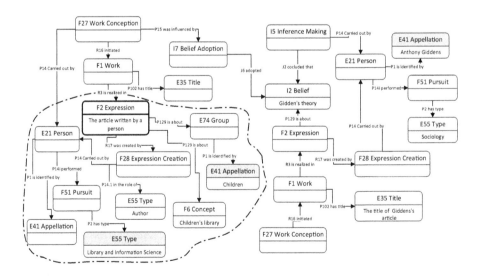

Fig. 2. Modelling information need in CIDOC CRM: example 2.

For this case, the main notion is the article that is an expression or kind of information object.

3.3   Example 3

The third example is about a second year PhD student in history. Her project is about history and fashion and she is studying a few particular fashion designers. As part of her project, she needs to find a photo of a specific garment that has been designed by one of the fashion designers. The garment has been counterfeited and has been the subject of a trial. The user would like to see an image of this garment and all the descriptive information about it, such as dimension, colour, material and texture, intended usage, date it was produced, where it is stored, etc. In order to find the answer to this information need, the user makes a query using the name of the designer. Another thing that this user is interested to find information about, besides the physical object, is the production process of the mentioned physical object and all its related information such as how it was produced; by whom it has been produced and etc.

As the interpretation of this case and its modelling in CRM shows (Fig. 3), there is a creation activity (E65) that has been carried out by the designer (E21). The creation activity has created the design of a garment and followed by a production activity (E12) that carried out by a person (E21) and results in the production of a specific garment (E22) at a specific time (E52) and place (E53). Additionally, there may be some information about the object and its current location.

The central notions in this case are the garment as a man-made object for the first need and production activity for the second need.

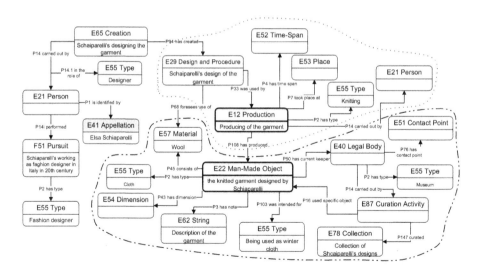

Fig. 3. Modelling information need in CIDOC CRM: example 3.

4 Discussion

The preliminary results of this study shows that although users have complex information needs, they are still basing their queries upon common and familiar entities such as Person or Place. Another common characteristic in our results is that the main entities in the information need often is different from the entities in the query. In our use cases the main entities for a result listing would have been of the types Event or Activity, and Information object or Physical object.

Finding common patterns in what entities users are querying and what entities they typically expect to be presented with in a result listing, is important input when designing and implementing information systems for accessing rich semantic information. Our research in this area is so far exploratory and further research in a larger sample in each of subject domains will be needed in order to generalize the result. So far, our case studies have been professionals with a somewhat explicit and known information need. Users having other motivation for the search, or less knowledge of the domain, may show different patterns.

Future work includes analysing a larger set of use cases as well as identifying other aspects of the patters such as characterizing the distance and paths between the query entities and the information need entities.

References

1. Bates, M.J., Wilde, D.N., Siegfried, S.: An analysis of search terminology used by humanities scholars: the Getty Online Searching Project Report Number 1. Lib. Q. Inf. Community Policy **63**(1), 1–39 (1993)
2. Collins, K.: Providing subject access to images: a study of user queries. Am. Arch. **61**(1), 36–55 (1998)
3. Crofts, N., Doerr, M., Gill, T., Stead, S., Stiff, M.: Definition of the CIDOC Conceptual Reference Model. Version 6.2. Technical report, ICOM/CIDOC Documentation Standards Group and CIDOC CRM Special Interest Group (2010)
4. Duff, W., Johnson, C.: Where is the list with all the names? Information-seeking behavior of genealogists. Am. Arch. **66**(1), 79–95 (2003)
5. Hennicke, S.: Representation of archival user needs using CIDOC CRM. In: Practical Experiences with CIDOC CRM and its Extensions (CRMEX 2013) Workshop, 17th International Conference on Theory and Practice of Digital Libraries (TPDL 2013), pp. 48–60. CEUR Workshop Proceedings (2013)
6. Tibbo, H.R. (ed.): Abstracting, Information Retrieval, and the Humanities: Providing Access to Historical Literature, No. 48. American Library Association, Chicago (1993)
7. Yi, K., Beheshti, J., Cole, C., Leide, J.E., Large, A.: User search behavior of domain-specific information retrieval systems: an analysis of the query logs from psycinfo and ABC-Clio's historical abstracts/America: history and life. J. Am. Soc. Inf. Sci. Technol. **57**(9), 1208–1220 (2006)

Mapping the Hierarchy of EAD to VRA Core 4.0 Through CIDOC CRM

Panorea Gaitanou[1]([✉]), Manolis Gergatsoulis[1], Dimitrios Spanoudakis[2],
Lina Bountouri[1], and Christos Papatheodorou[1,3]

[1] Laboratory on Digital Libraries and Electronic Publishing,
Database and Information Systems Group (DBIS), Department of Archives,
Library Science and Museology, Ionian University, Corfu, Greece
{rgaitanou,manolis,boudouri,papatheodor}@ionio.gr
[2] Pervasive Computing Group, Department of Informatics and Telecommunications,
University of Athens, Athens, Greece
dspanoud@di.uoa.gr
[3] Digital Curation Unit, Institute for the Management of Information Systems,
Athena R.C., Athens, Greece

Abstract. Metadata interoperability requires the development of mappings and crosswalks between different schemas. Crosswalks development is a laborious and complex task especially when metadata of compound objects include hierarchical structures. A typical case of such a complexity is finding aids that describe archives that include hierarchies of subordinate components. This paper makes a step forward to obtaining interoperability between two well-known metadata schemas, the EAD standard for the encoding of archival descriptions and the VRA Core 4.0 standard for describing compound visual resources. The paper presents an algorithm for mapping archival hierarchical structures expressed in the EAD standard to the VRA Core 4.0 standard. The input of the algorithm can be any EAD document as well as the crosswalk from EAD to VRA Core that has already been defined, by exploiting the mappings of both EAD and VRA Core 4.0 to the CIDOC Conceptual Reference Model. The output of the proposed process is a VRA Core 4.0 document that includes the information of the EAD document and represents its archival structure.

Keywords: Metadata interoperability · Crosswalks · Hierarchical structure · Encoded Archival Description · VRA Core 4.0 · CIDOC CRM

1 Introduction

Cultural heritage institutions regularly implement various interoperability techniques in order to homogeneously manage their heterogenous data. One of the

This research has been co-financed by the European Union (European Social Fund - ESF) and Greek national funds through the Operational Program "Education and Lifelong Learning" of the National Strategic Reference Framework (NSRF) - Research Funding Program: Heracleitus II. Investing in knowledge society through the European Social Fund.

E. Garoufallou et al. (Eds.): MTSR 2016, CCIS 672, pp. 193–204, 2016.
DOI: 10.1007/978-3-319-49157-8_17

most common interoperability techniques is *crosswalks*. A *crosswalk* defines the semantic mapping of the elements of a source metadata schema to the elements of a target metadata schema, so as to semantically translate the description of sources encoded in different schemas. A crosswalk is expressed through a table that shows the equivalent metadata fields of the metadata schemas involved.

In the framework of our research on ontology-based metadata interoperability, we proposed an algorithm [1] that generates crosswalks between metadata schemas that have already been mapped to the *CIDOC Conceptual Reference Model (CRM)* [7]. The algorithm gets as input two schemas, the source and the target schema and their mappings to the *CIDOC CRM*. It exploits the mappings and, based on their common semantics, produces a crosswalk from the source to the target schema. The algorithm has been evaluated on a crosswalk from Encoded Archival Description (EAD) [12] to VRA Core 4.0 [13].

EAD describes archives, which are compound and hierarchically structured objects/collections. An archive may consist of a set of sub-fonds or series, while the series of an archive consist of files or records and the files usually consist of records. The algorithm in [1] provides a mapping of the elements/attributes of the archival description to VRA Core 4.0. However, the mapping does not represent adequately the hierarchical structure of the archive.

This paper resolves this issue and proposes a new process that produces a VRA Core 4.0 document that represents the hierarchical structure and keeps the whole information of a given EAD document. Hence, the process gets as input an EAD document as well as the EAD to VRA Core 4.0 crosswalk and creates a semantically equivalent VRA Core 4.0 document that has the structure and carries the information of the EAD document.

The paper is structured as follows: In the next section, the algorithm for the generation of the crosswalk from EAD to VRA Core 4.0 is presented. In Sect. 3, the algorithm that maps content of an EAD document along with its structure to a VRA Core 4.0 document is demonstrated. Finally, Sect. 4 presents works related to the generation of crosswalks between metadata schemas and concludes the paper.

2   Mapping EAD to VRA Core 4.0 Through CIDOC CRM

2.1   EAD and VRA Core 4.0 Mappings to CIDOC CRM

An ontology-based integration architecture has been proposed in [4,15], which uses *CIDOC CRM*, a semantically rich model that conceptualizes the cultural heritage domain.

As part of this architecture, mappings have been defined between metadata schemas of the participating sources and *CIDOC CRM* that acts as the mediator, promoting the interoperability between the sources and the mediator. Two of such mappings are the mapping from EAD [4] to *CIDOC CRM*, as well as the mapping from VRA Core 4.0 [2] to *CIDOC CRM*. The mappings are expressed in a rule-based path oriented language named *Mapping Description Language* (*MDL*) [3,4]. It is path oriented because represents XML XPATH expressions as *CIDOC CRM* paths.

Definition 1. *A* CIDOC CRM path *is a sequence of the form:*

$$C_0 \rightarrow P_1 \rightarrow C_1 \rightarrow \ldots \rightarrow P_n \rightarrow C_n$$

with $n \geq 0$, such that C_i, with $0 \leq i \leq n$, are CIDOC CRM classes and P_i, with $1 \leq i \leq n$, are CIDOC CRM properties.

In this section, we shortly present EAD and VRA Core 4.0 as well as their mappings to *CIDOC CRM*.

EAD is an XML-based metadata schema used to encode the archival description, expressed in the form of *finding aids*. An EAD document starts from the ead root element and consists of three basic elements: the eadheader (including the meta- metadata); the frontmatter (carrying information for the printed finding aid); and the archdesc (providing information on the archives content and context).

EAD's documentation logic is based on the four fundamental multilevel description rules, as defined in the General International Standard Archival Description (ISAD (G)) [11]. The first of these rules, "*Description from the general to the specific*", mentions that the archival description must represent the context of the hierarchical structure of the fonds and its parts. In other words, the archival description must begin with the description of the fonds as a whole and then, at the next and subsequent levels gives information for the parts being described; hence, it must present the descriptions in a hierarchical part-to-whole relationship proceeding from the broadest (fonds) to the more specific.

As a consequence the archival description, expressed through EAD, follows a multi-level hierarchical structure, which represents the archive and its components (encoded in the dsc subelement of archdesc element through the component elements c01-c12, and c).

While mapping EAD to *CIDOC CRM*, we observed that an EAD document is conceptualized as an instance of the E31 Document class that includes (P106 is composed of) the archival description (E31 Document), which is the documentation (P70 documents) of a physical object that has been created by human activity (E22 Man-Made Object) and that carries (P128 carries) information, which is immaterial and can be carried by any physical medium (E73 Information Object). Finally, the information carried by the archive can be expressed (P67 refers to) in one or more languages (E33 Linguistic Object). Therefore, the archival description can be represented in *CIDOC CRM* by the following path:

E31 Document→P106 is composed of→E31 Document→P70 documents→E22 Man-Made Object→P128 carries→E73 Information Object→P67 refers to→E33 Linguistic Object.

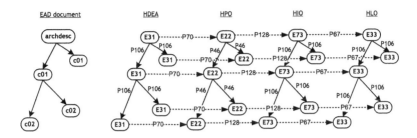

Fig. 1. EAD Semantic hierarchies and their relationships.

Given that an archive has a hierarchical structure, four isomorphic semantic hierarchies are defined in *CIDOC CRM* (see Fig. 1):

a. the *hierarchy of documentation elements and attributes (HDEA)*, where the ead, eadheader, frontmatter, archdesc, c01-c12 and c elements are mapped to instances of the E31 Document class, linked between them through the property P106 is composed of,

b. the *hierarchy of physical objects (HPO)*, where the archive and its components, namely archdesc, c01-c12 and c, are considered as physical objects and are mapped to instances of the E22 Man-Made Object class and linked between them via the P46 is composed of property,

c. the *hierarchy of information objects (HIO)*, where the archive and its components are considered as information objects and are mapped to instances of the E73 Information Object class, linked between them via the P106 is composed of property, and

d. the *hierarchy of linguistic objects (HLO)*, where the archive and its components as linguistic objects are mapped to instances of the E33 Linguistic Object class, linked between them via the P106 is composed of property.

On the other hand, VRA Core 4.0 is a metadata schema for the cultural heritage community that allows the description of three broad groups of entities: *works* (element work) and *collections* (element collection) of visual culture, as well as *images* (element image) that document them. It contains a set of top level elements (agent, culturalContext, date, description, inscription, location, material, measurements, relation, rights, source, stateEdition, stylePeriod, subject, technique, textref, title and worktype), and several optional global attributes (dataDate, extent, href, pref, refid, rules, source, vocab, xml:lang), which are applied to any element or subelement, when necessary. An essential feature of VRA Core 4.0 is that it provides mechanisms to define hierarchical relationships between VRA records, through the relation element and its type attribute (including its values partOf and largerContextFor).

The VRA Core 4.0 to CIDOC CRM mapping focuses on the restricted version of the schema, which imposes controlled vocabularies and type lists as values of the XML nodes of the schema. As a consequence, each attribute assigned to an element of the metadata schema may lead to the generation of different semantic paths in the ontology, depending on the values of that attribute, and produces a plethora of conceptual expressions corresponding to the same element. Furthermore, the use of several global attributes provided by the schema makes the mapping procedure even more complicated, by generating additional semantic paths in the ontology. In the VRA to CIDOC CRM mapping, the work element is associated with an instance of the E24 Physical Man-Made Thing class, which comprises all persistent physical items that are purposely created by human activity. The E24 class was selected since it is considered as a semantically broad class that comprises other more specialized classes, like E22 Man-Made Object or E25 Man-Made Feature.

2.2   Crosswalk Generation Algorithm

In this section, we briefly present the algorithm, presented in [1], that automatically generates crosswalks between metadata schemas. The algorithm applies to pairs of metadata schemas (source and target schema), provided that both the source and the target schema map a resource to an instance of a single class of *CIDOC CRM*. The algorithm proceeds in two phases. In Phase 1, the algorithm accepts the source and target metadata schema mappings to *CIDOC CRM*, expressed in MDL, and produces a two column table, which in its first column contains all possible paths of either the source or target schema, and in the second column the corresponding *CIDOC CRM* paths.

In Phase 2, the algorithm uses (a) the two column table, and (b) the *CIDOC CRM* classes and properties hierarchy. The algorithm compares the *CIDOC CRM* paths in the second column of the table. In case one *CIDOC CRM* path of the source schema *isa-subsumes* another *CIDOC CRM* path of the target schema, then the corresponding source path and the corresponding target path form a new mapping pair, which is added to the crosswalk table.

Definition 2. *Let A, B be two CIDOC CRM paths where A is of the form $C_0 \rightarrow P_1 \rightarrow C_1 \rightarrow \ldots \rightarrow P_n \rightarrow C_n$, and B is of the form $C_0' \rightarrow P_1' \rightarrow C_1' \rightarrow \ldots \rightarrow P_n' \rightarrow C_n'$, with $n \geq 0$. We say that A isa-subsumes B if for each i with $0 \leq i \leq n$, C_i is either the same class or a subclass of C_i' and for each j with $1 \leq j \leq n$, P_j is either the same property or a subproperty of P_j'.*

Example 1. Consider the CIDOC CRM paths A and B where
$A : E24 \rightarrow P108B \rightarrow E12 \rightarrow P14 \rightarrow E21 \rightarrow P131 \rightarrow E82$, and
$B : E24 \rightarrow P108B \rightarrow E12 \rightarrow P14 \rightarrow E39 \rightarrow P1 \rightarrow E41$.
Given that the E21 Person class is a subclass of E39 Actor, P131 is identified by is a subproperty of P1 is identified by and E82 Actor Appellation is a subclass of E41 Appellation, we can conclude that path A isa-subsumes path B.

It should be noticed that the algorithm is based on the following assumption:

Assumption 1. *The algorithm applies to pairs of metadata schemas (source and target schema), provided that the mapping of each participating schema maps a resource documented by a document of the schema to single class instance in CIDOC CRM.*

However, EAD violates this assumption as each archival object (the archive and its components) described in EAD is mapped to four different semantic hierarchies in *CIDOC CRM* (Fig. 1). Besides, its elements are mapped to *CIDOC CRM* paths that are associated with one or more of these hierarchies, according to their semantics. Therefore, EAD cannot be used by the present form of the algorithm.

We can tackle this problem by considering the set of the four classes as a new *joint class* i.e. we can consider the four classes E31 Document, E22 Man-Made Object, E73 Information Object and E33 Linguistic Object as a new class {E31, E22, E73, E33}. Thus the EAD path /ead/archdesc maps to the *CIDOC CRM* path E31→P106→{E31, E22, E73, E33}. Additionally the definition of the *isa-subsumes* relationship is revised as follows:

Definition 3. *Let A, B be two CIDOC CRM paths where A is of the form $C_0 \to P_1 \to C_1 \to \cdots \to P_n \to C_n$, and B is of the form $C_0' \to P_1' \to C_1' \to \cdots \to P_n' \to C_n'$, with $n \geq 0$. Assume now that (some of) the classes participating in these paths may be joint classes represented as a set of conventional CIDOC CRM classes. For simplicity reasons every conventional CIDOC CRM class C is considered to be a joint class represented by the singleton $\{C\}$. We say that A isa-subsumes B if for each i with $0 \leq i \leq n$, C_i is a v-subclass of C_i' and for each j with $1 \leq j \leq n$, P_j is either the same property or a subproperty of P_j'. We say that a joint class C is a v-subclass of a virtual class C' if there is a class $c \in C$ and a class $c' \in C'$ such that c is either the same class or a subclass c'.*

Table 1. Part of the Phase 1 results.

Source and target path	CIDOC CRM path
/ead/archdesc	{E31,E22,E73,E33}
/ead/archdesc/did/unitid	{E31,E22,E73,E33}→P1→E42
/ead/archdesc/did/origination/corpname	{E31,E22,E73,E33}→P108B→E12→P14 →E40→P131→E82
/ead/archdesc/did/unittitle	{E31,E22,E73,E33}→P102→E35
/ead/archdesc/did/physloc	{E31,E22,E73,E33}→P53→E53
/ead/archdesc/dsc/c01/@level	E31→P106→{E31, E22, E73, E33}→P2 →E55→P71B→E32{=''level''}
/vra/work	E24
/vra/work/@id	E24→P48→E42
/vra/work/titleSet/title	E24→P102→E35
/vra/work/agentSet/agent	E24→P108B→E12→P14→E40→
[name/@type="corporate"]/name	P131→E82
/vra/work/locationSet/location	E24→P53→E53
/vra/work/relationSet/relation [@type=partOf][@relids]	E24→P46→E24

Table 2. EAD to VRA Core 4.0 crosswalk table.

EAD path	VRA Core 4.0 path
`/ead/archdesc/@level="fonds"`	`/vra/work/`
`/ead/archdesc/did/unitid`	`/vra/work/@id`
`/ead/archdesc/did/unittitle`	`/vra/work/titleSet/title`
`/ead/archdesc/did/origination/corpname`	`/vra/work/agentSet/agent [name/@type="corporate"]/name`
`/ead/archdesc/did/physloc`	`/vra/work/locationSet/location`
`/ead/archdesc/dsc/c01/@level`	`/vra/work/relationSet/relation [@type=partOf][@relids]`

Table 1 presents the results of the first phase of the algorithm, where an indicative subset paths of the source (EAD) and target (VRA Core 4.0) schema are listed in the left column, while the corresponding *CIDOC CRM* paths are listed in the right column. Given that the *CIDOC CRM* class E22 is subclass of the class E24 and applying the revised *isa-subsumes* relationship, a fragment of the crosswalk from EAD to VRA Core 4.0 is presented in Table 2.

3   Mapping Hierarchies

The presented algorithm generates a crosswalk from EAD to VRA Core 4.0 based on their common semantics as represented in the terms of *CIDOC CRM* paths. However, the algorithm does not take into account the hierarchical structure of EAD and produces the mapping of a single archival component to VRA Core 4.0. Therefore, a process is required that exploits the produced crosswalk and transforms an archival description, including its subordinate components, expressed in EAD, to a VRA Core 4.0 document.

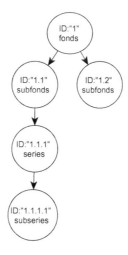

Fig. 2. Example of an archival hierarchy.

3.1 Problem Definition Through an Example

As mentioned, the hierarchical relationship between the archive and its components can be expressed by the elements c01-c12 and c, which are subelements of the dsc subelement of archdesc element. On the other side, VRA allows the encoding of relations between resources within the same XML document through the relation element and its attribute type. The relation element provides links to the IDs of related works, images or collections, while the attribute type identifies the type of the relationship. The values partOf and largerContextFor of the attribute type are used for the representation of hierarchical relationships between resources and their components.

Example 2. Consider the simplified EAD document presented in the left column of Table 3. The hierarchy of the archival description is presented in Fig. 2. The EAD document maps to the VRA document appearing in the right column of Table 3.

3.2 The Hierarchy Mapping Algorithm

The main idea behind the proposed algorithm is to produce interlinked VRA documents, one for each archival component described in an EAD document, which will encompass the multi-level hierarchical structure. The algorithm proceeds as follows:

Phase 1. In Phase 1, the input is an EAD document. The algorithm performs a depth-fist traversal in the EAD document and a unique tag is assigned for the archive and each component of it. The tags have the form $x_1.x_2.x_3 \ldots$ where each x_i is a sequential number that represents a level in the archival hierarchy. For instance, for the archival description of Example 2, the tag 1 is assigned to the archive, the tag 1.1 is assigned to the first c01 element, the tag 1.1.1 is assigned to the first c02 etc. The algorithm assigns tags incrementally until an EAD element denoting higher level of hierarchy appears. Then, the tag of the higher hierarchy is increased by one and the tag of the lower hierarchy turns back to 1. In our example, when a new c01 element appears, then the tag 1.2 will be assigned to it.

The output of this phase is a three-column table, called *EAD XPath Paths Table*, which stores the EAD XPath paths that encode the archive and its components in its first column, the corresponding unique tags in the second column and in the third column the corresponding VRA work elements that will be generated along with the values of their ID attributes. Table 4 presents the results of Phase 1 when applied to the archival structure of Example 2.

Phase 2. In this Phase, the input is the EAD XPath Paths Table produced in the previous phase. For each row of this table a new cell is generated that stores the IDs of the corresponding VRA 4.0 Core work element, as well as the included relation elements along with their relids attributes. Hence, a two column table named *VRA Component Table* is generated as the output of this Phase; the first column represents the EAD XPath of each archival structure and the

Table 3. Mapping EAD hierarchy to VRA Core 4.0.

EAD document	VRA Core 4.0 document

`<ead>`
`<eadheader/>`
`<archdesc ID=''1'' level=''fonds''>`
`<did/>`
`<dsc>`
`<c01 ID=''1.1'' level=''subfonds''>`
`<did/>`
`<c02 ID=''1.1.1'' level=''series''>`
`<did/>`
`<c03 ID=''1.1.1.1''`
`level=''subseries''>`
`<did/>`
`</c03>`
`</c02>`
`</c01>`
`<c01 ID=''1.2'' level=''subfonds''>`
`<did/>`
`</c01>`
`</dsc>`
`</archdesc>`
`<ead>`
	`<vra>`
	`<work ID=''1''>`
	`<titleSet/>`
	`<relationSet>`
	`<relation type=''largerContextFor''`
	`relids=''1.1''/>`
	`<relation type=''largerContextFor''`
	`relids=''1.2''/>`
	`</relationSet>`
	`</work>`
	
	`<work ID=''1.1''>`
	`<titleSet/>`
	`<relationSet>`
	`<relation type=''partOf'' relids=''1''/>`
	`<relation type=''largerContextFor''`
	`relids=''1.1.1''/>`
	`</relationSet>`
	`</work>`
	
	`<work ID=''1.2''>`
	`<titleSet/>`
	`<relationSet>`
	`<relation type=''partOf'' relids=''1''/>`
	`</relationSet>`
	`</work>`
	
	`<work ID=''1.1.1''>`
	`<titleSet/>`
	`<relationSet>`
	`<relation type=''partOf'' relids=''1.1''/>`
	`<relation type=''largerContextFor''`
	`relids=''1.1.1.1''/>`
	`</relationSet>`
	`</work>`
	
	`<work ID=''1.1.1.1''>`
	`<titleSet/>`
	`<relationSet>`
	`<relation type=''partOf'' relids=''1.1.1''/>`
	`</relationSet>`
	`</work>`
	`</vra>`

Table 4. XPaths for the archive and its components.

EAD XPath Paths	Tags	VRA work ID
/ead/archdesc/@level="fonds"	1	`<work ID=''1''>`
/ead/archdesc/dsc/c01/@level="subfonds" [1]	1.1	`<work ID=''1.1''>`
/ead/archdesc/dsc/c01/@level="subfonds" [2]	1.2	`<work ID=''1.2''>`
/ead/archdesc/dsc/c01/c02/@level="series"	1.1.1	`<work ID=''1.1.1''>`
/ead/archdesc/dsc/c01/c02/c03/ @level="subseries"	1.1.1.1	`<work ID=''1.1.1.1''>`

Table 5. EAD Paths describing the archive and its component and the equivalent VRA paths.

EAD XPath Paths	VRA work and relation elements
/ead/archdesc/@level=''fonds'' /ead/archdesc/dsc/c01/@level=''subfonds''[1] /ead/archdesc/dsc/c01/@level=''subfonds''[2]	<work ID=''1''> <relation type=''largerContextFor'' relids=''1.1''/> <relation type=''largerContextFor'' relids=''1.2''/>
/ead/archdesc/dsc/c01/@level=''subfonds''[1] /ead/archdesc/dsc/c01/c02/@level="series"	<work ID=''1.1''> <relation type=''largerContextFor''/ relids=''1.1.1''/> <relation type=''partOf'' relids=''1''/>
/ead/archdesc/dsc/c01/@level="subfonds"[2]	<work ID=''1.2''> <relation type=''partOf'' relids=''1''/>
/ead/archdesc/dsc/c01/c02/@level="series" /ead/archdesc/dsc/c01/c02/c03/ @level="subseries"	<work ID=''1.1.1''> <relation type=''largerContextFor'' relids=''1.1.1.1''/> <relation type=''partOf'' relids=''1.1''/>
/ead/archdesc/dsc/c01/c02/c03/ @level="subseries"	<work ID=''1.1.1.1''> <relation type=''partOf'' relids=''1.1.1''/>

second column is filled by the IDs of the corresponding VRA 4.0 Core work elements, and their relation subelements. Table 5 presents the results of Phase 1 when applied to the archival structure of Example 2.

Phase 3. In this Phase, for each EAD component the crosswalk generation algorithm presented in Sect. 2.2 is applied and the output is n work elements, where n is the number of the components of the archive. Then the work elements are appended to a new VRA Core 4.0 document. The right column of Table 3 presents the corresponding VRA Core 4.0 document for Example 2.

4   Discussion and Outlook

Developing metadata schema crosswalks is an open research issue in the last decades, especially due to the need for interoperability among various resources disseminated in the web. A significant effort on metadata crosswalks is presented in Godby et al. [6] where translation services between metadata schemas using crosswalks as part of their infrastructure have been developed.

Additionally, it is worth mentioning that related conversions to bibliographic standards have been implemented with metadata schemas coming from the educational domain. In [14] the process that Eisenhower National Clearinghouse (ENC) staff went through to develop crosswalks between metadata based on three different standards and the generation of the corresponding XML records is described.

Moreover, in [5] an effort to develop a software that performs metadata schema transformations is presented. The goal of this work is focused on the development of a usable prototype for translating among metadata schemas. This work concerns the development of a self-contained metadata translation service that automates routine processes.

The appearance of *CIDOC CRM* revealed the issue of the semantic interoperability between the metadata vocabularies. Actually, *CIDOC CRM* provides a reference model, which can reveal the semantics of the elements of metadata schemas and is very effective in correlating them via properties. The result is the development of coherent and semantically rich descriptions of the resources. Therefore, *CIDOC CRM* can be used as mediating schema, able to ensure interoperability between metadata schemas. The main requirement for implementing such an architecture is to map the elements of metadata schemas to *CIDOC CRM* paths [8]. Several initiatives deal with *CIDOC CRM*-based integration. [16] describes an effort to utilize *CIDOC CRM* in the core model of the BRICKS project. In addition, [9] presents an effort to combine MPEG-7 schema and *CIDOC CRM* into a single ontology for describing and managing multimedia in museums. Towards this direction, a methodology for aligning *CIDOC CRM* with MPEG-7 is also presented in [17].

In this paper, we proposed a process that transforms an archival description encoded by an EAD document to a VRA Core 4.0 document. The proposed process exploits the crosswalk between the two schemas, which has been resulted by an algorithm based on the mappings from EAD and VRA Core 4.0 to *CIDOC CRM*. The paper confronted the issue of mapping multi-level hierarchical archival structures to VRA and, actually, addresses the issue of structural heterogeneity of metadata schemas.

Given that EAD is one of the most widely used metadata schema in (digital) archival collections it would be a great challenge to provide an algorithm that maps an archival hierarchy to a well - known metadata schema, such as VRA Core 4.0. Although EAD and VRA are targeted to describe resources with different substance and structure, there exist and other efforts for generating mappings between these two schemas. For instance, the Getty Research Institute has published a crosswalk among various metadata schemas, which includes both EAD and VRA [10]. Nonetheless, the work presented in this paper differs from the Getty's crosswalk in the following points: (a) Getty's crosswalk does not take into account the multi-level hierarchical structure of EAD, while mapping it to VRA and (b) it proposes mappings for a limited number of EAD elements to VRA elements. In general, the crosswalks published until now, do not take into account the need to represent the existing hierarchical relationships between the archival components.

Concluding this work aligned semantically EAD and VRA Core 4.0 schemas. The next step of our research aims to optimise the proposed algorithm so as to transform efficiently very large archival descriptions to VRA Core 4.0 documents. For this purpose, we will test the scalability of the algorithm on huge volumes of archival descriptions.

References

1. Gaitanou, P., Bountouri, L., Gergatsoulis, M.: Automatic generation of crosswalks through CIDOC CRM. In: Dodero, J.M., Palomo-Duarte, M., Karampiperis, P. (eds.) MTSR 2012. CCIS, vol. 343, pp. 264–275. Springer, Heidelberg (2012). doi:10.1007/978-3-642-35233-1_26

2. Gaitanou, P., Gergatsoulis, M.: Defining a semantic mapping of VRA Core 4.0 to the CIDOC conceptual reference model. Int. J. Metadata Semant. Ontol. **7**(2), 140–156 (2012)

3. Gergatsoulis, M., Bountouri, L., Gaitanou, P., Papatheodorou, C.: Mapping cultural metadata schemas to CIDOC conceptual reference model. In: Konstantopoulos, S., Perantonis, S., Karkaletsis, V., Spyropoulos, C.D., Vouros, G. (eds.) SETN 2010. LNCS (LNAI), vol. 6040, pp. 321–326. Springer, Heidelberg (2010). doi:10.1007/978-3-642-12842-4_37

4. Gergatsoulis, M., Bountouri, L., Gaitanou, P., Papatheodorou, C.: Query transformation in a CIDOC CRM based cultural metadata integration environment. In: Lalmas, M., Jose, J., Rauber, A., Sebastiani, F., Frommholz, I. (eds.) ECDL 2010. LNCS, vol. 6273, pp. 38–45. Springer, Heidelberg (2010). doi:10.1007/978-3-642-15464-5_6

5. Godby, C.J., Smith, D., Childress, E.: Two paths to interoperable metadata. In: Proceedings of the International Conference on Dublin Core and Metadata Applications (DC). DCMI 2003, Seattle, Washington. Dublin Core Metadata Initiative (2003)

6. Godby, C.J., Smith, D., Childress, E.: Toward element-level interoperability in bibliographic metadata. Code4Lib J. (2) (2008)

7. ICOM/CIDOC CRM Special Interest Group. Definition of the CIDOC Conceptual Reference Model, Version 5.0.4, November 2011. http://www.cidoc-crm.org

8. Martin, D., Kondylakis, H., Plexousakis, D.: Mapping language for information integration. Technical report, Institute of Computer Science, Foundation of Research and Technology, December 2006

9. Hunter, J.: Combining the CIDOC CRM and MPEG-7 to describe multimedia in museums. In: Proceedings of the Museums and the Web International Conference. Museum Publications, Boston, April 2002

10. Getty Research Institute. Metadata standards crosswalk (research at the getty).   http://www.getty.edu/research/publications/electronic_publications/intrometadata/crosswalks.pdf

11. International Council on Archives. Committee on Descriptive Standards. ISAD(G): General International Standard Archival Description. ICA, 2nd edn. (2000)

12. Library of Congress. EAD 2002 Schema (2002). http://www.loc.gov/ead/eadschema.html

13. Library of Congress. VRA CORE - a data standard for the description of works of visual culture: Official Web Site (Library of Congress) (2011). http://www.loc.gov/standards/vracore/

14. Lightle, K.S., Ridgway, J.S.: Generation of XML records across multiple metadata standards. D Lib Mag. **9**(9) (2003)

15. Lourdi, I., Papatheodorou, C., Doerr, M.: Semantic integration of collection description: combining CIDOC/CRM and Dublin core collections application profile. D Lib Mag. **15**(7/8) (2009)

16. Meghini, C., Risse, T.: Bricks: a digital library management system for cultural heritage. ERCIM News (61) (2005)

17. Ntousias, A., Gioldasis, N., Tsinaraki, C., Christodoulakis, S.: Rich metadata and context capturing through CIDOC/CRM and MPEG-7 interoperability. In: Luo, J., Guan, L., Hanjalic, A., Kankanhalli, M.S., Lee, I. (eds.) Proceedings of the 7th ACM International Conference on Image and Video Retrieval, CIVR, Niagara Falls, 7–9 July 2008, pp. 151–160. ACM (2008)

Linking Named Entities in Dutch Historical Newspapers

Theo van Veen[✉], Juliette Lonij, and Willem Jan Faber

Koninklijke Bibliotheek, National Library of the Netherlands, The Hague, The Netherlands
{theo.vanveen,juliette.lonij,willemjan.faber}@kb.nl

Abstract. We improved access to the collection of Dutch historical newspapers of the Koninklijke Bibliotheek by linking named entities in the newspaper articles to corresponding Wikidata descriptions by means of machine learning techniques and crowdsourcing. Indexing the Wikidata identifiers for named entities together with the newspaper articles opens up new possibilities for retrieving articles that mention these resources and searching the newspaper collection using semantic relations from Wikidata. In this paper we describe our steps so far in setting up this combination of entity linking, machine learning and crowdsourcing in our research environment as well as our planned activities aimed at improving the quality of the links and extending the semantic search capabilities.

Keywords: Named entities · Linked data · Entity linking · Semantic enrichment · Semantic search · Machine learning · Classification · Crowdsourcing

1 Introduction

One of the strengths of the semantic web [1] is the possibility it offers to identify resources and link the mentions of these resources to relevant descriptions from external data sources. In the research environment of the Koninklijke Bibliotheek (KB) we started to enrich Dutch historical newspaper articles with named entities (i.e. names of persons, locations, organizations and others) linked to their resource descriptions in various knowledge bases, such as DBpedia [2], Wikidata [3] and VIAF [4]. We combine all relevant links for an entity into a single enrichment record, which is stored in a dedicated enrichment database. We discussed the data model for these enrichment records and the architecture of our enrichment infrastructure in a previous article [5]. This paper will focus on the process of entity linking and the application of the results in semantic search, as currently available in the KB research environment.

Entity linking has received much attention in recent years and often DBpedia plays a central role [6, 7]. In Sect. 2 we will describe the automatic process we developed for generating links to DBpedia based on machine learning techniques and entity context information. As we cannot train the machine-learning algorithm to give results with 100 % accuracy, we will consider crowdsourcing as an option to correct false and missing links in Sect. 3. Indexing generated links makes it possible to use them in various forms of semantic search, some of which we have implemented in a purpose-built research portal that we will present in Sect. 4. We will draw some provisional conclusions and

© Springer International Publishing AG 2016
E. Garoufallou et al. (Eds.): MTSR 2016, CCIS 672, pp. 205–210, 2016.
DOI: 10.1007/978-3-319-49157-8_18

point out the next steps we have planned for improving the quality of the links and extending the semantic search capabilities in Sects. 5 and 6 respectively.

2   The Automatic Linking Process

The development of the automatic entity linking process has gone through a number of different phases.

Our first attempt of searching all titles from DBpedia descriptions in our newspapers was not very successful. Thus, in the next phase, we did it the other way around: we first performed named entity recognition [8] on the newspaper articles and subsequently looked for matching descriptions in DBpedia. For this purpose we constructed a Solr [9] index out of DBpedia dumps, combining the relevant data, such as label, abstract, VIAF and Wikidata identifiers, for each resource into records. The most appropriate link candidate was selected with a simple, rule-based approach looking at various forms of string matching (e.g. exact match, partial match, last-part match) and a few additional features such as the number of inlinks. This stage is illustrated in Fig. 1.

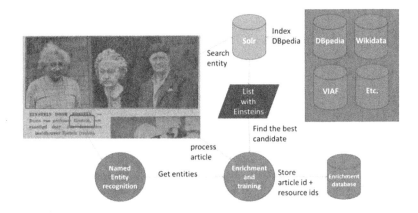

Fig. 1. Schematic overview of the process for entity linking

The third phase was speeding up the process by using the processing capacity of the HPC cloud infrastructure at SURFsara [10]. This has increased the throughput of the number of articles by more than an order of magnitude. Because of the large number of articles and the time it still takes to process them all, we keep the linking process running continuously, with each improved version starting from the point in the collection where the previous version left off. Once all articles have been processed we start at the beginning again. In this way we are able to enrich the entire collection of articles as quickly possible, improving the quality level with each iteration.

In the fourth phase we started applying machine-learning techniques to the linking process, training a Support Vector Machine classifier [11] on a labeled example set of several thousands of potential links. To the string matching features we added a number of features based on contextual information. These include the type and subtype match, e.g. "person" or "politician", between the named entity and the candidate description, the occurrence of other named entities from the article in the DBpedia abstract, and the compatibility of the publication year of the article and any known year of birth.

During this last phase the fact that DBpedia is language dependent became increasingly problematic. We wanted to include data from both Dutch and English descriptions in the index, because the English DBpedia dumps contain more names and the English descriptions often provide additional data about a resource. To avoid having to deal with different identifiers for the same entity in different languages, we decided to switch to the Wikidata identifiers as the main identifier. From Wikidata we can still obtain all the links to DBpedia, as well as many other databases.

We measured the quality of the results from the second phase onward by means of an accuracy score for a manually linked evaluation set of 349 named entities. With the rule-based approach we were able to obtain an accuracy of 74,50 % after the last iteration. For the machine-learning approach, while improving the features, the accuracy gradually increased to 83,09 %. Based on manual inspection of the results we estimate that approximately 5 % of the examples cannot be automatically linked because of serious OCR errors or because a significant amount of human knowledge about the entity is required. Thus, we expect 95 % to be the maximum accuracy possible and we hope to get closer to that number using a neural network approach.

3   Crowdsourcing

Since the accuracy of the results of the automatic linking process will never reach 100 % we need user feedback to make corrections and add missing links. This crowd-sourced data can be used to extend the training set for the machine-learning algorithm and user feedback is also useful for preventing the suggestion to the end user that the links are 100 % reliable.

Users can currently provide feedback through an enrichment page displaying a newspaper article with the linked named entities marked in the text. Clicking on a name shows the linked resources for that name and the option to remove incorrect links. Selecting the text of a name appearing in the article will enable the option to add a new link to that name, either by entering a URL or by choosing the most appropriate entry in DBpedia.

At the moment the enrichment page is only available to KB-employees. We expect a significant number of other users to be intrinsically motivated to contribute links in areas of their interest or expertise, but before we can offer this functionality to the general public we will need to take measures to minimize the chances of abuse and the introduction of errors. Options such as crowdsourced moderation are being considered, but the exact measures have yet to be decided upon.

4   Semantic Search

The links resulting from the automatic linking process and from crowdsourcing can be used to improve accessibility and usability of the newspaper collection, which is the actual goal of our project.

The links function as identifiers and, by indexing them along with the newspaper articles, knowledge bases can play the role of thesaurus for the named entities. We believe that Wikidata is the most promising candidate for this purpose, since Wikidata is not restricted to a specific domain, like VIAF, and it is language independent, as opposed to DBpedia. It already has quite a large number of descriptions (over 19 million at the moment [12]) and can easily be extended with entries for entities that are not yet included.

Using the indexed Wikidata identifiers it also becomes possible to search the newspaper articles for resources based on semantic relations present in Wikidata. This concept, as shown in Fig. 2, is very simple: a list of Wikidata identifiers resulting from a SPARQL [13] query in Wikidata is used as input for a conventional SRU [14] query in the enriched newspaper article index. As Wikidata resource descriptions contain many other existing resource identifiers, this approach can even be applied to library catalogues and other databases by replacing the obtained Wikidata identifiers with the local identifiers listed in Wikidata.

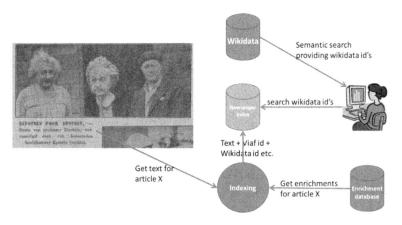

Fig. 2. Overview of the usage of the Wikidata identifier

For demonstration purposes we have developed a research portal [15] for searching and viewing the KB collections and their various enrichments, a screenshot of which is shown in Fig. 3. The following semantic query functionality is available for the enriched newspaper index ("Newspapers +"), the last three options for the library catalogue ("KB Catalogue") as well:

- Each resource identified in an article is provided with an infobox that has the option to search articles mentioning this specific resource using its Wikidata identifier.
- The infobox also contains contextual information about the resource in the form of properties from Wikidata and allows searching for all resources in the newspaper articles with the same value for a particular property by clicking on that value. The actual query that is generated looks like [property = value] using the Wikidata identifiers of both property and value.

- Using square brackets in a query, e.g. "[Beatles]", the application tries a few SPARQL queries using different properties as a first guess, thus avoiding the need for user knowledge on Wikidata property names. In this example the property that can have "Beatles" as its value turns out to be "is member of" and the query results are the articles enriched with the Wikidata identifier of any of five the members of the Beatles (Pete Best being the fifth Beatle).
- For advanced users it is possible to enter a property-value combination between square brackets, such as "[P737 = Q1203]", which will search for articles mentioning resources that are influenced (P737) by John Lennon (Q1203). Here, the user must know the Wikidata identifiers of both the property and the value but in a future version we might give the user some help in finding these identifiers.
- The even more skilled user might enter a very complex SPARQL query on the Wikidata website directly and use the resulting Wikidata identifiers in a query on the research portal. In this case the query is treated as a conventional query, but still the added value of our approach is that the Wikidata identifiers are available in the index.

Fig. 3. Screenshot of the research portal

5   Conclusion

In this paper we have described the combination of building blocks making up the entity linking process for Dutch historical newspapers currently running in the KB research environment, as well as the ways in which the resulting links can be used to improve access and usability. The linking quality achieved so far is quite promising, as we expect to be able to further increase it. Our approach of using Wikidata as both thesaurus and data source for semantic search is conceptually simple and makes available new search functionality. Adopting a "release early, release often" strategy has ensured that we can already use and assess these results in our research portal. New links and functionality are added there as soon as they become available.

6 Next Steps

We have several next steps planned for further improving the quality of the links and the possibilities for semantic search. The first step is replacing the conventional classification algorithm with a neural network approach. Moreover, we want to start using the data obtained from crowdsourcing as additional input for training the network. Another, somewhat more distant goal is to extract new relations between named entities occurring in the newspaper collection that are not (yet) part of knowledge bases. We hope to be able to present some results of these upcoming steps at the conference.

References

1. Semantic Web. http://www.w3.org/standards/semanticweb/
2. DBpedia. http://dbpedia.org/
3. Wikidata. https://www.wikidata.org/
4. VIAF, Virtual International Authority File. http://viaf.org/
5. Van Veen, T., Lonij, J., Koppelaar, H.: Semantic enrichment: a low-barrier infrastructure and proposal for alignment. D-Lib Mag. (2015). doi:10.1045/july2015-vanveen
6. Odijk, D., Meij, E., de Rijke, M.: Feeding the second screen: semantic linking based on subtitles. In: Open Research Areas in Information Retrieval (OAIR 2013), Lisbon (2013)
7. Sil, A., Croning, E., et al.: Linking named entities in any database. In: EMNLP-CoNLL 2012 Proceedings of the 2012 Joint Conference on Empirical Methods in Natural Language Processing and Computational Natural Language Learning, Jeju Island, Korea (2012)
8. Stanford Named Entity Recognizer. http://nlp.stanford.edu/software/CRF-NER.shtml
9. Apache Solr. http://lucene.apache.org/solr/
10. SURFsara. https://www.surf.nl/en/services-and-products/hpc-cloud/
11. mySVM. http://www-ai.cs.uni-dortmund.de/SOFTWARE/MYSVM/index.html
12. Wikidata statistics. https://www.wikidata.org/wiki/Wikidata:Statistics
13. SPARQL, query language for RDF. http://www.w3.org/TR/rdf-sparql-query/
14. SRU, Search and Retrieval via URL's. http://www.loc.gov/standards/sru/
15. KB research portal. http://www.kbresearch.nl/xportal/

Exploring Audiovisual Archives Through Aligned Thesauri

Victor de Boer[1,2(✉)], Matthias Priem[3], Michiel Hildebrand[4], Nico Verplancke[3], Arjen de Vries[4], and Johan Oomen[1]

[1] Netherlands Institute for Sound and Vision, Hilversum, The Netherlands
{vdboer,joomen}@beeldengeluid.nl
[2] Department of Computer Science, VU University Amsterdam, Amsterdam, The Netherlands
[3] Flemish Institute for Archiving, Ghent, Belgium
{matthias.priem,nico.verplancke}@viaa.be
[4] Spinque B.V., Utrecht, The Netherlands
{michiel,arjen}@spinque.com

Abstract. As audiovisual archives are digitizing their collections and making these collections available online, the need arises to also establish connections between different collections and to allow for cross-collection search and browsing. Structured vocabularies, made available as Linked Data, can be used as connecting points by aligning thesauri from different institutions. In this paper, we present a case study where partial collections of two audiovisual archives are connected by aligning their thesauri. We report on the conversion of one of the thesauri to SKOS and on the subsequent application of an interactive alignment tool "CultuurLINK". Finally, we introduce an cross-collection browser which uses the produced alignment to allow users to explore connections between the two collections.

Keywords: Audiovisual archives · Thesaurus alignment · Cross-collection browsing

1 Introduction

The task of audiovisual archives is to store audiovisual heritage from various sources and to make this material available to media professionals, researchers and the general public. In recent years, audiovisual archives have initiated large-scale digitisation projects and started ingesting born-digital material. Collections are being made available on the Web so they can be accessed by various user-groups and making them available on the Web. However, these collections are mostly still only available through their own Web interfaces and are rarely connected to outside collections and other information sources.

At the same time, end users are more and more expecting to be able to access, browse and search across different collections. Especially for media researchers,

© Springer International Publishing AG 2016
E. Garoufallou et al. (Eds.): MTSR 2016, CCIS 672, pp. 211–222, 2016.
DOI: 10.1007/978-3-319-49157-8_19

such cross-collection exploration is extremely valuable [1]. Structured vocabularies, published online as SKOS [2] and made available as Linked Data [3] offer excellent opportunities to provide this type of integration.

In this document, we describe a case study where parts of two national audiovisual collections are described using SKOS thesauri. The collections are archived by the Flemish Institute for Archiving (VIAA)[1] and the Netherlands Institute for Sound and Vision (NISV)[2]. Both institutions manage large digital archives, composed of a variety of sources. The material comes from the public broadcaster(s), regional broadcasters and/or cultural heritage institutions. This archive material is accessible to diverse audiences, such as customers themselves, research or education. The thesauri of the two institutions are aligned using an interactive and transparent alignment tool. We finally introduce a cross-collection browser which uses the produced alignment to allow users to explore connections between the two collections. The contributions of this paper are the following:

- We describe the entire pipeline of a real-world, international use case that illustrates the end-user benefit of aligned SKOS thesauri;
- We present a method and tools for converting XML thesauri to SKOS;
- We introduce CultuurLINK, an interactive tool for thesaurus alignment;
- We present an application that enables cross-collection search and browsing using the aligned thesauri.

2   The Two Institutions and Their Data

2.1   Netherlands Institute for Sound and Vision and the GTAA

NISV is the largest audiovisual archive in the Netherlands, with more than 800,000 h of radio, television, film and music archived. The archive makes its collection accessible to diverse audiences, including media professionals, the creative industries, education and the general public. Through research and innovation, the institute has developed into a broad cultural institution that plays a central role through his knowledge and infrastructure within the archive and media sectors. NISV makes its collection available online through various enduser services, including services for the creative industry, education and research.

GTAA. The Common Thesaurus for Audiovisual Archives[3] (GTAA). The GTAA is used by NISV to annotate the different collections. The GTAA closely follows the ISO-2788 standard for thesaurus structures and consists of several facets for describing TV programs: subjects, people mentioned, named entities (Corporation names, music bands etc.), locations, genres, producers and presenters. The GTAA, available as SKOS, contains approximately 180.000 terms and is actively maintained, being updated as new concepts emerge on television.

---

[1] http://viaa.be.

[2] http://beeldengeluid.nl.

[3] http://datahub.io/dataset/gemeenschappelijke-thesaurus-audiovisuele-archieven.

Approximately 20,000 terms have broader or narrower relationships. Nine concept schemes divide the thesaurus terms of content in geographical terms, persons, genres, etc. The thesaurus includes about $90,000$ scope notes, and 33,542 terms are related to each other.

OpenImages. The GTAA thesaurus is used to annotate the entire NISV collection, however, due to licensing issues, only a small part of this collection is made publicly and freely available using Creative Commons licenses. This "open images" collection is a set of $1,700$ video items freely available on the Web, mostly consisting of Dutch public news items from the mid 20th century. The Open Images dataset can be accessed through a web portal[4] and a set of APIs. There also is a version available in the Resource Description Framework (RDF). For the research described here, we use this RDF version.

2.2 VIAA and the VRT Thesaurus

VIAA is the Flemish institute for archiving and was founded in 2012. VIAA digitizes, archives and makes available material from more than 80 organizations. Among these organizations are the Flemish public broadcaster VRT, regional broadcasters, archives and cultural heritage institutions. The digitized and archived material is made available to education, research and the public (through public libraries).

Conversion of the VRT Thesaurus. The VRT archive is managed using a digital system, and since 1986, the media items were already annotated using a central keyword list. Recently (in 2014), this thesaurus was greatly downsized and imported to the media management system. The thesaurus currently consists of 102,172 terms. To enable reuse of the thesaurus and linking within the project, we converted the thesaurus to the SKOS format. First, the thesaurus was exported in XML format, which provided insight into the structure and other features of the thesaurus. Unlike GTAA, the VRT thesaurus does not separate different types of terms in different concept schemes. The 'what', 'where' and 'who' terms are all be found in the same list. Terms have alternative labels and relations to other terms. Each term, moreover, has been given a unique ID, which we can use for the assignment of URIs for the $102,172$ `skos:Concept` instances. The original relations were mapped to SKOS relations as follows:

- 'ParentID' attributes, indicating hierarchical relations are mapped to `skos:broader` and `skos:narrower` relations. Examples of this include geographical part-of relations as well as subclass relations. In total $97,744$ concepts have a broader or narrower relation.
- Terms without parentID indications are modeled as `skos:TopConcepts`. In total there are $4,429$ top concepts.
- Preferred and alternative labels are mapped to `skos:prefLabel` and `skos:altLabel`, respectively. The labels receive RDF language tags specifically indicating the Flemish dialect of Dutch ('`@nl-be`').

---

[4] http://openimages.eu.

- 'Relation' attributes are mapped to `skos:related` RDF triples. Examples include cities and their football club ('Amsterdam' - 'Ajax').
- 'Explanation' attributes are mapped to `skos:scopeNotes`. Only a small fraction of terms have this attribute (212 terms).

As, at the time of conversion, the SKOS thesaurus has not yet received an official status and is not yet published by either VIAA or VRT, the URI namespace is left unspecified in the converted SKOS thesaurus[5]. The conversion script is based on NodeJS and uses the Skosify library[6], and is available as open source code[7]. It is well-documented and can be reused to convert similar thesauri or to be re-run in case adaptations or additions are made to the thesaurus in the original management system.

Collection Subset. Unfortunately, for the Flemish audiovisual data, neither the entire archive, nor any subset are at the moment openly licensed. They are only available for research or educational purposes. Nevertheless, we selected a subset of the VRT video collection of 35,000 items. Like Open Images, this is only a small subset of an archive containing more than 1 million records. These items are contemporary television broadcasts and are annotated with thesaurus terms.

3 Thesaurus Alignment

3.1 CultuurLINK

To align the two SKOS thesauri, we used the functionalities of the CultuurLINK tool[8]. CultuurLINK was based on research on interactive alignment and the Amalgame tool described in [4]. CultuurLINK extends that tool with a more efficient backend implementation and an improved end-used interface. CultuurLINK is an interactive web-based vocabulary alignment tool in which strategies can be constructed to optimally produce correspondences (links) between concepts of two or more thesauri. It features an intuitive end-user interface where collection managers arrange different work-flow elements using a drag-and-drop interface. These elements include different filters and word-matching techniques. The interface also allows for inspection and evaluation of intermediary or final results. This highly interactive alignment allows the collection managers - who know the different features and peculiarities of the source and target thesauri - to develop a specialized and transparent workflow which leads to high quality alignment between the two vocabularies. CultuurLINK features (fuzzy) string matching strategy elements, regular expression operations, and basic Natural Language Processing options (e.g. stemming). Additional strategy elements can be applied to select concepts based on structural properties.

---

[5] We use http://example.org as temporary namespace in the produced SKOS files.
[6] https://github.com/NatLibFi/Skosify/.
[7] https://github.com/viaacode/skoscreator.
[8] http://cultuurlink.beeldengeluid.nl.

3.2   Aligning the Two Thesauri

Several strategies have been designed for the alignment of the VRT and GTAA thesaurus corresponding to different types of terms. The reason for this division is that different types of terms have different properties. For example, person names consist of at least two parts (In GTAA "Last-Name, First-Name", in the VRT thesaurus "First-Name Last-Name") requiring the use of a regular expression and topic terms might take a singular or plural form, calling for word stemming. For optimal matching a division into four strategies corresponding to different types of terms was made. These categories share some characteristics in terms of the way the actual terms are constructed. Therefore, by isolating them, we can use specific workflows of string matching techniques for each of the term types. There are four such substrategies for:

- **Topics.** Topic terms are generic concepts. (e.g. "transportation").
- **Locations.** Concepts denoting geographical names (e.g. "Amsterdam")
- **Entities.** For example organization names (e.g. football club "Ajax")
- **Persons.** For example: media producers or persons appearing in news footage.

The CultuurLINK screenshot in Fig. 1 shows the entire strategy for Topics visually. It shows the different filters to isolate the topics from the two vocabularies. Subsequently, string-matchers and other building blocks are used to identify matching terms. The bottom half of the screen shows the inspection part of the tool, where the user can inspect and evaluate intermediate or final mappings. The strategies can be explored at http://cultuurlink.beeldengeluid.nl[9].

Table 1. Number of established links between the thesauri per strategy

Term type	Links
Topics	4,167
Entities	2,197
Locations	4,011
Persons	11,265
Total	21,640

Table 1 shows how many links are eventually found between the thesauri. A total of $21,640$ links are found, which indidates that 21% of the VRT terms are mapped to a corresponding GTAA term. The percentage might seem low, however, the two thesauri each have a different focus (Dutch vs. Flemish). GTAA, for example, contains many names of Dutch media producers or actors who do not appear in any Flemish media items, and therefore absent from the VRT

---

[9] A strategy can be revisited by entering a session identifier "vrt_onderwerpen", "vrt_plaatsen", "vrt_namen" and "vrt_personen".

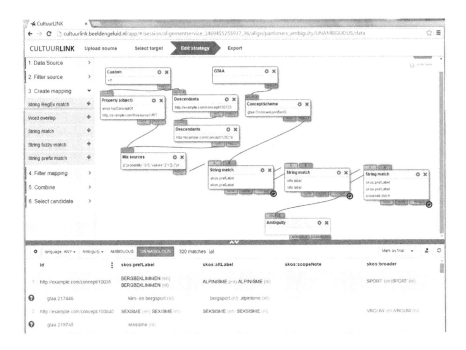

Fig. 1. Screenshot of the CultuurLINK tool, showing the strategy for Topics

thesaurus. The same holds for geographical terms. For the more neutral 'Topics', the overlap is in fact significant. GTAA includes 4, 683 topics ("subjects") while the VRT thesaurus contains 25, 155 potential subjects (identified by the exclusion of people and geographical concepts). In total, 4, 167 mappings are found, which corresponds to 89 % of the GTAA terms. 4, 011 out of 8, 617 locations are matched, corresponding to 47 %. Inspection of the unmatched location terms shows that these are mostly smaller places, appearing in one but not in the other thesaurus. A formal analysis of the unmatched terms would give more insight into the exact quality of the alignment and whether it matches expectations.

The exported links, together with the SKOS thesauri themselves are published on github[10]; they are also accessible through an online triple store[11] allowing for browsing, downloading and querying using the SPARQL protocol.

4   The Demonstrator

4.1   User Interface

The demonstrator can be reached at http://link.spinque.com/VIAA-1.0/. As part of the collections cannot be made available to the general public, a password

---

[10] http://github.com/biktorrr/gtou_taalunie.
[11] http://semanticweb.cs.vu.nl/test/.

is needed[12]. To visualize the functionality of the demonstrator, a screencast of the demonstrator is publicly available at https://youtu.be/iOJvcHRfvDY. Figure 2 shows two annotated screenshots of the application.

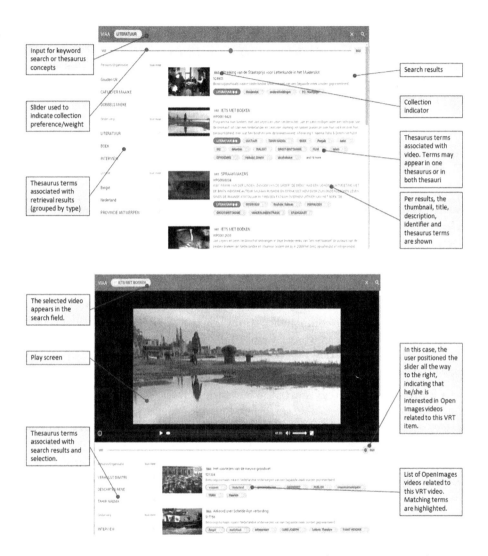

Fig. 2. Annotated screenshot of the demonstrator (Color figure online)

After logging in, a user starts by using a search term. The interface displays matching results based on titles and descriptions. The concepts with which videos are annotated are presented. For these concepts, the interface shows whether

---

[12] Available upon request.

they occur in a single, or in both thesauri (yellow/blue dot). The concepts can be selected, after which they are used as search terms. The top screenshot shows an example where the concept "Literature" is selected[13]. This concept is present in both thesauri and a link was established in the alignment (both a blue and yellow marker are shown). The demonstrator shows results from both collections. At the top of the screen a slider can be found, which allows users to adjust the weight given to one of the two collections. This is an important feature of the demonstrator, as it allows users to actively find relations between items *across* the collections, even when initially one collection might have many more "hits". When a user has selected a video (bottom screenshot), this video is presented to the user (full screen if preferred). Below the video, related videos are shown. These are determined through their corresponding thesaurus concepts.

4.2   Demonstrator Backend

The demonstrator takes as input a query, which is any combination of keywords, selected thesaurus terms or selected items. The application then presents as search results the best matching videos of the two collections, based on their annotations with concepts from the two thesauri.

The application can be considered an information retrieval application (a search engine), where the desired results would be evaluated on their relatedness to the user query. The demonstrator is constructed using the "search by strategy" approach [5], in which the back end developer can connect visually a variety of search-related components to design a *search strategy* that specifies how to retrieve relevant videos[14]. Figure 3 shows the search strategy designed for the search and suggestion functionality for VIAA application. We explain the elements below.

Data Source. The strategy starts top left with a 'Data Source' block, that represents the entire database. The corresponding data consists of the two thesauri, the two sub-collections, and the link sets resulting from the alignment. In the application, we only search for videos, in our case identified as instances of the class http://schema.org/VideoObject. The search strategy uses a filter to only retrieve objects of this type Video.

Inputs. In the application, the user can create a complex query consisting of a combination of keywords, thesaurus concepts, as well as an example video. In the strategy these inputs are represented by the green-labeled blocks at the top.

Search. The three types of inputs from the search request contribute to the search results in different ways. The keywords are used to search the titles and descriptions of the videos (four blocks on the left side of Fig. 3). Also, the keywords are used to locate thesaurus concepts, which then lead through the subject

---

[13] Note that the thesaurus labels are in Dutch which we translate for this paper.

[14] Specifically, the application is implemented using the Spinque Core platform http://www.spinque.com/spinque-core.

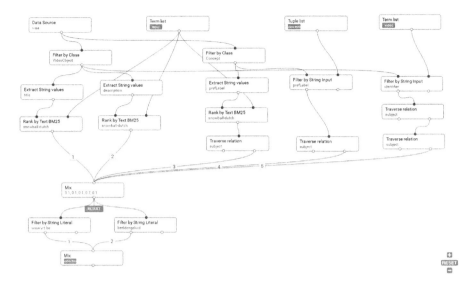

Fig. 3. Screenshot of the search strategy used in the backend of the demonstrator.

relation to relevant videos. The concepts in a search query may lead directly to a video related through the subject relationships.

If a user-selected video is a part of the search query, then the goal is to find videos related to that video. In order to achieve this, the thesaurus concepts are used. This is done with the blocks on the right side of the strategy shown in Fig. 3. Concepts with which the selected video is annotated are selected, and videos described with the same concepts are retrieved from the collection. The related videos are weighted by the number of concepts matches: the more overlap, the more relevant the results.

Combining Results. The sub-results of the various sub-strategies are combined in a "mix block". This block has 5 inputs, each with a weight that indicates how much should be included. In our application, results found through matching concepts are considered more important than the results using keywords and thus receive a higher weight.

Collection Weighing. The results contain videos of the VRT collection as well as the OpenImages collection. The last three blocks determine the priority of each collection in the results. In the search application, the user can specify (with a slider) which collection should be emphasized more for that specific search query. In the last mix block this user input is used to set the weights. Weights can be set between 0–1 with the total weight of the two collections adding up to 1.

5   Discussion

Converting the VRT Thesaurus. The conversion of the VRT thesaurus was done by IT specialists. However, knowledge of the internal structure of the source thesaurus provides significant time savings and quality assurance. Combining "technical know-how" (about XML, RDF, SKOS) and "content know-how", we estimate that a conversion project such as this can be completed in about 5 working days. This indicates that it is a viable option for institutions to work together with specialists to convert their thesaurus. Using the produced scripts, any updates on the datamodel or actual content are easily carried through to the SKOS version. For completely new thesauri, the scripts can be used as starting points, saving time. The work performed may thus serve as a basis for publishing a linked thesaurus. The thesaurus could be published online as Linked Open Data using a specific solution, such as the OpenSKOS platform[15], which is used by the GTAA. The experience gained during this project shows us not only some possible further steps, but can also be used as a use case to convince other organizations to disclose current non-standardized and/or published thesauri as linked data.

Mapping Strategies and Generated Links. In general, aligning thesauri is still a daunting task, even when the labels of the terms are available in the same language (in our case Dutch). A variety of reasons may underly the way different structures are arranged in the thesauri [4]. In our specific case, it was non-trivial to identify the corresponding parts between two thesauri (Persons, Places etc.) as both thesauri took different modeling decisions and have a different structure. The fairly 'flat' GTAA is at the highest level divided in a number of concept schemes whereas the VRT thesaurus has a lot or hierarchical structure, but no concept schemes. However, by allowing the application of different filtering options and fine-tuning these interactively, CultuurLINK makes it possible to identify these corresponding parts. For each of the parts, specific sub-strategies could be made using different label matchers. Even though the thesauri have different origins, structure and usage, we still find large numbers of links between the two. In addition, it is likely that the four strategies which developed within this project serve as blueprints for mapping strategies between other thesauri.

Here too, a mixed team is in the best position to find the links. Ideally combining content managers and IT people, aware of the (dis)advantages and workings of strings matchers, structural matchers, fuzzy matching, etc. Even so, an interactive, user-friendly tool such as CultuurLINK allows for users with less technical knowledge to still produce good alignments.

The links produced should be further explored to determine and possibly expand the coverage and quality. This process is guided by revisiting unmapped concepts to determine whether these are terms that have no counterpart in the other thesaurus, or they should be classified as errors.

The links between the VRT thesaurus and GTAA act as a possible bridge to the larger Web of Data [3]. In previous projects, the GTAA thesaurus has been linked to other thesauri and datasets, including Wordnet[16] and DBpedia[17] (cf. [6,7]). Through the alignment described here, the VRT thesaurus and collection

---

[15] http://openskos.beeldengeluid.nl/.

[16] https://wordnet.princeton.edu/.

[17] http://dbpedia.org.

are now linked to these sources as well. Furthermore, these existing links can also be exploited to either add or verify the links between VRT thesaurus and GTAA.

In the current form, we only produce and show exact matches. However, we could also consider close matches, based on near-exact string matches for example. These could also be used in the demonstrator when no exact matches are found.

Extension and Reusability of the Demonstrator. At this time, the demonstrator covers two subsets of the much larger Flemish and Dutch collections guarded by the institutions. Both VIAA and NISV work towards making the metadata of a much larger set of items available as Linked Data, using the thesauri described here. As soon as this metadata becomes available, it can be added to the dataset of the demonstrator to encourage wider usage and retrieval results.

6 Related Work

Prior descriptions of tools and use cases for linking cultural heritage thesauri exist. Van Assem et al. describe a method for converting thesauri to SKOS [8], while the museum use case described in [9] discusses a similar approach to conversion. Cross-collection browsers that exploit linked thesauri have been previously explored in the MultimediaN Eculture [10] and MuseoFinland [11], but with less emphasis on exploring links between two collections. The DIVE linked media browser [12] partially overlaps in terms of collections, also enabling browsing of the Open Images collection and the links to other collections. Here, also the user has limited control over the retrieval of related objects from two collections. The level of user control in our demonstrator allows for more effective exploration of cross-collection links.

7 Conclusions

The case study shows how we combine existing and new tools to provide integrated browsing and search across different (sub)collections from two national audiovisual archives. We describe the conversion of a legacy thesaurus and present the reusable conversion algorithms. We illustrate the benefits of interactive alignment using the CultuurLINK tool; specifically, we show how filters can be used to isolate different sections of each thesaurus, so that *section-specific* matchers can be employed. This alignment strategy uses a combination of filtering and matching techniques that work for these two specific thesauri and the exact same strategy will not work (as well) for two different thesauri. This is precisely why a transparent, interactive alignment method (and tool) is needed. The converted thesaurus and the links produced are represented as SKOS RDF files that can be accessed online for easy reuse.

We finally introduce a cross-collection browser which uses the produced alignment to allow users to explore connections between the two collections. This application uses a flexible search strategy to retrieve relevant items from the two collections based on user search queries in the form of keywords, thesaurus terms or selected videos. The tool features a slider to allow users to put more emphasis on one or the other collection.

Acknowledgments. This research was funded by Taalunie project "Gemeenschap-pelijke Thesaurus voor Uniforme Ontsluiting".

References

1. Bron, M., van Gorp, J., Nack, F., de Rijke, M., Vishneuski, A., de Leeuw, S.: A subjunctive exploratory search interface to support media studies researchers. In: Proceedings of SIGIR12 (2012)
2. Miles, A., Matthews, B., Wilson, M., Brickley, D.: SKOS core: simple knowledge organisation for the web. In: International Conference on Dublin Core and Metadata Applications, p. 3 (2005)
3. Bizer, C., Heath, T., Berners-Lee, T.: Linked data - the story so far. Int. J. Semant. Web Inf. Syst. **5**(3), 1–22 (2009)
4. van Ossenbruggen, J., Hildebrand, M., de Boer, V.: Interactive vocabulary alignment. In: Gradmann, S., Borri, F., Meghini, C., Schuldt, H. (eds.) TPDL 2011. LNCS, vol. 6966, pp. 296–307. Springer, Heidelberg (2011). doi:10.1007/978-3-642-24469-8_31
5. de Vries, A.P., Alink, W., Cornacchia, R.: Search by strategy. In: Proceedings of the Third Workshop on Exploiting Semantic Annotations in Information Retrieval, pp. 27–28. ACM (2010)
6. Bouma, G.: Cross-lingual ontology alignment using EuroWordNet and Wikipedia. In: Proceedings of the 7th Conference on International Language Resources and Evaluation (LREC 2010), pp. 1023–1028 (2010)
7. Malaisé, V., Isaac, A., Gazendam, L., Brugman, H.: Anchoring Dutch cultural heritage thesauri to WordNet: two case studies. In: ACL 2007, p. 57 (2007)
8. van Assem, M., Malaisé, V., Miles, A., Schreiber, G.: A method to convert thesauri to SKOS. In: Sure, Y., Domingue, J. (eds.) ESWC 2006. LNCS, vol. 4011, pp. 95–109. Springer, Heidelberg (2006). doi:10.1007/11762256_10
9. Boer, V., Wielemaker, J., Gent, J., Hildebrand, M., Isaac, A., Ossenbruggen, J., Schreiber, G.: Supporting linked data production for cultural heritage institutes: the Amsterdam museum case study. In: Simperl, E., Cimiano, P., Polleres, A., Corcho, O., Presutti, V. (eds.) ESWC 2012. LNCS, vol. 7295, pp. 733–747. Springer, Heidelberg (2012). doi:10.1007/978-3-642-30284-8_56
10. Schreiber, G., et al.: MultimediaN E-culture demonstrator. In: Cruz, I., et al. (eds.) ISWC 2006. LNCS, vol. 4273, pp. 951–958. Springer, Heidelberg (2006). doi:10.1007/11926078_70
11. Hyvönen, E., et al.: CultureSampo: A National Publication System of Cultural Heritage on the Semantic Web 2.0. In: Aroyo, L., et al. (eds.) ESWC 2009. LNCS, vol. 5554, pp. 851–856. Springer, Heidelberg (2009). doi:10.1007/978-3-642-02121-3_69
12. de Boer, V., Oomen, J., Inel, O., Aroyo, L., Van Staveren, E., Helmich, W., De Beurs, D.: Dive into the event-based browsing of linked historical media. Web Semant. Sci. Serv. Agents World Wide Web **35**, 152–158 (2015)

Natural History in Europeana - Accessing Scientific Collection Objects via LOD

Jörg Holetschek[1(✉)], Gisela Baumann[1], Gerda Koch[2], and Walter G. Berendsohn[1]

[1] Botanic Garden and Botanical Museum Berlin-Dahlem, Berlin, Germany
{j.holetschek,g.baumann,w.berendsohn}@bgbm.org
[2] AIT Angewandte Informationstechnik Forschungsgesellschaft mbH, Graz, Austria
kochg@ait.co.at

Abstract. Millions of specimens housed in collections of natural history institutions document our planet's biodiversity over centuries and represent both an indispensable knowledge base for today's biological research as well as a cultural heritage. Digitization efforts of the past years have produced a substantial amount of digital assets: high-resolution images, videos, sound files, 3D imagery and 3D models. The OpenUp! Natural History Aggregator draws together these virtual representations of specimens from á multitude of institutions and feeds them into Europeana, the cross-domain portal for Europe's digitized cultural heritage. Enriching their metadata with data drawn from additional resources such as common names, taxonomic literature and geographic terms helps to increase discoverability und usability. The assignment of stable uniform resource locators and the application of standard vocabularies, existing ontologies and frameworks like RDF allow effective linking of web resources from different knowledge domains, thus creating linked open data.

Keywords: ABCD · BioCASe · XML · GBIF · DarwinCore · IPT · B-HIT · Natural history collections · Specimens · Open data · OpenUp! · Europeana · EDM · Linked Open Data · LOD · Semantic web · RDF · Aggregator · Primary biodiversity data

1 Collections in the Natural History Community

Natural history collections worldwide house huge amounts of specimens (i.e. preserved organisms or parts of organisms) that have been gathered by collectors over the past centuries. Their stock is the result of innumerable working years of professional and volunteer botanists, zoologists and mycologists collecting specimens in all regions of the world [1].

The objects in those collections present an astonishing variety. To name some of the more common ones: stuffed and mounted animals and paleontological specimens as known to visitors of natural history museums; pinned insects; pressed plants, dried seeds, fruits and wood samples in herbarium collections; entire animals or parts of organisms stored in alcohol or other preservatives; bones and sculls; environmental samples with water and soil organisms; microscopic slides of microorganisms or tissues; and substance collections ranging from plant exudates to DNA [2]. They provide rich and

© Springer International Publishing AG 2016
E. Garoufallou et al. (Eds.): MTSR 2016, CCIS 672, pp. 223–234, 2016.
DOI: 10.1007/978-3-319-49157-8_20

verifiable documentation of the planet's flora and fauna throughout the centuries. Together with annotations attached to them and literature documenting the species, they and their metadata serve today's biological research, for example as a source of material for research using new analytical methods and for finding evidence for past and ongoing changes and losses of our biodiversity. Apart from these uses in natural sciences, they also represent cultural objects, for example in documenting the voyages of the European explorers of the world, such as Humboldt and Livingstone.

In the traditional work with collections, physical inspection of specimens plays an important role. This requires that either the scientist travels to the storage place of the collection, or the specimen is sent to where the scientist is located. The first approach is time-consuming and cost-intensive and only pays if work with many specimens is planned. The latter approach is cheaper if only a single or a few specimens are of interest, but it comes with the risk of damaging specimens during transportation, so that historic and valuable material usually won't be sent via mail.

Therefore, high-quality digitization of specimens is getting increasingly used to replace the need for physically sending specimens to scientists and to advertise the availability of specific specimens for research. If a specimen loan is requested, a digital representation of the specimen is produced and sent to the requestor instead of the physical specimen itself. Moreover, whole collections are being digitized and made available on the Internet as virtual collections, often eliminating the need for a scientist to request a loan at all [3, 4]. If the lack of resources hinders whole collections from being digitized, priorities for partial digitization can be set based on the needs of the scientific community [2].

Digitization efforts result in a multitude of digital assets, depending on the collection type, ranging from images of two-dimensional specimens (e.g. herbarium sheets) and 2D-like objects (e.g. insect drawers) to three-dimensional imagery and 3D models of preserved specimens to videos and sound recordings of living specimens [5–7].

The metadata describing natural history specimens are rather complex, including collector(s), date and locality of the gathering event, person(s) identifying the organism, information about collection methods used (e.g. trapping equipment), preservation methods, and historic identification results, potential nomenclatural type designations (fixing the scientific name to a specific specimen), ownership information and intellectual property rights statements. The two common standards used for exchanging metadata in the natural history domain diverge considerably in their complexity: Access to Biological Collections Data (ABCD) is a multipart, hierarchical and fine-granular XML schema reflecting the richness of natural history specimen metadata [8]. In contrast, the Darwin Core standard is a set of terms and a specification on how to use these terms to describe specimens in a flat, row-oriented format [9].

For each of these two data standards, additional specifications and software packages implementing these specifications exist. The BioCASe protocol (Biological Collection Access Service) is used by networks (data portal, aggregators or other data consumers) to retrieve data from collections offered in the ABCD data format [10]. The BioCASe Provider Software is a middleware that implements this protocol and can be used to publish natural history collection data stored in a relational database management system or row-like format like spreadsheets or CSV files [11]. Once configured, collection data will be available as ABCD documents through a BioCASe web service. The Integrated Publishing Toolkit (IPT) can be used for the Darwin Core standard (DwC) [12]. It allows extracting collection data from a relational database or other tabular file formats and publishing them as Darwin Core archives (zip-compressed files containing the DwC data table and a metadata document).

In the past decade, several networks and aggregators have been established that draw data from a huge number of different data sources. Some focus on certain taxonomic groups and/or geographic regions, others have a global scope and deal with all types of primary biodiversity data. The Global Biodiversity Information Facility (GBIF) is the most comprehensive network, covering both natural history collection and observation data on a global scale. Currently (July 28th 2016), the GBIF data portal offers access to 639 million records of 29,331 datasets provided by 822 data publishers [13]. Australia's Virtual Herbarium (AVH) links together 5 million specimen records from 16 Australian herbaria, acting as an aggregator for the Atlas of Living Australia network and offering its own data portal [14], apart from providing data to GBIF. Numerous smaller networks exist, like the Geosciences Collections Access Service (GeoCASe) providing a network for paleontological and mineralogical collections [15] or the biodiversity data network of the German institutions of the Humboldt-Ring (BiNHum) [16].

As more and more datasets are published, the role of stable identifiers becomes increasingly important. Objects made available on the web – both physical specimens and digital assets resulting from digitization efforts – need to be uniquely referenceable through permanent Uniform Resource Identifiers (URIs) in order to allow proper quoting and interlinking with other online resources. For natural history specimens, the Consortium of Taxonomic Facilities (CETAF) has developed a standard for stable identifiers based on the hypertext transfer protocol (HTTP) that is being implemented by an increasing number of institutions [17]. For datasets lacking unique permanent identifiers, aggregators are used to creating their own identifiers for addressing resources uniquely within the networks. In the natural history community, usually a combination of institution code, collection code and catalogue number – the so-called triple ID – is applied.

2 A Cross-Domain Portal for Europe's Cultural Heritage

Europeana is the cross-domain central portal and single entry point to Europe's digitized cultural heritage; its latest version links to over 53 million digital objects such as images, texts, sounds, and videos [18]. Some of these are world famous, others are hidden treasures from more than 3,500 museums, galleries, archives, libraries, and audio-visual collections all over Europe. Europeana's vision is to make cultural heritage as easily accessible and as freely re-usable as possible. Every digital object in Europeana is provided with internationally compatible and machine-readable rights statements giving explicit information about conditions for re-use [19]. All metadata published by Europeana is available free of restriction under the Creative Commons Universal Public Domain Dedication (CC0 1.0) [20].

Reflecting a natural history contribution to Europeana, prospective data providers had to face significant obstacles. Most of the providers run their own collection management system and store multimedia objects and corresponding metadata in a different way. Having providers to map their data models individually to the Europeana metadata standard (EDM) would have meant a lot of parallel effort, i.e. familiarizing with EDM, data field mapping, transforming and providing the collection's metadata. Individual mappings would have increased the likelihood of errors which would have resulted in lengthy publishing and communication processes. Once published, each data provider would have to check for technical adaptations at the Europeana side repeatedly and might have to adjust the way of metadata provision, leading to increased maintenance efforts. Finally, every provider would have had to determine and manage its own metadata enrichment processes to boost the accessibility of scientifically described multimedia objects by culturally interested users.

It was soon obvious that an aggregation mechanism was needed that would merge metadata from different data models into a common metadata standard, which could then be enriched, transformed and fed into Europeana. Without such an effective workflow, only a minority of natural history institutions would have had the technical know-how and necessary resources to open up their collections to Europeana.

3   OpenUp! - The Natural History Aggregator

Until 2011, Europeana was mainly dedicated to cultural content and focused on artwork, texts and audio-visual material. Although being clearly within the scope of Europeana, multimedia objects belonging to the natural history domain were dramatically underrepresented. As a consequence, members of the Consortium of European Taxonomic Facilities (CETAF) and several European nodes of the Global Biodiversity Information Facility initiated the OpenUp! project. The initiators aimed at "opening up the natural history heritage for Europeana" in order to raise public awareness of the scientific and cultural importance of these collections.

OpenUp! started in 2011 as a 3-year-project funded by the European Commission. Its central idea was to provide online access to a wide range of collection objects and connect the cultural domain to the natural history domain [21]. OpenUp! complemented the Biodiversity Heritage Library Europe project [22], which mobilized digitized literature in the natural history domain in Europe.

Today OpenUp! is a growing network and Europeana's aggregator for the natural history domain. So far, 25 partners from 13 European countries joined the network and contribute 2.9 million records with links to about 3 million multimedia objects. Some of the most outstanding natural history museums and botanical gardens in Europe are part of the network. Content served includes specimen images, images of fossils and geological objects, movies, and animal sound files, normally with a reference to an observation or collection event. Europe's natural history collections cover most of the world's described organisms, ranging from common and famous species to those that have already gone extinct. Some were collected during historical expeditions by well-known epochal explorers and scientists like Darwin or Humboldt. OpenUp! makes multimedia representations of these treasures available to the general public, often for the first time.

The OpenUp! Natural History Aggregator is coordinated by the Freie Universität Berlin, Botanic Garden and Botanical Museum Berlin. OpenUp! created a software suite and workflow for the harvesting of community standard data, their transformation, enrichment and provision in EDM format (Europeana Data Model). Harvests are repeated at regular intervals to ensure availability of up-to-date information on the Europeana data portal [23]. The OpenUp! Natural History Aggregator is contained in a virtual machine and can be operated at different places. For a sustainable operation of the whole process, outsourcing has proven to be the most cost-effective option. Currently, the OpenUp! Natural History Aggregator platform is operated by the company AIT (Angewandte Informationstechnik Forschungsgesellschaft mbH, Graz, Austria). The operation is financed by the data providers under a Service Level Agreement with the operator of the platform. For new data providers, the aggregation fee is waived for the first two years.

At present, OpenUp! is Europeana's 4th largest content provider.

4   OpenUp! Data Flow

Figure 1 shows the basic data flow established for the OpenUp! project: The Natural History Aggregator at the center draws data from the institutions holding natural history collections (left), enriches them with supplemental information provided by OpenUp! partners or third parties (right), and finally feeds them into Europeana (top). The dotted boxes mark the different domains, the natural history world at the left, Europeana at the top, and the supplemental resources at the right stemming from different domains. The task of the Natural History Aggregator is to draw data from different sources and merge them, to bridge the different access protocols and methods required for retrieving these sources, and to transform and map the diverse data standards used in the different domains.

As described before, accepted standards and methods exist for storing and transferring metadata of natural history collections. OpenUp accepts both methods of publication by using the Berlin Harvesting Toolkit (B-HIT), a software product allowing the harvest of both BioCASe data sources and Darwin Core archives [24]. Data providers that do not want to use either of these tools can also offer their data as Darwin Core archives produced through custom export procedures or packages. The open source data management and distribution platform CKAN (Comprehensive Knowledge Archive Network) [25], for example, supports exporting Darwin Core archives through the CKAN Packager [26].

B-HIT stores metadata as XML files (extensible markup language) for BioCASe data sources and CSV files (comma separated values) for IPT providers or custom-made DarwinCore archives. In a first step, these files are parsed into a relational metadata database. This database is not bound to a specific data standard; it can accommodate both ABCD and Darwin Core data. In particular, it contains the URLs for the multimedia objects that Europeana is centered around. In a subsequent step, the natural history metadata is enriched with information drawn from diverse resources and transformed into the Europeana data model, EDM (described in Sect. 4).

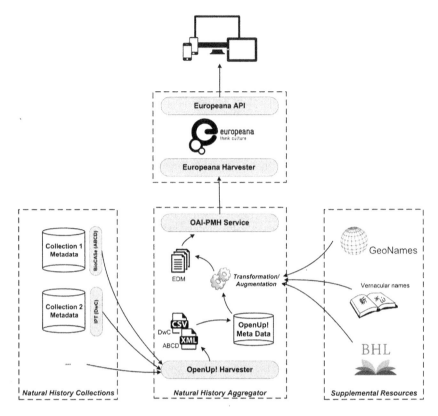

Fig. 1. Dataflow through the OpenUp! Natural History Aggregator

Currently, the Natural History Aggregator uses information of three kinds to augment the original collection data and to make it easier accessible by non-scientific users:

1. The coordinates of the specimen's gathering site are used to add links to GeoNames, a geographic thesaurus database available through web services [27]. These links can be used to visualize the gathering site on data portals; moreover, since GeoNames maintains hierarchies of place names, more geographic information can be inferred and added, like superordinate administration areas, country names or continents.
2. Most specimens are identified by Latin names which are the basis of all scientific work, but rarely used in everyday life. In order to allow non-scientists to discover and understand the names of the specimen's species, vernacular names in 753 different languages and dialects are added. Currently, the Natural History Aggregator draws 635,020 common names for 259,300 species from 30 different taxonomic thesauri.
3. The Biodiversity Heritage Library (BHL) is a consortium of natural history libraries cooperating in digitizing legacy literature on biodiversity and making it publicly available as a part of a global "biodiversity commons". Currently, BHL provides open access to 183,385 volumes of biodiversity literature [22]. The Natural History

Aggregator uses the specimens' scientific species names to find volumes in BHL, which are then added to the specimen's metadata.

After processing, both the original specimen metadata and the enrichments are stored as EDM documents and published through an OAI-PMH server (Open Archive Initiative, Protocol for Metadata Harvesting) [28].

This server is visited by the Europeana harvester monthly; the documents published are ingested and become accessible on the Europeana portal. In addition, all records available on the portal can be retrieved through the web services defined in the Europeana API [29]. This allows a wider dissemination of the data; it can be used by other portals or by apps on mobile devices.

5   Linking Data in a Semantic Web

Over the past years, the idea of the semantic web has been filled with life by real-world applications. Numerous research projects are dedicated to training computers, robots and databases to better understand users' search requests and to provide information in a semantically linked way. The foundation is the availability of semantically rich and open data sets.

Linked open data (LOD) is an approach for interrelating freely available data on the Internet by way of semantic techniques and for publishing this data in a machine readable way to facilitate searching and analyzing the data. It follows Tim Berners-Lee's vision that in semantic networks associative relationships can be established between completely independent objects, because the data and schemas are well described and understandable [30]. User-friendly integrated platforms for Linked Open Data Consumption that will meet the technical user requirements for LOD retrieval and presentation are currently under development [31]. A key requirement for LOD is the assignment of stable Uniform Resource Identifiers (URIs).

The Resource Description Framework (RDF) provides the syntax and rules that form the basis of LOD [32], structures information as triples of subjects, predicates and objects, and can be expressed using various formats such as RDF-XML, JSON-LD or Turtle. The Europeana Data Model, the metadata model used for aggregating and ingesting the diverse European cultural and natural heritage metadata into the joint digital library, describes data using RDF [33]. EDM is an open, cross-domain semantic web-based framework that accommodates a variety of heritage community standards and re-uses existing namespaces, for examples from OAI ORE [34], Dublin Core [35], Creative Commons [36], or the Simple Knowledge Organization System framework (SKOS) [37] for the integration of vocabulary concept terms. The reuse of existing standards within EDM provides the basis to integrate a multitude of different cultural and natural heritage data into the single repository formed by Europeana.

Ontologies and vocabularies reduce the complexity of the world to a manageable extent and help to put vast amounts of data into structured forms. A basic premise in the LOD approach is to reuse terms from existing standard vocabularies wherever possible, rather than to reinvent them. This maximizes the probability that the data can be used without additional modifications by applications that are tuned to well-known vocabularies [38]. OpenUp! is applying international vocabularies like the Catalogue of Life [39] or PESI (The Pan-European Species-directories Infrastructure, [40]), and about 25 different national vernacular names lists for data enrichment.

Standards on creating and publishing vocabularies on the web offer valuable solutions for interconnecting isolated data silos and support cross search and data comparison (ISO 25964 [41], SKOS). SKOS is the accepted standard model for expressing the structure and content of concept schemas like vocabularies, authority lists and thesauri on the web using RDF. The SKOS framework is promoted by the W3C initiative and is a lightweight approach with a limited set of properties describing the relations between concepts.

6   Semantic Enrichment for Europeana

Semantic enrichment of metadata plays a key role in Europeana as it improves access to the objects, defines relations among them and allows multilingual retrieval of documents. There are three hierarchical levels at which metadata can be enriched during the provision process: on data provider level, on aggregator level, and on Europeana level directly before publishing the data on the portal. Due to the lack of resources at the institutions hosting the collections, the data providers often do not have the capacity to enrich their metadata at the source. Europeana, on the other hand, does a variety of metadata enrichments, including concepts (GEMET Thesaurus [42, 43]). The following focuses on enrichment processes at aggregator level.

Metadata enrichment is a core activity and one of the value adding services of the OpenUp! Natural History Aggregator. It opens scientific collections to non-scientific users and interlinks natural history objects to other cultural heritage in Europeana. Although all of the additional information is added during the transformation process, the three types of enrichment (see Sect. 4) are executed in different ways:

1. For specimen records with geographic coordinates, a GeoNames stable URI is added. This link refers to the GeoNames website, which visualizes the locality and displays textual information; moreover, it links to any higher-ranking localities, such as administrative units, countries or continents. All additional textual information is displayed in the Location section on the Europeana portal (Fig. 2), and a link to GeoNames is added to the References and Relations section.
2. During the OpenUp! project, a vocabulary web service providing vernacular names of organisms for scientific names was created [44]. The service is hosted at the Natural History Museum in Vienna and is continually improved both technically and regarding to the amount of underlying data sources. It returns matching vernacular names in an EDM-specific SKOS format. As an alternative to using the web service, aggregators can also request a bulk format of the results for a given list of scientific names. The common names added are displayed in the Subject section of the Europeana portal.
3. For easy access to taxonomic literature available at the Biodiversity Heritage Library, a link to a BHL web service is added that will be displayed in the Relations

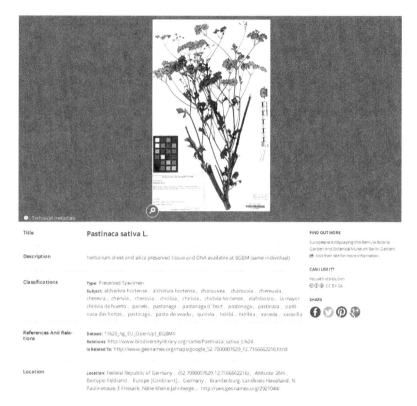

Fig. 2. An OpenUp! record on the Europeana portal reflecting three kinds of metadata enrichment: common names in different languages, a BHL link and a GeoNames link

section. When used, this web service retrieves a list of all publications in the online digital library for the organism of the specimen record.

Modern business intelligence (BI) tools assist data managers in the collection, transformation and processing of massive amounts of data from various sources. They provide a set of methodologies, processes, architectures, and technologies that load data from heterogeneous operational databases, transform and merge them into a single integrated data warehouse [45]. Data sources and sinks as well as transformation steps can be assembled into data flows using simple drag and drop, sparing the need of writing code manually in most cases. This reduces development costs and maintenance efforts considerably, as both data sources and sinks might change and evolve during the life cycle of a system.

In OpenUp!, the open source BI tool Pentaho Kettle is applied for data integration and ETL (Extract, Transform, Load) [46]. The complete process in Pentaho is divided into three steps: transform, validate, and OAI import. Within the transformation step, the native ABCD and DarwinCore data standards are mapped to the Europeana Data Model. These mappings can be extended for other standards whenever necessary, for example LIDO, an exchange format widely used in the museum community (Lightweight Information Describing Objects, [47]).

7 Outlook

The OpenUp! Natural History Aggregator has paved the way for natural history institutions to feed their collections into Europeana. Even after the initial EU-funded project ended, organizations are willing to pay an annual fee for the aggregation and data enrichment services provided. New technical developments like the support of Darwin-Core archives allow new providers to join in, resulting in a constantly rising number of published objects.

Especially the data enrichment is as very sought-after component. Some institutions cannot join OpenUp! because they are obliged to use national aggregators. They are aware of the added value they're missing by not providing common names, for example. Separating data enrichment from aggregation and offering it as a separate service may be a way forward.

The list of data enrichments currently done should be extended in the future. Even though common names allow non-scientists to search for data, they can still be very restrictive. The inclusion of generic search terms that include a group of organisms would yield more complete search results for users that don't know exactly what they're looking for. For example, search for *mouse* should also retrieve hits for the different vole species; more generic search terms could include even several genera.

As described before, LOD heavily relies on stable identifiers. Even though some institutions provide them for their specimens and OpenUp! adds GeoNames identifiers for localities, this list should be extended. Authority files exist for other data items relevant for natural history collections, for example person names [48]. The collector of a specimen is an often used search criterion for specimens. Since most specimens stem from the analogue age, their names often exist in numerous variations in respect to spelling, abbreviation or word order, making it hard to get complete hit lists. Enriching specimen data with stable identifiers for collectors would allow finding variants of a person name and yielding reliable search results.

The latest version of the Europeana data portal supports state-of-the-art search features like facets and full-text search. However, as more items are added to the metadata and the number of objects available grows, the need for more detailed search options gets more important to allow precise searches and finding the relevant objects. Facets should include the different ontologies included by the Europeana Data Model, which could be made available to advanced users in an expert-mode.

In order to create a shared, dynamic, efficient and cost-effective metadata aggregation for the Europeana DSI, semantic enrichment and Linked Open Data play a key role to interlink collection objects provided by diverse aggregators and thus build bridges between knowledge domains.

Acknowledgements. Funding for the OpenUp! Natural History Aggregator was provided by the Information and Communication Technologies Policy Support Programme (ICT PSP, grant agreement 270890) and the Connecting Europe Facility, Telecommunications sector (CEF-TC, grant agreements CEF-TC-2014-2-01 and CEF-TC-2015-1-01) of the European Commission. The development of the BioCASe Provider Software and the BioCASe protocol has been funded under grant number EVR1-2001-00003.

References

1. Holetschek, J.: Approaches for involving volunteers into the process of metadata capture from specimens. Report for the SYNTHESYS II project, Network Activity 3, Deliverable 3.1 (2011). http://www.biocase.org/synthesys/del_3.1_part1.pdf
2. Berendsohn, W.G., Seltmann, P.: Using geographical and taxonomic metadata to set priorities in specimen digitization. Biodivers. Inform. **7**, 120–129 (2010)
3. Berendsohn, W.G., Einsiedel, B., Merkel, U., Nowak-Krawietz, H., Röpert, D., Will, I.: Digital imaging at the Herbarium Berolinense. In: Häuser, C.L., Steiner, A., Holstein, J., Scoble, M.J. (eds.). Digital Imaging of Biological Type Specimens - A Manual of Current Best Practice. Stuttgart (2005)
4. Pignal, M., Michiels, H.: Switching to the fast track: rapid digitization of the world's largest herbarium. Botany 2011, Columbus, Ohio (2011). http://digitarium.fi/sites/digitarium.fi/files/Mnhn-Herbarium-digitization.pdf
5. The EoS project (Erschließung objektreicher Spezialsammlungen). http://eos.naturkunde museum-berlin.de/singleSpecimens
6. The virtual zooarchaeology of the arctic project (VZAP). http://vzap.iri.isu.edu
7. The sound animal archive at the museum für Naturkunde Berlin. http://www.tierstimmenarchiv.de
8. Holetschek, J., Dröge, G., Güntsch, A., Berendsohn, W.G.: The ABCD of primary biodiversity data access. Plant Biosyst. Int. J. Dealing Aspects Plant Biol. Off. J. Soc. Bot. Ital. **146**(4), 771–779 (2012)
9. Wieczorek, J., Bloom, D., Guralnick, R., Blum, S., Döring, M., et al.: Darwin Core: an evolving community-developed biodiversity data standard. PLoS ONE **7**(1), e29715 (2012). doi:10.1371/journal.pone.0029715
10. The BioCASe protocol. http://www.biocase.org/products/protocols
11. The BioCASe provider software. http://www.biocase.org/products/provider_software
12. Robertson, T., Döring, M., Guralnick, R., Bloom, D., Wieczorek, J., Braak, K., et al.: The GBIF integrated publishing toolkit: facilitating the efficient publishing of biodiversity data on the internet. PLoS ONE **9**(8), e102623 (2014). doi:10.1371/journal.pone.0102623
13. The GBIF data portal. http://www.gbif.org
14. Australia's virtual herbarium. http://avh.chah.org.au
15. Geosciences access service (GeoCASe). http://geocase.eu
16. BiNHum – Biodiversitätsnetzwerk des Humboldt-Rings. http://wiki.binhum.net
17. CETAF stable identifiers. http://cetaf.org/cetaf-stable-identifiers
18. The Europeana data portal. http://www.europeana.eu
19. The rights statements initiative. http://rightsstatements.org
20. Creative commons universal public domain dedication. https://creativecommons.org/publicdomain/zero/1.0/

21. Berendsohn, W.G., Güntsch, A.: OpenUp! - creating a cross-domain pipeline for natural history data. ZooKeys **209**, 47–54 (2012). doi:10.3897/zookeys.209.3179. In: Blagoderov, V., Smith, V.S. (eds.) No specimen left behind: mass digitization of natural history collections

22. Biodiversity heritage library. http://www.biodiversitylibrary.org

23. Europeana data portal. http://www.europeana.eu/portal

24. Kelbert, P., Droege, G., Barker, K., Braak, K., Cawsey, E.M., Coddington, J., et al.: B-HIT - a tool for harvesting and indexing biodiversity data. PLoS ONE **10**(11), e0142240 (2015). doi:10.1371/journal.pone.0142240

25. CKAN – the open source data portal software. http://ckan.org

26. CKAN packager. https://github.com/NaturalHistoryMuseum/ckanpackager

27. GeoNames geographical database. http://www.geonames.org

28. Open archives initiative protocol for metadata harvesting. https://www.openarchives.org/pmh/

29. Europeana API. http://labs.europeana.eu/api

30. Berners-Lee, T.: Linked data. https://www.w3.org/DesignIssues/LinkedData.html

31. Klímek, J., Škoda, P., Necaský, M.: Requirements on linked data consumption platform. In: WWW 2016 Workshop: Linked Data on the Web (LDOW2016)

32. Resource description framework. https://www.w3.org/RDF/

33. Europeana data model. http://pro.europeana.eu/page/edm-documentation

34. OAI object reuse and exchange (ORE). http://www.openarchives.org/ore/terms/

35. Dublin core. http://dublincore.org/

36. Creative commons. https://creativecommons.org/

37. Simple knowledge organization system. https://www.w3.org/TR/skos-reference/

38. Heath, T., Bizer, C.: Linked data: evolving the web into a global data space. Synth. Lect. Semant. Web Theor. Technol. **1**(1), 1–136 (2011)

39. Species 2000 & ITIS catalogue of life. http://www.sp2000.org/

40. PESI Pan-European species-directories infrastructure. http://www.eu-nomen.eu/pesi/

41. ISO 25964 – the international standard for thesauri and interoperability with other vocabularies. http://www.niso.org/schemas/iso25964/

42. Eionet GEMET thesaurus. http://www.eionet.europa.eu/gemet

43. DBpedia: structured content of wikipedia. http://www.dbpedia.org

44. OpenUp component report. http://open-up.eu/en/content/deliverables-and-components-pu

45. Obeidat, M., North, M., Richardson, R., Rattanak, V., North, S.: Business intelligence technology, applications, and trends. Int. Manage. Rev. **11**(2), 47–56 (2015)

46. Pentaho data integration. http://www.pentaho.com/product/data-integration

47. Lightweight information describing objects. http://network.icom.museum/cidoc/working-groups/lido/what-is-lido/

48. Virtual international authority file. http://viaf.org

Track on Digital Libraries, Information Retrieval, Big, Linked and Social Data

Enabling Multilingual Search Through Controlled Vocabularies: The AGRIS Approach

Fabrizio Celli[✉] and Johannes Keizer[✉]

Food and Agriculture Organization of the United Nations, Rome, Italy
{Fabrizio.celli,Johannes.keizer}@fao.org

Abstract. AGRIS is a bibliographic database of scientific publications in the food and agricultural domain. The AGRIS web portal is highly visited, reaching peaks of 350,000 visits/month from more than 200 countries and territories. Considering the variety of AGRIS users, the possibility to support cross-language information retrieval is crucial to improve the usefulness of the website. This paper describes a lightweight approach adopted to enable the aforementioned feature in the AGRIS system. The proposed approach relies on the adoption of a controlled vocabulary. Furthermore, we discuss how expanding user queries with synonyms increases the sensitivity of a search engine and how we can use a controlled vocabulary to achieve this result.

Keywords: Cross-language information retrieval · Controlled vocabulary · Query expansion · Search engine · Digital repository · Agriculture

1 Introduction

The debate on the usefulness of controlled vocabularies has been carried out for more than two decades [11, 14, 16, 17, 20], and there are still controversial opinions. On the one hand, there are supporters of the theory of abandoning controlled vocabularies [2, 5], but on the other, there are those who argue that controlled vocabularies are essentials to ensure the right recall when searching in bibliographic databases [7, 21]. The former base their assertion on the evidence that keyword-based searching has become the preferred method of searching in online information systems [11]. Thus, according to them, a textual search is everything users need; there is no value in using controlled vocabularies, but free keywords are enough to help users in retrieving resources from bibliographic databases. However, several studies emphasize that many resources returned in a keyword-based search would be lost without controlled vocabularies. Gross and Taylor [10] argue that 35.9 % of results would not be found if subject headings were removed from catalog records. In fact, subject fields very often contain terms that are not available in titles and abstracts, since expert cataloguers avoid repetitions [13]. In addition to that, controlled vocabularies can mediate the implementation of advanced features, like semantic search in information retrieval systems, as in the case of the European project INSEARCH [1].

© FAO, 2016
E. Garoufallou et al. (Eds.): MTSR 2016, CCIS 672, pp. 237–248, 2016.
DOI: 10.1007/978-3-319-49157-8_21

In this paper, we show how the adoption of a controlled vocabulary helps in implementing the multilingual search functionality in the AGRIS information system, in order to retrieve multilingual content whose language may be different from the language of the query. In that way, this functionality refers to cross-language information retrieval. Section 2 introduces AGRIS and AGROVOC multilingual controlled vocabulary. Section 3 presents the problem of enabling multilingual search in AGRIS. We discuss a methodology that relies on AGROVOC to expand user queries in order to retrieve resources in different languages. This methodology can be generalized and applied to other systems that make use of a multilingual controlled vocabulary. In Sect. 4, we analyze how expanding user queries with synonyms may help in improving the recall of a search engine. Section 4 is only analytical, since we have not implemented the proposed solution yet. Finally, in the last section we draw our conclusions.

2 An Overview of AGRIS and AGROVOC

Over the last few years, AGRIS has dramatically changed its shape. AGRIS is the International Information System of Agricultural Science and Technology. It was set up in 1974 as an initiative of around 180 member countries of the Food and Agriculture Organization of the United Nations (FAO). The main objective was to improve access and exchange of information on agricultural research serving the information needs of developed and developing countries on a partnership basis. Now, AGRIS ambition is to be a global hub to agricultural research and technology information.

AGRIS is a collection of more than 8 million multilingual bibliographic references, mainly accessible through the AGRIS website[1]. On the data acquisition side, the AGRIS team collects and publishes data from more than 150 partners all over the world. The data ingestion process includes disambiguation of AGRIS entities, de-duplication, and semantic enrichment. Then, data are published as machine-readable RDF triples and become freely downloadable through a SPARQL endpoint or FTP. On the data dissemination side, since 2013 the AGRIS website has been completely revamped as a semantic mash-up that uses formal alignments across many systems to provide a universe of data around each bibliographic record. Users can browse the AGRIS core database, looking for information about a topic in the AGRIS domain. When users select a bibliographic resource, the system shows its associated *mashup page*. A mashup page is a web page that displays an AGRIS resource together with relevant knowledge extracted from external data sources (as the World Bank[2], DBPedia[3], and Nature[4]). The availability of external data sources is not under AGRIS control. Thus, if an external data source is temporary unreachable, it is not displayed in AGRIS mashup pages.

---

[1] http://agris.fao.org.

[2] http://data.worldbank.org/.

[3] http://wiki.dbpedia.org/.

[4] http://api.nature.com/.

The mediation of AGROVOC[5] makes the generation of mashup pages possible. AGROVOC is a thirty years old multilingual controlled vocabulary containing over 32,000 concepts in 23 languages, and covering all areas of interest of FAO. A community of experts maintains AGROVOC and edits it through VocBench [19], an open source web application for editing SKOS and SKOS-XL thesauri. AGROVOC is aligned with 16 multilingual knowledge organization systems related to agriculture. The AGRIS system relies on those alignments and on the high quality of AGROVOC content to query external web services and interlink AGRIS bibliographic resources to relevant content. In fact, AGRIS records are indexed with AGROVOC descriptors. Sometimes data providers produce records where AGROVOC is already available in their metadata; other times, AGROVOC descriptors are added to AGRIS metadata as a result of the semantic enrichment process. The availability of AGROVOC descriptors in AGRIS metadata represents the backbone for the generation of mashup pages [4].

The AGRIS audience is mainly composed of domain experts, researchers, librarians, information managers, and everyone with an interest in agricultural subjects. According to Google Analytics, every month hundreds of thousands of users from about 200 countries and territories access the system. Considering the high variety of AGRIS users, their needs, and the uniqueness of the AGRIS content, we have the duty to explore new possibilities of usage of AGRIS data. We want to provide AGRIS users with additional features that derive from intrinsic characteristics of AGRIS data. The mediation of AGROVOC controlled vocabulary can be the key of our exploitation of AGRIS data. In another work [4], we have explored the possibility to crawl unstructured web resources, use an automatic indexer to assign AGROVOC descriptors to crawled web resources, and interlink them with AGRIS bibliographic data. In this paper, we show how we can enable multilingual search and other searching features through the usage of AGROVOC controlled vocabulary.

3   Enabling Multilingual Search Using a Controlled Vocabulary

Xian is a Chinese researcher and he wants to retrieve some scientific publications from the AGRIS database. His main research interest is about "rice". Xian performs a keyword-based search using the Chinese keyword 稻米 (which means "rice" in English), but the AGRIS system returns only 14 documents. This result looks strange to Xian, since "rice" is the agricultural commodity with the third-highest worldwide production according to 2013 FAOSTAT data [6]. Thus, Xian is expecting to retrieve a lot of scientific material about this important cereal. He analyzes results and discovers that all of them have Chinese metadata. Xian realizes that he has to query the system in English (and maybe in other languages) to access the international literature. Xian is quite unhappy with AGRIS. He would like to query the system in his native language,

---

[5] http://aims.fao.org/agrovoc.

which would simplify the choice of further keywords to refine his query, but he would also like to access the international literature.

The above paragraph reflects a typical scenario of cross-language information retrieval. In order to understand Xian's problem better, we should provide some background information about the AGRIS default search. When a user queries the system, their query refers to metadata available in the AGRIS database. Thus, if a user searches for 稻米, by default the system returns all bibliographic references containing the word 稻米 in the title, in the abstract, or as a keyword. The problem with this behavior is that the user may be interested in results in all languages or in a subset of them, thus not only in results whose metadata are available in the language of their query. As we show in Sects. 3.2 and 3.3, a multilingual controlled vocabulary is a valid tool to deal with this scenario. In fact, it can be used to expand user queries by translating keywords in all languages available in the vocabulary. What we want to achieve is to let users searching in their native languages and retrieving scientific publications in all languages.

3.1   Related Work

Several authors have observed that the development of methodologies and tools supporting multilingual information discovery is essential to make non-English content available to end users [8, 12, 15]. Gohrab [8] proposes a framework that performs on-the-fly machine translation of queries and documents. This framework does not rely on controlled vocabularies, but supports automatic translation of queries using external services like "Google Translate[6]" or "Microsoft Translator[7]". It also adopts "Open-MaTrEx[8]" for domain-specific translations. We believe that the usage of a controlled vocabulary for the translation of user queries is important when searching the scientific literature. In fact, it allows searching mediated by concepts, overcoming problems related to synonyms, scientific names, and abbreviations, and increasing the level of precision of the translations. By the way, searching mediated by concepts is still based on words and, in case of polysemy, there is the risk of wrong translations of user queries. Using a domain-specific controlled vocabulary like AGROVOC reduces the impact of polysemy. There is "one sense per discourse" [9]; given a context, there is a high probability that polysemous words are used in a single sense.

Kaplan [12] describes a methodology that uses different lexical resources. The proposed query translator module tries to perform the translation using first term networks, then domain specific controlled vocabularies, and finally a general-purpose query translation service. The software component allows querying in English, French, German, or Swedish, and retrieving results in one or more of those languages. Our approach is based on AGROVOC, a multilingual thesaurus covering 23 languages. We are not only interested in translating the source query, but also in extending the query making use of synonyms in the available languages.

---

[6] https://cloud.google.com/translate/.

[7] https://www.microsoft.com/translator/.

[8] http://www.openmatrex.org/.

3.2   The AGRIS Approach to Multilingual Search

AGRIS approach to multilingual search is based on the adoption of AGROVOC as an instrument to translate user search keywords. In this way, we demonstrate that a controlled vocabulary is not only good for document indexing, but it can be applied to other aspects of information retrieval, as enabling multilingual search through automatic query expansion (AQE). AQE has a 50-year history but, as the survey [3] states, only in recent years it has reached a good level of scientific maturity to lose the status of experimental technique.

We have developed a software component for AGRIS that implements the following algorithm. We call this component the *multilingual query expansion module*. This module is responsible for translations of user keywords, but it does not translate titles or phrases. When a user performs keyword searching in the AGRIS database, the system:

- Identifies the query pattern;
- Uses AGROVOC to translate keywords;
- Expand the user query, boosting keywords provided by the user;
- Returns results in all available languages.

The identification of the query pattern is needed to allow the system to expand the query. In fact, users may perform keyword searching or they may perform structured searching. In the second case, the query presents controlled keywords that must not be translated. As an example, if a user wants to search only in the subject field, they can use the query `subject:rice`, where `subject` is the controlled keyword that tells the system in which bibliographic field the user wants to look for the keyword `rice`. In the same example, `rice` is the keyword that the system has to translate. In addition to that, special characters like '*' and '-' have to be discarded, since they are used by the system to build negative and wildcard queries. The special character "+" can be used to define mandatory keywords.

In the current implementation of the algorithm, we have considered a limited set of query patterns. The system expands the source query if:

- The query contains 4 terms or pictograms, without the Boolean operators "AND" and "OR". We have identified this threshold to distinguish keywords searching from title, serials, and author searching. In fact, a study conducted in 2001 [18] affirms that the average length of a search query is 2.4 terms. Furthermore, in March 2016, the average length of a search query in AGRIS was 4.7, but longest queries referred to titles or authors. This parameter affects the retrieval performance, since it can cause very long expanded queries.
- The query has pattern `subject:($keywords)`, `+subject:($keywords)`, `subject:$keywords`, or `+subject:$keywords`[9]. It is the case of searching only in the subject field.

---

[9] `$keywords` stands for a list of terms or pictograms satisfying constraints expressed in the previous bullet point.

The implementation relies on two Apache Solr indexes:

- AGROVOC label index. This index contains all concepts available in the AGROVOC thesaurus. For each concept identified by a URI, the index stores preferred and alternative labels in all languages.
- AGRIS core index, which contains all AGRIS resources. This is the main index used by the AGRIS website to retrieve records after the submission of a user query.

Once the system has identified the query pattern, the *multilingual query expansion module* queries the AGROVOC label index to obtain translations of source keywords. The module matches source keywords against both preferred and alternative labels to identify the AGROVOC concept, but it considers only preferred labels as output of the translation process. In fact, as we show in Sect. 4, alternative labels can mediate query expansion with synonyms. After that, the module expands the source query by building a union of source keywords and their translations. The system boosts source keywords by a factor of 50, since we think that it is important to return to users results of their original query first, and then results of the multilingual query. As an example, if the source query is +subject:rice, the system builds the query:

```
+(subject:"rice"^50 OR subject:("चावल" OR "Reis" OR "рис
(з ерно)" OR "ເຂົ້າ" OR "벼" OR "Arroz" OR "Riso" OR "Riz"
OR "rizs" OR "稻米" OR "rýže" OR "أرز" OR "ข้าว" OR "米" OR
"ryža" OR "جنرب" OR "Ryż (ziarno)" OR "pirinç"))
```

As depicted in Fig. 1, after query expansion, the system queries the AGRIS core index using the expanded query Q1. The AGRIS website displays results in all languages, boosting results coming from the original query. Overall, the user is not aware that the system has modified their query. In fact, the system never shows the expanded query to the user.

3.3 Analysis of Results

Here follows a sequel to the scenario introduced at the beginning of Sect. 3. *The Chinese researcher Xian has just discovered that AGRIS has implemented the multilingual search functionality. Xian queries the system using the keyword 稻米. The system returns only 14 results, since only 14 AGRIS records contain the keyword 稻米 in title, abstract, or subject field. Xian clicks on the button to enable the multilingual search, and the system returns 166,639 results*[10]*. The new set of results is composed of bibliographic references about the concept "rice", but only 14 of them contains the Chinese word 稻米 actually. Now Xian has a lot of material to analyze and he can apply filters to make his query more specific.*

---

[10] http://agris.fao.org/agris-search/searchIndex.do?enableField=Enable&query=%E7%A8%BB%E7%B1%B3.

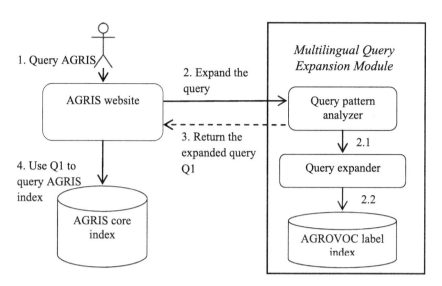

Fig. 1. Workflow of the multilingual query expansion module

There is also another interesting scenario to take into account. It concerns the absence of results after searching in a specific language. Let us consider an Indian user who queries the AGRIS system using the Hindi keyword "फसलें" (which means "crops" in English). The system returns zero results. It means that there are no resources in the AGRIS database containing the word "फसलें" in title, abstract, or subject field. Enabling the multilingual search, the user gets access to 474,854 scientific papers in several languages. Unfortunately, the result set does not contain metadata in Hindi, since the AGRIS Indian data provider only produces metadata in English. Be that as it may, our user is now able to query the system in their native language and access scientific literature even when there are no publications in that language.

Table 1 shows the effectiveness of the multilingual search feature, comparing the number of results before and after enabling this functionality in the AGRIS website. Correctness of results depends on the correctness of the AGROVOC thesaurus and AGRIS metadata. A community of domain experts from different countries contributes to the quality and correctness of labels available in AGROVOC. Thus, the multilingual translation based on AGROVOC is highly reliable as far as the agricultural domain concerns. Results of a multilingual search on AGRIS are composed of the union of results of several monolingual searches. The main disadvantage is that there could be too many results; this is why advanced filters are essential to allow AGRIS users to refine their queries reducing the number of results.

Table 2 compares the execution time of the default search with the execution time of the multilingual search. The execution time of the multilingual search is composed of two parts:

- The time to expand the user query with translations. This is the time that the multilingual query expansion module needs to translate the source query.

Table 1. Comparison of number of results before and after enabling the multilingual search in the AGRIS website

Query ID	Source query	English translation	Number of results	Number of multilingual search results
Q1	稻米	rice	14	166,639
Q2	फसले	crops	0	474,854
Q3	latte	milk	8,019	189,475
Q4	Klimaänderung	climate change	23	31,028
Q5	"su muhafazası"	water conservation	22	15,285
Q6	انتظام حرارى للتربة	soil thermal regimes	21	368
Q7	"forest mensuration"	forest mensuration	3,679	3,930

- The time to execute the expanded query. This is longer than the time to execute the source query, since the expanded query contains more terms.

On average, the execution of multilingual search requires 68.75 ms more than the default search in our implementation. This delay is highly acceptable.

An analysis of usage of this functionality show that 2 % of AGRIS active users enable the multilingual search. Let us focus on the expression "active users" to define it better. According to Google Analytics, in March 2016 AGRIS received around 513,000 unique page views. 80 % of them come from Google.com and Google Scholar, while 20 % of them represent activity of users in the AGRIS website. The latter percentage is about users who rely on AGRIS to actively search for scientific literature, i.e. the "active users"; on the other hand, users coming from Google access a bibliographic record directly, without using the AGRIS website search feature. On average, among active users, 2 % of them enable multilingual search. This number is quite satisfying, since multilingual search is an advanced functionality and we expect a small percentage of usage. In addition to that, the multilingual search is a new AGRIS functionality and it needs time to reach the public. It is highly likely that the percentage will increase over the time and after we will promote the multilingual search in webinars and events.

In order to improve the multilingual search usefulness, we have to explore possible extensions. First, we should allow users to select a subset of languages when enabling the functionality. In fact, it may well be that a user wants to retrieve results only in a couple of languages and not in all possible languages. Second, we need to solve the problem of singular/plurals, abbreviations, and misspellings. Currently the translation of the user query relies on exact match of strings; if AGROVOC contains a keyword in the exact way the user has written it, the system can translate the keyword in all available languages. At present, the system can manage singular/plural variations only for English terms. Finally, we have to explore the possibility to expand user queries to synonyms. In this way, the system can increase the recall, including all resources about the same topic in a specific language. The combination of this functionality with the multilingual search option is a valuable tool for end users, as we argue in the next section.

Table 2. Comparison of execution time (in milliseconds) before and after enabling the multilingual search in the AGRIS website

Query ID	Execution time	Time for query expansion	Execution time of the expanded query	Total multilingual search execution time
Q1	350 ms	25 ms	390 ms	415 ms
Q2	340 ms	30 ms	370 ms	400 ms
Q3	360 ms	20 ms	400 ms	420 ms
Q4	400 ms	30 ms	430 ms	460 ms
Q5	390 ms	25 ms	440 ms	465 ms
Q6	370 ms	30 ms	430 ms	460 ms
Q7	360 ms	30 ms	390 ms	420 ms

4   The Synonyms Problem with Recall

Methodologies for multilingual information retrieval through controlled vocabularies can be applied to different scenarios. In the previous section, we have described the implementation of the translation of user keywords through AGROVOC. In this way, users can access the AGRIS scientific literature in any languages, querying the system in any languages too. There are other situations where we can adopt the same methodology. For instance, it can be used to implement the expansion of user queries with synonyms in a given language, or even in all languages.

We can consider the following example. Peanut is a crop that mainly grows in the tropics and subtropics. People with a general background usually know this crop by the name of *Peanut*, but people in the field know it as *Groundnut* (this is the technical name, while the scientific name is *Arachis hypogaea*). If an AGRIS user queries the system with the keyword *Peanut*, the system returns only results containing such keyword in their metadata, but not results containing the keyword *Groundnut*. A search mediated by concepts returns all results related to the crop Peanut, containing both *Peanut* and *Groundnut* in the metadata.

We can apply the methodology described in Sect. 3.2 to implement this behavior. The *Synonyms Query Expansion Module* works exactly as the Multilingual Query Expansion Module described in Fig. 1. The new module identifies the query pattern and then uses the AGROVOC index to retrieve all synonyms of a keyword in a given language. The difference is that now the expansion module includes both preferred and alternative labels in a specific language as output of the translation process, while the Multilingual Query Expansion Module considers only preferred labels in all languages. In fact, the union of preferred and alternative labels in a language compose the set of available synonyms for that language.

We can further extend this process by combining the Synonyms Query Expansion Module and the Multilingual Query Expansion Module. If the user looking for *Peanut* enables the synonyms retrieval, they obtain results including also the keyword *Groundnut*. If the user also enables the multilingual retrieval, they obtain results in all languages, including synonyms for each different language. This is a very important step, since it is not a mere translation of strings. Different languages may have different synonyms, which are not the direct translation of one another. Relying on a controlled vocabulary like AGROVOC solves this issue. In fact, we do not translate the main keyword and all its synonyms in other languages, but we search for the AGROVOC concept, and we extract all preferred and alternative labels of the concept for all the available languages.

Even if we have not implemented the Synonyms Query Expansion Module yet, we can provide some numbers to demonstrate its power. We can use the AGRIS website to simulate the synonyms expansion manually. We start with three fulltext queries, using the keywords *Peanuts*, *Groundnuts*, and their combination. This is the number of results:

1. Groundnuts: 2,824 results
2. Peanuts: 6,750 results
3. Peanuts OR Groundnuts: 9,222 results

The third query is exactly the query that the *Synonyms Query Expansion Module* would generate. As we can see, enabling the synonyms retrieval improves the recall. Another observation is that the sum of results of the first two queries is 9,574 while the synonyms expansion returns 352 results less; this means that results sets 1 and 2 have a small overlapping, because 352 AGRIS records contain both *Peanut* and *Groundnut* in their metadata.

Now, we simulate the combination of the synonyms expansion with the multilingual search. We enable the multilingual search for the query Groundnuts, and then for the synonyms expanded query Peanuts OR Groundnuts:

4. Groundnuts (multilingual query): 4,713 results
5. Peanut OR Groundnut[11] (multilingual query): 10,842 results

The fourth query shows that the multilingual retrieval allows obtaining 1,889 records more than the default query with the keyword *Groundnuts* (query number 1). On the other hand, the combination of synonyms expansion and multilingual query (query number 5) returns 6,129 results more than the multilingual expansion of the keyword *Groundnuts* (query number 4) and only 1,620 results more than the synonyms expansion performed by the third query.

The major impact of the synonyms expansion with respect to the multilingual one is due to the fact the AGRIS has an high coverage of English metadata, thus synonyms has more impact than translations when the source keyword is in English. AGRIS resources cover 64 languages, and there is a coverage of at least 10,000 resources for 28 languages. However, many data providers translate metadata also into English. For example, Chinese resources have titles, abstracts, and keywords both in English and in Chinese.

---

[11] We have manually built the multilingual expansion for this query. In fact, the current system does not recognize a query pattern including the OR operator, as we have explained in Sect. 3.2. We have executed a union of the two expanded queries generated for the keywords *Peanuts* and *Groundnuts*.

5 Conclusions

Multilingual search and synonyms expansion have a profound impact on searching in online repositories. While multilingual search allows users to search in their native language and to retrieve documents in several languages, expanding user queries with synonyms allows retrieving more resources about the given topic. In this paper, we have proposed a methodology that relies on a controlled vocabulary to implement the aforementioned features. We have implemented the methodology in the AGRIS website to enable the multilingual search; our implementation has required AGROVOC controlled vocabulary and a software component that detects query patterns and translates a query through AGROVOC. We have also discussed how the same methodology can be adopted to expand user queries with synonyms. Experimental results demonstrate significant improvements of recall in both cases. The high amount of retrieved resources can be reduced by an effective advanced search that helps users in refining and filtering out results.

The current implementation in the AGRIS system can be improved. First, we have to provide the possibility to select a subset of languages when enabling the multilingual search. This feature would allow users to retrieve results only in their favorite languages, reducing the number of undesired results. Second, we have to implement the Synonyms Query Expansion Module. Finally, we have to work on homonyms and variations of keywords, like abbreviations and misspellings, especially for non-Latin characters.

There are also further scenarios to explore. As future work, it would be useful to study which additional expansions of queries can be useful to users. For instance, a controlled vocabulary like AGROVOC allows generalizing or restricting the topic of a query by navigating the hierarchy of concepts. Even more useful would be a system that automatically performs different query expansions and combinations of them, presenting to end users alternative subsets of results. In this case, users can select the desired result set by considering the number of results and the specific mechanism under the retrieval of different result sets.

Acknowledgement. The views expressed in this information product are those of the author(s) and do not necessarily reflect the views or policies of FAO.

References

1. Basili, R., Stellato, A., Daniele, P., Salvatore, P., Wurzer, J.: Innovation-related enterprise semantic search: the INSEARCH experience. In: 2012 IEEE Sixth International Conference on Semantic Computing (ICSC), pp. 194–201. IEEE (2012)
2. Bibliographic Services Task Force of the University of California Libraries: Final Report, Rethinking How We Provide Bibliographic Services for the University of California (2005)
3. Carpineto, C., Romano, G.: A survey of automatic query expansion in information retrieval. ACM Comput. Surv. (CSUR) **44**(1), 1 (2012)

4. Celli, F., Keizer, J., Jaques, Y., Konstantopoulos, S., Vudragović, D.: Discovering, indexing and interlinking information resources. F1000Research **4**, 432 (2015). doi:10.12688/f1000research.6848.2

5. Marcum, D.B.: The future of cataloging: address to the ebsco leadership seminar, Boston, Massachusetts (2005)

6. FAOSTAT Food and Agriculture commodities production. http://faostat3.fao.org/browse/rankings/commodities_by_regions/E

7. Gardner, S.A.: The changing landscape of contemporary cataloging. Cataloging Classif. Q. **45**(4), 81–99 (2008)

8. Ghorab, M.R., Leveling, J., Lawless, S., O'Connor, A., Zhou, D., Jones, G.J.F., Wade, V.: Multilingual adaptive search for digital libraries. In: Gradmann, S., Borri, F., Meghini, C., Schuldt, H. (eds.) TPDL 2011. LNCS, vol. 6966, pp. 244–251. Springer, Heidelberg (2011). doi:10.1007/978-3-642-24469-8_26

9. Gale, W.A., Church, K.W., Yarowsky, D.: One sense per discourse. In: Proceedings of the Workshop on Speech and Natural Language, pp. 233–237. Association for Computational Linguistics

10. Gross, T., Taylor, A.G.: What have we got to lose? the effect of controlled vocabulary on keyword searching results. Coll. Res. Libr. **66**(3), 212–230 (2005)

11. Gross, T., Taylor, A.G., Joudrey, D.N.: Still a lot to lose: the role of controlled vocabulary in keyword searching. Cataloging Classif. Q. **53**(1), 1–39 (2015)

12. Kaplan, A., Sándor, Á., Severiens, T., Vorndran, A.: Finding quality: a multilingual search engine for educational research. In: Gogolin, I., Åström, F., Hansen, A. (eds.) Assessing Quality in European Educational Research, pp. 22–30. Springer Fachmedien, Wiesbaden (2014)

13. Lu, C., Park, J.R., Hu, X.: User tags versus expert-assigned subject terms: a comparison of library thing tags and library of congress subject headings. J. Inf. Sci. **36**(6), 763–779 (2010)

14. McCutcheon, S.: Keyword vs controlled vocabulary searching: the one with the most tools wins. Indexer **27**(2), 62–65 (2009)

15. Peters, C., Braschler, M., Clough, P.: Cross-language information retrieval. In: Peters, C., Braschler, M., Clough, P. (eds.) Multilingual Information Retrieval, pp. 57–84. Springer, Berlin, Heidelberg (2012)

16. Rowley, J.: The controlled versus natural indexing languages debate revisited: a perspective on information retrieval practice and research. J. Inf. Sci. **20**(2), 108–118 (1994)

17. Scicluna, R.: Should libraries discontinue using and maintaining controlled subject vocabularies? (2015)

18. Spink, A., Wolfram, D., Jansen, M.B., Saracevic, T.: Searching the web: the public and their queries. J. Am. Soc. Inf. Sci. Technol. **52**(3), 226–234 (2001)

19. Stellato, A., Rajbhandari, S., Turbati, A., Fiorelli, M., Caracciolo, C., Lorenzetti, T., Keizer, J., Pazienza, M.T.: VocBench: a web application for collaborative development of multilingual thesauri. In: Gandon, F., Sabou, M., Sack, H., d'Amato, C., Cudré-Mauroux, P., Zimmermann, A. (eds.) ESWC 2015. LNCS, vol. 9088, pp. 38–53. Springer, Heidelberg (2015). doi:10.1007/978-3-319-18818-8_3

20. Voorbij, H.J.: Title keywords and subject descriptors: a comparison of subject search entries of books in the humanities and social sciences. J. Documentation **54**(4), 466–476 (1998)

21. Zavalina, O.L.: Collection-level subject access in aggregations of digital collections: metadata application and use. Ph.D. dissertation, University of Illinois at Urbana-Champaign (2010)

Exploring Term Networks for Semantic Search over RDF Knowledge Graphs

Edgard Marx[1,3]([✉]), Konrad Höffner[1], Saeedeh Shekarpour[4],
Axel-Cyrille Ngonga Ngomo[1], Jens Lehmann[2], and Sören Auer[2]

[1] AKSW, University of Leipzig, Leipzig, Germany
marx@infomatik.uni-leipzig.de, konrad.hoeffner@uni-leipzig.de
[2] Computer Science Institute, University of Bonn, Bonn, Germany
{lehman,auer}@cs.uni-bonn.de
[3] Instituto de Pesquisa e Desenvolvimento Albert Schirmer, Teófilo Otoni, Brazil
[4] Knoesis Research Center, Fairborn, OH, USA
saeedeh@knoesis.org

Abstract. Information retrieval approaches are considered as a key technology to empower lay users to access the Web of Data. A large number of related approaches such as Question Answering and Semantic Search have been developed to address this problem. While Question Answering promises more accurate results by returning a specific answer, Semantic Search engines are designed to retrieve the best top-K ranked resources. In this work, we propose *path, a Semantic Search approach that explores term networks for querying RDF knowledge graphs. The adequacy of the approach is evaluated employing benchmark datasets against state-of-the-art Question Answering as well as Semantic Search systems. The results show that *path achieves better F_1-score than the currently best performing Semantic Search system.

1 Introduction

The growth of Semantic Web technologies has led to the publication of large volumes of data. Approximately 10000 *Resource Description Framework (RDF)*[1] datasets are available via public data portals.[2] However, retrieving desired information from datasets still poses a significant challenge. Lay users cannot be expected to make themselves familiar with the underlying query languages and modeling structures.

A major challenge is the efficient retrieval of the resource that best represents the user's intent via natural language (NL) keyword queries. Relying solely on off-the-shelf triple stores or document retrieval may lead to poor performance or precision (see Sect. 5). To address this problem, we propose an approach for Semantic Search RDF knowledge graphs by exploring its Term Network. A Term Network (see Sect. 4) is a graph whose vertices are labeled terms. Overall, our contributions are as follows:

---

[1] http://www.w3.org/RDF.

[2] http://lodstats.aksw.org/.

© Springer International Publishing AG 2016
E. Garoufallou et al. (Eds.): MTSR 2016, CCIS 672, pp. 249–261, 2016.
DOI: 10.1007/978-3-319-49157-8_22

- We develop a new formal model for Semantic Search (SemS) based on Term Networks;
- We present a ranking method that increases the precision on retrieving RDF data;
- We compare our approach with state of the art SemS techniques on the QALD-4 [17] benchmark and show that we achieve a higher F_1-score.

The rest of this paper is organized as follows: The related work is reviewed in Sect. 2. Section 3 defines the preliminaries. Section 4 describes the *path model. Section 5 outlines the evaluation and discusses the results. Finally, Sect. 6 concludes giving an outlook of potential future work.

2   Related Work

Information retrieval (IR) over Linked Data is an active and diverse research field with many existing related work focusing on designed for different environments, diverging in complexity and precision. The related work can be mainly categorized in two types of approaches that recover information from Linked Data Knowledge Graphs (KGs, see Definition 1): (1) by using conventional IR techniques and (2) by answering natural language questions. While the use of time efficient traditional IR systems lacks the ability to deal with complex queries, they are usually faster. Wang et al. [19] shows that pure traditional IR engines are faster than the combination of a triple store with a full-text index. However, both models explore the semantics of an NL query for delivering the response by applying statistics measures and heuristics in the KG.

Semantic Search (SemS) approaches aim to retrieve the top-k ranked resources for a given NL input query. Swoogle [3], introduces a modified version of PageRank that takes into account the types of the links between ontologies. Sindice [10], Falcons [2] and Sig.ma [16] explores traditional document retrieval to index and locate relevant sources and/or resources. Sindice is a search engine that can retrieve documents containing a given statement. Falcon, uses a built-in ranking mechanism for entity ranking while Sig.ma allows the use of constraints to query for particular classes and/or properties. In all cases, the structure and semantics are not taken into account during the matching phase.

YAHOO! BNC [4] used a local, per property, term frequency as well as a global term frequency. It also applied a boost based on the number of matched query terms. Umass [4] explored existing ranking functions applied to four field types: (1) title; (2) name; (3) dbo:title, and; (4) all others. The fields were weighted separately with a specific boost applied to each of them.

Later, Blanco et al. [1] proposed a modified version of BM25F ranking function adapted for RDF data. The function was applied to a horizontal pairwise index structure composed of the subject and its property values. However, the most important feature in the proposed structure is the possibility to assign different weights to predicates. The proposed adaptation is implemented in the $Glimmer_{Y!}$ engine and is shown to be time efficient as well as outperforms other state-of-the-art methods in ranking RDF resources.

Recently, Virgilio et al. [18] introduced a distributed technique for SemS on RDF data using MapReduce. The method uses a distributed index of RDF paths. The proposed strategy returns *the best top-k answers in the first k generated results.* The retrieval is done by evaluating the paths containing the terms of the query using two strategies: (1) Linear and (2) Monotonic. (1) The Linear strategy uses only the high ranked path(s). As a consequence, it does not produce an optimum solution but has linear complexity with respect to the size of matched entities. (2) The Monotonic strategy uses all matched paths and, thus, produces better results. Intuitively, measuring all suitable paths from all entities is less time efficient. Please refer to the work of Mangold et al. [8] for a more detailed analysis of SemS approaches.

One of the biggest challenges in SemS method lies in evaluating the relatedness between the terms in a KG and an NL query. Document retrieval engines rely on term frequency weighting, which is based on the assumption, that the more frequently a term occurs, the more related it is to the topic of the document [7]. While good retrieval performance needs to take the frequency into account, it suffers from frequent yet unspecific words such as "the", "a" or "in". Inverse document frequency corrects this by diminishing the weight of words that are frequently occurring in the corpus, leading to the combined term frequency–inverse document frequency (tf-idf) [15] to score documents for a query.

3   Preliminaries

We begin by introducing a formal definition of the RDF model. Thereafter, we introduce fundamental concepts that are required for full understanding of the rest of the paper.

RDF[3] is a standard for describing Web resources. A resource can refer to any physical or conceptual thing, such as a Web site, a person or a device. The RDF data model expresses statements about resources in the form of subject-predicate-object triples. The subject denotes a resource; the predicate expresses a property (of the subject) or a relationship (between subject and object); the object is either a resource or literal. Resources are identified with IRIs, a generalization of URIs, while literals are used to identify values such as numbers and dates by means of a lexical representation.

Definition 1 (RDF knowledge Graph, KG). *Formally, let K be a finite RDF knowledge graph (KG). K can be regarded as a set of triples $(s, p, o) \in (\mathcal{I} \cup \mathcal{B}) \times \mathcal{P} \times (\mathcal{I} \cup \mathcal{L} \cup \mathcal{B})$, where $\mathcal{R} = \mathcal{I} \cup \mathcal{B}$ is the set of all RDF resources $r \in \mathcal{R}$ in the KG, $\mathcal{I}$ is the set of all IRIs, $\mathcal{B}$ is the set of all blank nodes, $\mathcal{B} \cap \mathcal{I} = \emptyset$. $\mathcal{P}$ is the set of all predicates, $\mathcal{P} \subseteq \mathcal{I}$. $\mathcal{L}$ is the set of all literals, $\mathcal{L} \subset \Sigma^*$ and $\mathcal{L} \cap \mathcal{I} = \emptyset$, where Σ is the unicode alphabet. $\mathcal{E}$ is the set of all entities, $\mathcal{E} = \mathcal{I} \cup \mathcal{B} \setminus \mathcal{P}$. An RDFTerm φ refers to any edge label $p \in \mathcal{P}$ or vertex in the KG $\varphi \in (\mathcal{I} \cup \mathcal{B} \cup \mathcal{L})$. A KG is modeled as a directed labeled graph $\mathcal{G} = (\mathcal{V}, \mathcal{D})$, where $\mathcal{V} = \mathcal{E} \cup \mathcal{L}$, $\mathcal{D} \subseteq \mathcal{E} \times (\mathcal{E} \cup \mathcal{L})$ and the labeling function[4] of the edges is a mapping $\lambda : \mathcal{D} \mapsto \mathcal{P}$. We disregard literal language tags and data types.*

---

[3] https://www.w3.org/TR/REC-rdf-syntax/.
[4] Not to be confused with rdfs:label.

Figure 1 shows an excerpt of a KG where a literal vertex $v_i \in \mathcal{L}$ (respectively a resource vertex $v_i \in \mathcal{R}$) is illustrated by a rectangle, respectively an oval. Each edge between two vertices corresponds to a triple, where the first vertex is called the subject, the labeled edge the predicate and the second vertex the object. For example, $\textcircled{e2} \xrightarrow{\texttt{rdfs:label}} \boxed{\text{Mona Lisa}}$ corresponds to the triple <e2, rdfs:label, "Mona Lisa">.

In this work, we address the problem of SemS systems that aim to retrieve the top-k ranked entities representing the intention behind an NL user query.

Definition 2 (Natural Language Query). *A NL query $q \in \Sigma^*$ is a user given keyword string expressing a factual information needed.*

4   Approach

For many years, scientists from the most diverse fields of cognitive science have tried to explain and reproduce the human cognition system, including psychology, neuroscience, philosophy, linguistics and artificial intelligence. While diverse theories have been developed, a commonly shared idea is that knowledge is organized as a network [12]. Hudson et al. [6] go further and states that grammar is organized as a network as well. According to Hudson's work, the syntactic structure of a sentence consists of a network of dependencies between single terms. Thus, everything that needs to be said about the syntactic structure of a sentence can be represented in such a network. Hudson explores Saussure's [13] idea that *"language is a system of interdependent terms in which the value of each term results solely from the simultaneous presence of the others"*. He also argues about the psycholinguistic evidence for the use of *spreading activation* in supporting knowledge reasoning. However, according to Hudson et al., the main challenge is finding out how the activation occurs in mathematical terms [6]. Our intuition is that as the KG contains a network of terms formed by the label (e.g. rdfs:label) of the RDFTerms—properties, classes and entities—they can be used to query.

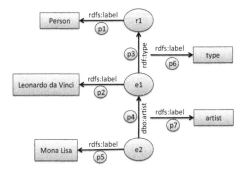

Fig. 1. An excerpt of a KG. The label of rdfs:label properties were omitted for simplification.

Definition 3 (Term). *A term[5] can be a word or a phrase used to describe a thing or to express a concept* [11]. *In this work, we consider as term any literal $(l \in L)$ in a KG.*

Definition 4 (RDFTerm Label). *A term associated with an RDFTerm φ, denoted by $L(\varphi)$, is the literal respectively the label of φ. Considering the* rdfs:label[6] *as labeling property:*

$$\text{label}(r) := \{l \in L \mid (r, \textit{rdfs:label}, l) \in K\}$$

$$L(\varphi) := \left\{ \begin{array}{ll} \{\varphi\} & \textit{if } \varphi \in L, \\ \text{label}(\varphi) & \textit{otherwise.} \end{array} \right\}$$

Although there is no evidence that the previous works were influenced by Hudson's theory, there are models that make use of the KG in order to evaluate the answer [14,20]. Figure 1 shows a set of literals associated with the resources in the KG sample. Each resource contains a set of terms $LR(r)$. This terms are called *Resource-Associated Terms* and are defined as follows:

Definition 5 (Resource-Associated Terms). *The set of terms associated with a resource r denoted by $LR(r)$ is the union of all literals as well as labels of each property and object in the triples in which r is the subject.*

$$LR(r) := \{l \in L \mid \exists (r, p, o) \in K : \\ \exists \varphi \in \{p, o\} : l = L(\varphi)\}$$

Example 1 (Resource-Associated Terms).   Considering the KG depicted in Fig. 1, the triples having the entity e_2 as subject are as follows:

```
1.  e2 rdfs:label "Mona Lisa".
2.  e2 dbo:artist e1.
```

The associated terms for e_2 are: $LR(e_2) = \{$ "label", "Mona Lisa", "artist", "Leonardo da Vinci"$\}$

Definition 6 (Term Network). *A Term Network is a graph whose vertices are labeled with terms.*

A KG can be converted to a TN by visiting all vertices and edges executing the following operations (Fig. 2 shows the TN for Example 1):

1. Labeling edges and non-literal vertices by a copy of their respective labels defined by the labeling property rdfs:label;
2. Converting edges to vertices.

The TN of a KG is connected and its paths can have cycles as well as an arbitrary length. In order to simplify the TN and eliminate its ambiguity, the proposed model works on a simplified version of the TN extracted from a structure called Semantic Connected Component (SCC), defined as follows:

---

[5] Not to be confused with an RDFTerm.
[6] Other labeling properties may also be used.

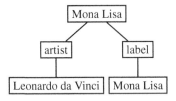

Fig. 2. Representation of a TN extracted from the triples that have e_2 as subject from the KG depicted in Fig. 1.

Definition 7 (Semantic Connected Component). *The Semantic Connected Component (SCC) of an entity e in an RDF graph G under a consequence relation $\models$ is defined as $SCC_{G,\models}(e) := \{(e,p,o) \mid G \models \{(e,p,o)\}\} \cup \{(p, rdfs:label, l) \in G\} \cup \{(o, rdfs:label, l) \in G\}\}$. If the graph and consequence relation is clear from the context, we use the shorter notation $SCC(e)$. Within this paper, we use the RDFS entailment consequence relation as defined in its specification[7].*

Example 2 (Semantic Connected Component). For instance, by RDFS entailment, the entity dbr:Australia is a dbo:PopulatedPlace. The inference is due to dbr:Australia being typed as dbo:Country which is a subclass of dbo:PopulatedPlace. Considering the running example, the SCC of the entity e_2 is $SCC(e_2) = (\{e_2, e_1, \text{"Mona Lisa"}\}, \{p_5, p_4\})$.

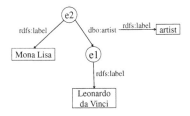

Fig. 3. Representation of the SCC of the entity e_2 extracted from the KG depicted in Fig. 1.

The structure used for ranking is called Semantic Unit (SU). The SU is a tree, where the nodes starting from its root node are labeled with tokens and have only one child. Tokens are sub-strings extracted from another string, they are formally defined as follows.

Definition 8 (Token). *A token $t \in \mathcal{T}$ is the result from a tokenizing function $\mathcal{T} : \Sigma^* \to \Sigma^{**}$, which converts a string to a set of tokens.*

---

[7] http://www.w3.org/TR/rdf-mt/.

The root node sub-trees of the SU form a set of paths starting from the resource to which the SCC is associated, see Fig. 4. The SU is defined as follows:

Definition 9 (Semantic Unit (SU)). *The Semantic Unit is a tree where:*

- *The root node is an entity;*
- *All vertices in the root node sub-trees only have one child, and;*
- *Vertices in the root node sub-trees are labeled with tokens.*

Example 3 (Semantic Unit (SU)). Considering the running example, the SU of the entity e_2 is $SU(e_2) = (\{\ e_2, v_1, v_2, v_3, v_4, v_5, v_6, v_7\}, \{(e_2, v_1), (e_2, v_5), (v_1, v_2),$ $(v_2, v_3), (v_3, v_4), (v_5, v_6), (v_6, v_7)\})^8$ and is depicted in Fig. 4.

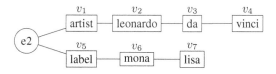

Fig. 4. Representation of the SU of the entity e_2 extracted from the KG depicted in Fig. 1.

An SCC can be converted into an SU as follows:

1. Converting the sub-trees starting from the root node of the SCC into TN;
2. Converting the literal vertices to a graph where there is an edge starting from each token to its subsequent one, defined as follows:

$$\mathcal{G}(l) := (\mathcal{T}(l), \mathcal{D}(l))$$
$$\mathcal{D}(l) := \{(t_1, t_2) \in \mathcal{T}(l) \mid \exists i \in \mathbb{N} : (\pi_i(\mathcal{T}(l)) = t_1) \wedge (\pi_{i+1}(\mathcal{T}(l)) = t_2)\}$$

Example 4 (Literal to graph). Converting the term `"mona lisa"` to a graph.

$$\mathcal{G}(\texttt{"mona lisa"}) = (\{\texttt{"mona"}, \texttt{"lisa"}\}, \{(\texttt{"mona"}, \texttt{"lisa"})\})$$

In the following sections, we start by describing how we retrieve SU in the KG using the query terms. Later, we discuss how we can efficiently rank it.

4.1 Retrieving

The idea is to perform the selection of SUs which have a term in intersection with the query terms. For instance, one possible solution for $\{$`"mona"`, `"lisa"`, `"artist"`$\}$ is the co-occurrence of all terms in a SU. The next possible solution is the co-occurrence of two of the three terms and so on. Thus, it is necessary to check for the existence of the query terms in different paths. For example, one SU may contain the token `"artist"` and another with the tokens (`"mona"`, `"lisa"`), see Example 5.

---

[8] The output of the tokenizer used in this example are lowercase lexemes from a literal.

Example 5 (Retrieving "Mona Lisa artist"). In the KG in Fig. 1, the SCC containing the answer for the query {"mona", "lisa", "artist"} is SCC(e_2) and can be retrieved by a simple lookup with a SPARQL query.

Query and Resource Labels Analysis. Information retrieval systems for RDF are commonly designed to support full or keyword NL queries. However, converting keywords to full queries is a more challenging task. The *path query approach is designed to deal with keyword or full queries by converting the latter into keyword queries. The process of conversion of a NL input query to a tuple of keywords consists of applying known techniques, in order: (1) lowercase and (2) lemmatization. In order to increase the number of matched SUs, the same analysis is applied to the SU labels.

After extracting the SUs, the SCC of the SU's entity is used for ranking.

4.2   Ranking

Document retrieval approaches are not suitable for RDF because the most important feature of RDF is not the terms, but the relation of the concepts underlying its graph structure. The challenge of adapting the ranking method is measuring the relatedness between the resources in the target KG and the input query terms. As a query rarely exactly matches the resource associated terms, both are first converted into tokens. Thereafter, the proposed ranking assumes that the probability of a resource being part of an answer correlates with the number of matched tokens between the query and the resource associated terms. For instance, a query containing *birth date* should be more related to the property dbo:birthDate than to the property dbo:deathDate or dbpprop:date. The strength is measured by the number of query tokens matching with the resource tokens.

Definition 10 (Resource Matching). *A resource matching is a function* $MT : \mathcal{T} \rightarrow 2^{\mathcal{R}}$ *that maps query tokens* $\mathcal{T} = \{t_1, t_2, t_3...t_n\}$ *to resources, formally defined by* $MT(t)$, *where* δ *is a string dissimilarity function and* $\theta \in [0,1] \subset \mathbb{R}$:

$$MT(t) := \{r \in \mathcal{R} \mid \exists t' \in \mathcal{T}(LR(r)) : \delta(t,t') < \theta\}$$

Example 6 (Resource Matching). Let $\mathcal{T}(q) = $ {"mona", "lisa", "artist"}. According to Fig. 1, the tokens are mapped to: $MT("mona")=\{e_2\}$, $MT("lisa")= \{e_2\}$, $MT("artist")=\{p_4\}$.

As the knowledge base is a graph, the resources and literal values are connected by paths formed by edges and vertices, see Fig. 3.

Example 7 (Path). In the SCC shown in Fig. 3, there are two paths starting from the entity e_2 as follows: $\gamma_1= ((e_2, "Mona\ Lisa"))$ and $\gamma_2 = ((e_2, e_1))$.

Furthermore, resources belonging to a path between one resource to another are labeled (e.g. rdfs:label). Therefore, it is possible to explore the terms associated to the entity's paths to determine its relevance.

Definition 11 (Path terms). *Path terms are the set of all literals in the path* γ, *defined as follows:*

$$LP(\gamma) := \{l \mid \exists \varphi \in \gamma : l \in L(\varphi)\}$$

Example 8 (Path terms). For Example 7, the set of associated terms for the two given paths are as follows: $LP(\gamma_1)=$ {"label","Mona Lisa"} and $LP(\gamma_2) =$ {"artist","Leonardo da Vinci"}.

Thus, the relevance score of an entity depends on the number of matched terms in its associated paths. The higher the number of matched terms, the higher the relevance of the entity. Furthermore, if a term matches multiple paths of an entity, it is only attributed to the path with the highest number of matched terms. The relevance score of an entity is the sum of all individual path scores; it is measured by the *Semantic Weight Model (SWM)*, which is formally defined as follows.

Definition 12 (Semantic Weight Model (SWM)). *Each token* t *in* $\mathcal{T}(q)$ *is first mapped to the* paths *of the* SCC S. *The set of matched tokens from a path* γ *is returned by the function* $TP(\gamma, q)$. *A path match of an* SCC S *is evaluated by the function* $\mathrm{MTP}(\gamma, q, S)$ *using a path weighting function* $\mathrm{w} : D^+ \to R$.

$$TP(\gamma, q) := \{t \in \mathcal{T}(LP(\gamma)) \mid \exists t' \in \mathcal{T}(q) : \delta(t, t') < \theta\}$$
$$\mathrm{MTP}(\gamma, q, S) := \{t \in TP(\gamma, q) \mid \forall \gamma' \in D(S)^+ : \mathrm{w}(\gamma)|TP(\gamma, q)| \geq \mathrm{w}(\gamma')|TP(\gamma', q)|\}$$

The final score of an SCC S *is a sum of its* n *path-scores and is measured by the function* score(S), *as follows:*

$$\mathrm{score}(S) = \sum_{\gamma \in D(S)^+} \begin{cases} \mathrm{w}(\gamma)|TP(\gamma, q)| & \text{if } \mathrm{MTP}(\gamma, q, S) \neq \emptyset, \\ 0 & \text{otherwise.} \end{cases}$$

In case there are terms matching multiple paths and the paths have equal number of matched terms and equal score, only one of the path scores is added to the SCC *score.*

The SWM assigns different weights based on the RDF properties on the path. This means that the weight of a term in a path is determined by the type of the properties (label, is-a relation, other) on that path and it acts as a tiebreaker for the paths with equal number of tokens. The weight hierarchy of paths is constructed to allow the exploration of the KG by querying entities by type, label, predicates and objects. Since terms extracted from resources can have overlaps, there is a need for providing a disambiguation method.

Weighing. Following we start explaining the rationality behind the defined weights, later we use examples to better illustrate it.

Is-a Relation. The problem is that tokens can exist in different paths of an SCC. Thereafter, a token in an is-a relation property can also exists in other properties. However, a property as an entity label references the entity itself while an is-a relation references classes of entities. In this case, if a query intends to select a specific class of entities, other entities can be retrieved by mistake. Thus, it is important to provide an efficient method to disambiguate between classes and entities. To alleviate this problem, the weight of the paths containing an is-a relation property are set higher than other paths. Thereafter, the selection of a specific entity can be done by building a more precise query. The reason is that beside the entity's label, other properties can be used to disambiguate. For instance, in the case of a class and an entity have the same label, the user can use other entity property's term. Therefore, the highest weight is assigned to paths with an is-a relation property γ_t—i.e. the paths containing rdf:type.

Entity Label. The second highest weight is assigned to labeling property paths γ_l—i.e. the paths containing the rdfs:label property—and those are assigned higher values than other property paths γ_o. Entities can be referenced multiple times in a KG, but when a query contains an entity label, it is more likely that it is looking for the entity than for its references—an object instance. Therefore, to prevent entities with references to be higher ranked than the entity itself, the weight of the path with an labeling property is set higher than a path with another property.

Despite the different weights, we still want a higher number of matched tokens to score higher in practical cases, i.e. $n + 1$ matched tokens should score higher than n matched tokens for reasonably low n:

$$(n + 1)\,\mathrm{w}(\gamma_t) > (n + 1)\,\mathrm{w}(\gamma_l) >$$
$$(n + 1)\,\mathrm{w}(\gamma_o) > n\,\mathrm{w}(\gamma_t) > \tag{1}$$
$$n\,\mathrm{w}(\gamma_l) > n\,\mathrm{w}(\gamma_o)$$

Following, the model is explained using examples.

Case 1: Querying by Entity label. For the query "Rio de Janeiro", the SWM should consider the DBpedia entity dbpedia:Rio_de_Janeiro as the best answer although the DBpedia entity dbpedia:Tom_Jobim has the DBpedia property dbpprop:birthPlace referencing the entity dbpedia:Rio_de_Janeiro. For the term "The" in a query, the model will consider as a possible answer the entities dbpedia:The_Simpsons and dbpedia:The_Beatles rather than the DBpedia property dbpprop:The_GIP.

Case 2: Querying by Is-a Relation. Considering the query "place", the implemented *SWM* will prefer the data type dbo:Place instead of the property dbo:Place.

Case 3: Querying by Another Properties. Let us consider the case that the query is "birth place" rather than "place" as in the previous example. As the number of matching terms in the property dbo:birthPlace is higher than for the data type dbo:Place, consequently the weight of dbo:birthPlace will be higher than the *data type*.

5   Experimental Evaluation

We evaluate the performance of *path in comparison to the state-of-the-art SemS system as well as QA in terms of Precision, Recall and F-measure. To the best of our knowledge is the first time that the precision of both approaches are measured in the same benchmark.

Benchmark. Several benchmarks can be used to measure the precision of our approach, including benchmarks from the initiatives *SemSearch* [4][9] and *QA Over Linked Data (QALD)*[10]. *SemSearch* is based on user queries extracted from the YAHOO! search log, with an average distribution of 2.2 words per-query. *QALD* provides both QA and keyword search benchmarks for RDF data that aim to evaluate the extrinsic behavior of systems. The QALD benchmarks are the most suitable for our evaluation due to the wide type of queries they contain and also because it makes use of *DBpedia*, a very large and diverse dataset. In this work, we use openQA framework [9] over the newest version of the QALD benchmark compatible with the framework—QALD version 4 (QALD-4) benchmark [17]. The proposed approach was compared with respect to the performance of Glimmer$_{Y!}$ because it is the best performing SemS system and it is open-source, which allows to evaluates its performance.

Results. Table 1 shows the performance of *path in comparison to Glimmer$_{Y!}$, the state-of-the-art SemS system [1], and all participating QA systems in the multilingual challenge of the *QALD-4* benchmark.

Discussion. The proposed approach is faster than the best SemS participating in SemSearch'10. The main reason is that Glimmer$_{Y!}$ build an an index without reasoning which imposes constraints on the precision (Table 1). The index without reasoning is a core limitation of Glimmer$_{Y!}$, since the user cannot query by using terms from properties as well as from entity objects. For instance, in *Case* 1 in Sect. 4.2, Glimmer$_{Y!}$ fails to retrieve dbpedia:Tom_Jobim because the terms of the entity dbpedia:Rio_de_Janeiro belonging to the property dbpprop:birthPlace are not indexed. The same occurs for the data type in *Case* 2 where the type is also given by a non-literal object. However, the F-measure of *path decreases sensitively (0.42) in comparison with the best performing QA system in QALD-4. The drawback is due to *path does not target the treatment of complex queries—i.e., queries that require the use of aggregations, restrictions as well as solution modifiers to be answered.

---

[9] http://km.aifb.kit.edu/ws/semsearch10/.
[10] http://greententacle.techfak.uni-bielefeld.de/~cunger/qald/.

Table 1. *Precision* (**P**), *Recall* (**R**) and *F-measure* (**F$_1$**) achieved by different SemS and QA systems in QALD-4 *Multilingual Challenge*. The systems are Glimmer$_{Y!}$, *path*, SINA, TBSL and all QALD-4 participating systems.

System	P	R	F_1	Approach
Xser	0.71	0.72	0.72	QA
gAnswer	0.37	0.37	0.37	QA
CASIA	0.40	0.32	0.36	QA
*path	0.30	0.30	0.30	SemS
Intui3	0.25	0.23	0.24	QA
ISOFT	0.26	0.21	0.23	QA
SINA	0.15	0.15	0.15	QA
RO FII	0.12	0.12	0.12	QA
TBSL	0.10	0.10	0.10	QA
Glimmer$_{Y!}$	0.07	0.07	0.07	SemS

6   Conclusion, Limitations and Future Work

We have presented a novel ranking method for SemS over KGs. The results of an experimental study show a significant improvement in comparison to the state-of-the-art SemS. Furthermore, the approach achieves comparable precision when compared with QA systems.

There are a few challenges not addressed in the current implementation as complex queries [5]. In future work, we plan to extend the precision of this approach by addressing the mentioned challenges. Furthermore, we plan to investigate indexing techniques. We see this work as the first step of a larger research agenda for SemS over Linked Data.

Acknowledgements. This work was supported by a grant from the EU H2020 Framework Programme provided for the projects Big Data Europe (GA no. 644564), HOBBIT (GA no. 688227), and CNPq under the program Ciências Sem Fronteiras.

References

1. Blanco, R., Mika, P., Vigna, S.: Effective and efficient entity search in RDF data. In: Aroyo, L., Welty, C., Alani, H., Taylor, J., Bernstein, A., Kagal, L., Noy, N., Blomqvist, E. (eds.) ISWC 2011. LNCS, vol. 7031, pp. 83–97. Springer, Heidelberg (2011). doi:10.1007/978-3-642-25073-6_6
2. Cheng, G., Qu, Y.: Searching linked objects with Falcons: approach, implementation and evaluation. Int. J. Semant. Web Inf. Syst. **5**(3), 49–70 (2009)
3. Ding, L., Finin, T., Joshi, A., Pan, R., Cost, R.S., Peng, Y., Reddivari, P., Doshi, V.C., Sachs, J.: Swoogle: a search and metadata engine for the semantic web. In: Proceedings of the Thirteenth ACM Conference on Information and Knowledge Management (CIKM), pp. 652–659. ACM (2004)

4. Halpin, H., Herzig, D.M., Mika, P., Blanco, R., Pound, J., Thompson, H.S., Tran, D.T.: Evaluating ad-hoc object retrieval. In: Proceedings of the International Workshop on Evaluation of Semantic Technologies (IWEST 2010), 9th International Semantic Web Conference (ISWC 2010), Shanghai, PR China, November 2010

5. Höffner, K., Walter, S., Marx, E., Usbeck, R., Lehmann, J., Ngonga Ngomo, A.C.: Survey on challenges of Question Answering in the Semantic Web. Submitted to the Semant. Web J. (2016). http://www.semantic-web-journal.net/content/survey-challenges-question-answering-semantic-web

6. Hudson, R.A.: Language Networks: The New Word Grammar. Oxford Linguistics, Oxford University Press, Oxford (2007)

7. Luhn, H.P.: A statistical approach to mechanized encoding and searching of literary information. IBM J. Res. Dev. **1**(4), 309–317 (1957)

8. Mangold, C.: A survey and classification of semantic search approaches. Int. J. Metadata Semant. Ontol. **2**(1), 23–34 (2007)

9. Marx, E., Usbeck, R., Ngonga Ngomo, A.C., Höffner, K., Lehmann, J., Auer, S.: Towards an open question answering architecture. In: SEMANTiCS (2014)

10. Oren, E., Delbru, R., Catasta, M., Cyganiak, R., Stenzhorn, H., Tummarello, G.: Sindice.com: a document-oriented lookup index for open linked data. IJMSO **3**(1), 37–52 (2008)

11. Pearsall, J., Hanks, P., Soanes, C., Stevenson, A. (eds.): Oxford Dictionary of English (Kindle Edition) (2010)

12. Reisburg, D.: Cognition: Exploring the Science of the Mind. Norton, New York (1997)

13. de Saussure, F.: Course in General Linguistics. McGraw-Hill, New York (1959). (Translated by Wade Baskin)

14. Shekarpour, S., Marx, E., Ngomo, A.C.N., Auer, S.: SINA: semantic interpretation of user queries for question answering on interlinked data. J. Web Semant. **30**, 39–51 (2015)

15. Sparck Jones, K.: A statistical interpretation of term specificity and its application in retrieval. J. Documentation **28**(1), 11–21 (1972)

16. Tummarello, G., Cyganiak, R., Catasta, M., Danielczyk, S., Delbru, R., Decker, S.: Sig.ma: live views on the web of data. J. Web Semant. **8**(4), 355–364 (2010)

17. Unger, C., Forascu, C., Lopez, V., Ngomo, A.C.N., Cabrio, E., Cimiano, P., Walter, S.: Question answering over linked data (QALD-4). In: Working Notes for CLEF 2014 Conference (2014)

18. Virgilio, R., Maccioni, A.: Distributed keyword search over RDF via MapReduce. In: Presutti, V., d'Amato, C., Gandon, F., d'Aquin, M., Staab, S., Tordai, A. (eds.) ESWC 2014. LNCS, vol. 8465, pp. 208–223. Springer, Heidelberg (2014). doi:10.1007/978-3-319-07443-6_15

19. Wang, H., Liu, Q., Penin, T., Fu, L., Zhang, L., Tran, T., Yu, Y., Pan, Y.: Semplore: a scalable IR approach to search the web of data. J. Web Semant. **7**(3), 177 (2009)

20. Zhang, L., Liu, Q.L., Zhang, J., Wang, H.F., Pan, Y., Yu, Y.: Semplore: an IR approach to scalable hybrid query of semantic web data. In: Aberer, K., et al. (eds.) ASWC/ISWC 2007. LNCS, vol. 4825, pp. 652–665. Springer, Heidelberg (2007). doi:10.1007/978-3-540-76298-0_47

Development of a Classification Server to Support Metadata Harmonization in a Long Term Preservation System

Sándor Kopácsi[✉], José Luis Preza, and Rastislav Hudak

Computer Center, University of Vienna, Vienna, Austria
{sandor.kopacsi,rastislav.hudak}@univie.ac.at,
jl@preza.org

Abstract. In this paper we will describe the implementation of a Classification Server to assist with metadata harmonization when implementing a long term Preservation System of digital objects. After a short introduction to classifications and knowledge organization we will set up the requirements of the system to be implemented. We will describe several SKOS (Simple Knowledge Organization System) management tools that we have examined, including Skosmos, the solution we have selected for our internal use. Skosmos is an open source, web-based SKOS browser based on Jena Fuseki SPARQL server. We will also discuss some crucial steps that occurred during the installation of the selected tools, and finally we will show potential problems with the classifications to be used, as well as possible solutions.

Keywords: Long term preservation · Metadata · Classification · SKOS · Skosmos · Jena Fuseki

1 Introduction

Long term preservation of digital objects is today a key issue for libraries and research institutes, because they need to ensure that the digital content of books, documents, pictures, research data, etc. remains accessible and usable within a required period of time [1]. Digital preservation includes the activities of planning, resource allocation, and application of preservation methods and technologies [2].

When we store digital objects in an archiving system, it is very important to assign well-defined metadata, to make the discoverability of the object easier. Metadata can provide the title, the authors and keywords, and other important information about a document. Metadata can also store technical details on the format and structure, the ownership and access rights information, as well as the history of the preservation activities of the digital object.

When the data provider of the digital object has to add standardized values as metadata, it can be challenging to find the appropriate keywords, a process that at times requires guessing. If we want to avoid ambiguities, misspellings, etc. it is better to select the terms from pre-defined controlled vocabularies.

© Springer International Publishing AG 2016
E. Garoufallou et al. (Eds.): MTSR 2016, CCIS 672, pp. 262–277, 2016.
DOI: 10.1007/978-3-319-49157-8_23

Controlled vocabularies, or rather classifications in multiple topics are available in several data sources, among which we can select. If we want to provide all relevant classifications to our archiving system and make them accessible to the users for adding metadata information so they can select terms during upload or search, a Classification Server that handles our relevant vocabularies and classifications seems to be the ideal solution.

In a Classification Server the information should be stored according to classification- or knowledge organization schemas, usually in the structure of Resource Description Framework (RDF) and/or as Simple Knowledge Organization System (SKOS), and should be organized as Linked Data.

The University of Vienna has developed its own solution, called Phaidra[1], for archiving digital objects, that is also in use at several other institutions. To make the services of Phaidra more comfortable and more reliable, we have developed a Classification Server from which the user can select terms by accessing controlled vocabularies and classifications.

The main goal of the described research was to collect the available methods and tools for classifications, and based on the gained knowledge to develop a standalone Classification Server.

In the next sections we will distinguish the different types of classifications, including controlled vocabularies, taxonomies, thesauri and ontologies. Following that we will introduce the Simple Knowledge Organization Systems (SKOS) that are usually based on the Resource Description Framework (RDF). After that we will define the goal of the Classification Server, and then we set up the general and technical requirements of the system. Before the implementation, we have evaluated some available tools for classifications that we have compared in the next section. By describing the implementation of the Classification Server, we will share some technical details that can be useful for the developers of similar systems. Finally we will show the available services and the usage of the Classification Server, as well as the available connections to our Preservation System.

2   State of the Art of Classification Systems

2.1   Classification

Classification is a form of categorization that collects objects or items according to their subjects usually arranged in a hierarchical tree structure. This knowledge organization technique can take many different forms, such as controlled vocabularies, taxonomies, thesauri, ontologies, and some others.

Controlled vocabulary is a closed list of words or terms that have been included explicitly, and that can be used for classification. It is *controlled* because only terms from the list may be used, and because there is control over who can add terms to the list, when and how.

---

[1] https://phaidra.univie.ac.at/.

Taxonomy is a collection of controlled vocabulary terms organized into a *hierarchical* structure by applying parent-child (broader/narrower) relationship. Each term in taxonomies is in one or more relationships (e.g. whole-part, type-instance) to other terms in the taxonomy.

Thesaurus is more structured, much richer taxonomy, that uses associative relationships (like "related term") in addition to parent-child relationships.

Ontology is a more complex type of thesaurus usually expressed in an ontology representation language that consists of a set of types, properties and relationship types. In ontology instead of simply having "related term" relationship, there are various customized relationship pairs that contain specific meaning, such as "owns" and its reciprocal "is owned by".

2.2   Knowledge Organization Systems and Linked Data

Classifications can be considered as a collection of organized knowledge, therefore the technical background of classification is based on Knowledge Organization Systems (KOS). In knowledge organization systems we usually store the knowledge in form of triplets, as object-predicate-subject, or object-attribute-value.

Classifications can be represented in Simple Knowledge Organization Systems (SKOS) [3] as a Resource Description Framework (RDF) vocabulary. Simple Knowledge Organization System is a W3C recommendation designed for representation of thesauri, classification schemes, taxonomies, subject-heading systems, or any other type of structured and controlled vocabulary.

Using RDF allows knowledge organization systems to be used in distributed, decentralized metadata applications. Decentralized metadata is becoming a typical scenario, where service providers want to add value to metadata harvested from multiple sources [4].

Each SKOS concept is defined as an RDF resource, and each concept can have RDF properties attached, which include one or more preferred terms, alternative terms or synonyms, and language specific definitions and notes. Established semantic relationships are expressed in SKOS and intended to emphasize concepts rather than terms/labels [5].

A special query language, called SPARQL, can be used to query and update data sources stored as RDF. SPARQL contains capabilities for querying required and optional graph patterns along with their conjunctions and disjunctions. SPARQL also supports extensible value testing and constraining queries by source RDF graph [6].

It is clear, that SKOS - as a modern, well established standard - can (potentially) support formal alignments and hierarchical grouping of concepts using different SKOS relations (e.g. *skos:exactMatch, skos:closeMatch, skos:narrower, skos:broader, skos:related),* translation of concept labels, and URI-based mapping to similar concepts in other KOS.

3   Application of the Classification Server

The Classification Server that we have integrated from available tools is an independent component of Phaidra, the Digital Asset Management Platform with long-term archiving functionality developed by the University of Vienna.

Phaidra is an acronym for Permanent Hosting, Archiving and Indexing of Digital Resources and Assets. Phaidra is implemented at several local Austrian institutions and also internationally, including universities in Serbia, Montenegro and Italy. Phaidra provides academic, research- and management staff the possibility to archive digital objects for an unlimited period of time, to permanently secure them, to supplement them with metadata, as well as to archive objects - and to provide world-wide access to them.

We are going to apply the Classification Server during the ingestion phase, when the user of Phaidra uploads new items to the archiving system and wants to assign metadata to it from controlled vocabularies, and also when the user searches for items supplying terms from existing classifications. We also need it for resolving the terms saved in objects when displaying them.

4   Requirements of the Classification Server

We are developing a Classification Server for Phaidra that supports classifications and controlled vocabularies. The requirements were grouped into the categories of General Requirements and Technical Requirements. At this level of development we haven't explicitly distinguished functional and non-functional requirements, but among the General Requirements and Technical Requirements we can discover constraints that are either more functional or more administrative feature of the system. Each requirements were prioritized between 1 and 3, where 1 means the highest priority (=most important), while 3 means the lowest priority (=least important).

4.1   General Requirements

The General Requirements (see Table 1) are related to the main goals of the system that we were going to achieve by the implementation of the Classification Server. Some of them (GR-1, GR-2, GR-3, GR-4) are functional requirements, but others (GR-5 and GR-7) are more likely administrative issues.

4.2   Technical Requirements

All the Technical Requirements (see Table 2) can be considered as functional requirement, and some of them (TR-1, TR-2, TR-3) are related to the input and output format of the system. TR-4 was a rather important requirement, because we definitely

Table 1. General requirements

Number	Requirement	Priority
		1: highest
	The Classification Server	3: lowest
GR-1	should resolve the URIs of the different terms	3
GR-2	should support multiple languages (the "official" languages in Phaidra are English, German, Italian and Serbian)	2
GR-3	should support multiple versions of classifications	1
GR-4	should return the list of sub terms (narrower concepts)	1
GR-5	should be Phaidra independent	1
GR-6	should have no assumptions about the contents, which means that the set of classifications can differ on instances that are locally managed	2
GR-7	does not require too much development efforts and have lower costs	2

Table 2. Technical requirements

Number	Requirement	Priority
		1: highest
	The Classification Server	3: lowest
TR-1	should return the terms in multiple formats (such as XML, JSON, RDF, TTL)	2
TR-2	it should support standard import formats for vocabularies (e.g. SKOS/RDF, TTL, N-Triples)	1
TR-3	should support Linked Data (in SKOS/RDF/XML formats)	1
TR-4	should provide a SPARQL endpoint	1
TR-5	should provide a comprehensive search needed for Phaidra	1
TR-6	should also support classifications/vocabularies that do not yet support linked data (do not have URIs)	2
TR-7	should be able to use external terminology services, e.g. dewey.info, so that we do not necessarily have to import it locally	3

wanted our system to provide a SPARQL endpoint, through which other system can access our Classification Server by using SPARQL queries[2].

5 Testing Some Available Tools for Classification

In this section we describe several relevant solutions that we have evaluated for implementing our Classification Server. We have also collected information about other tools (like HIVE, iQvoc, CATCH), but they did not fit our requirements, thus they are not described in this document.

---

[2] https://www.w3.org/TR/rdf-sparql-query/.

5.1   ThManager

ThManager[3] is an open source tool for creating and visualizing SKOS RDF vocabularies. ThManager was developed by the Advanced Information Systems Laboratory of the University of Zaragoza. It was implemented in Java using Apache Jena, and facilitates the management of thesauri and other types of controlled vocabularies such as taxonomies or classification schemes. ThManager allows for selecting and filtering the thesauri stored in the local repository. Description of thesauri by means of metadata is in compliance with a Dublin Core based application profile for thesaurus.

ThManager runs on Windows and UNIX, and only requires having a Java Virtual Machine installed on the system. The application is multilingual. The application supports out of the box Spanish and English languages, but with little effort other languages can be implemented.

The main features include the visualization of thesaurus concepts (alphabetically, in hierarchical structure, properties of selected concepts), ability to search concepts ("equals", "starts with" and "contains"), editing thesaurus content (creation of concepts, deletion of concepts, and update of concept properties), export of thesauri (including thesaurus metadata) in SKOS format.

Available vocabularies in ThManager include AGROVOC, DCType, GEMET, ISO639, and UNESCO.

Unfortunately, the latest version of ThManager was launched in 2006, and we cannot expect any updates. Another drawback of ThManager is that it does not provide a SPARQL endpoint for accessing the managed vocabularies.

5.2   TemaTres

TemaTres[4] is an open source vocabulary server developed in Argentina. It includes a web application to manage and exploit vocabularies, thesauri, taxonomies and other formal representations of knowledge stored in a MySQL database, and provides the created thesauri in SKOS format. TemaTres requires PHP, MySQL and a HTTP Web server.

TemaTres provides a SPARQL endpoint. Exporting and publishing controlled vocabularies is possible in many metadata schemas (SKOS-Core, Dublin Core, MADS, JSON, etc.). It can import data in SKOS-Core format and has a utility to import thesauri from tabulated text files.

It has an advanced search with search terms suggestions, and a systematic or alphabetic navigation. TemaTres has a special vocabulary harmonization feature where it can find equivalent, no equivalent, and partial terms against other vocabularies.

It supports multilingual thesaurus, multilingual terminology mapping, and includes a multilingual interface. It exposes vocabularies with powerful web services. TemaTres displays terms in multiple deep levels in the same screen. It also provides quality

---

[3] http://thmanager.sourceforge.net/.

[4] http://www.vocabularyserver.com/.

assurance functions (duplicates and free terms, illegal relations). The main drawback of TemaTres is that not all documentation is available in English.

5.3 SKOS Shuttle

SKOS Shuttle[5] is a multi-user/multi-tenant online Thesaurus Service developed by Semweb LLC (Switzerland). It supports building, maintaining and operating of SKOS thesauri. SKOS Shuttle allows operating on an internal own RDF repository and on any external SESAME compliant RDF repositories and it easily allows direct editing of RDF statements (triples) without restrictions.

The user interface is intuitive. SKOS Shuttle also integrates a full REST API to create, manage and navigate thesauri. It accesses securely all information through SSL transported authentication. It provides industrial security (Rights, Groups, User and Project Management) and a smart "Orphan Concept Analysis" together with an assistant for direct concept "deorphanization" without using one single line of SPARQL code.

With its "systematics assistant", several base URI's can be used inside one single thesaurus. RDF Import/Export and whole RDF snapshots are possible in 6 different formats (N3, N-Triples, TRIG, Turtle, NQuads, RDF/XML).

SKOS Shuttles allows downloading or uploading a full RDF snapshot preserving versioning of each thesaurus. SKOS language tags can be added or removed "on the fly" while editing the thesaurus, speeding up maintenance tasks. SKOS Shuttle allows quick filtering on any thesaurus language, and also during concepts navigation, this permits to find out missing labels during navigation.

The SKOS Shuttle REST API provides a full range of selections/commands that are embeddable into any application using three output formats: JSON, XML and YAML. The API access requires the same secured authentication as the application to provide online services.

SKOS Shuttle seems to be a very promising tool. SKOS Shuttle is available as a service and is already being used by several Universities. SKOS Shuttle is not an open source product. Pricing is not yet known but SKOS Shuttle will be provided as a commercial service for small thesauri and as a free service for universities (up to a larger extent).

5.4 PoolParty

PoolParty[6] is a commercial semantic technology suite, developed by Semantic Web Company that offers solutions to knowledge organization and content business problems.

---

[5] https://skosshuttle.ch/.

[6] https://www.poolparty.biz/.

The PoolParty Taxonomy & Thesaurus Manager is a powerful tool to build and maintain information architectures. The PoolParty thesaurus manager enables practitioners to start their work with limited training. Subject matter experts can model their fields of expertise without IT support.

PoolParty taxonomy management software applies SKOS knowledge graphs. With PoolParty, the import of existing taxonomies and thesauri (e.g. from Excel) and export them in different standard formats are possible. In addition to basic SKOS querying, the API also supports the import of RDF data, SPARQL update and a service to push candidate terms into a thesaurus.

5.5 Protégé

Protégé[7] is a free, open-source ontology editor and framework for building intelligent systems. Protégé was developed by the Stanford Center for Biomedical Informatics Research at the Stanford University School of Medicine. Protégé is supported by a strong community of academic, government, and corporate users, who use Protégé to build knowledge-based solutions in areas as diverse as biomedicine, e-commerce, and organizational modelling.

With the web-based ontology development environment of Protégé, called Web-Protégé, it is easy to create, upload, modify, and share ontologies for collaborative viewing and editing. The highly configurable user interface provides suitable environment for beginners and experts. Collaboration features abound, including sharing and permissions, threaded notes and discussions, watches and e-mail notifications. RDF/XML, Turtle, OWL/XML, OBO, and other formats are available for ontology upload and download.

Although it is a very good tool, it is too complex for editing and visualizing such a simple model as SKOS, and provides too many options not specifically adapted for the type of relationships used in SKOS.

5.6 Skosmos with Jena Fuseki

Skosmos[8], developed by the National Library of Finland, is an open source web application for browsing controlled vocabularies. Skosmos was built on the basis of prior development (ONKI, ONKI Light) for developing vocabulary publishing tools in the FinnONTO (2003–2012) research initiative from the Semantic Computing Research Group.

Skosmos is a web-based tool for accessing controlled vocabularies used by indexers describing documents, and by users searching for suitable keywords. Vocabularies are accessed via SPARQL endpoints containing SKOS vocabularies.

---

[7] http://protege.stanford.edu/.

[8] http://skosmos.org/.

Skosmos provides a multilingual user interface for browsing vocabularies. The languages currently supported in the user interface are English, Finnish, German, Norwegian, and Swedish. However, vocabularies in any language can be searched, browsed and visualized, as long as proper language tags for labels and documentation properties have been provided in the data.

Skosmos provides an easy to use REST API for read only access to the vocabulary data. The return format is mostly JSON-LD, but some methods return RDF/XML, Turtle, RDF/JSON with the appropriate MIME type. These methods can be used to publish the vocabulary data as Linked Data. The API can also be used to integrate vocabularies into third party software. For example, the `search` method can be used to provide autocomplete support and the `lookup` method can be used to convert term references to concept URIs [7].

The developers of Skosmos recommend using the Jena Fuseki[9] triple store with the Jena text index for large vocabularies. In addition to using a text index, caching of requests to the SPARQL endpoint with a standard HTTP proxy cache such as Varnish can be used to achieve better performance for repeated queries, such as those used to generate index view.

5.7   Overall Evaluation and Tool Selection

All of the evaluated tools had advantages and disadvantages, but the most important selection criteria for us were to find an open source tool that can provide a SPARQL Endpoint. A comparison of the evaluated tools can be seen in Table 3, where we have included the open source products only. Other important selection criteria was to find tool that is based on the stable and widespread Apache Jena technology and which can be accessed via REST API[10].

For the above mentioned selection criteria, Skosmos with Jena Fuseki seemed to be the best solution; therefore we have selected it for implementing our Classification Server.

Table 3.   Comparison of the evaluated tools

	Implemented in	Input	Multilingual	Backend	SPARQL Endpoint	REST API	Last update
ThManager	Apcahe Jena	SKOS RDF	yes	SPARQL	available	N/A	2006
TemaTres	N/A	SKOS/tabulated text	yes	MySQL	available	available	2016
Protégé	Java Swing	OWL, Excel, CSV	yes	SPARQL	available	N/A	2016
Skosmos	Apcahe Jena	SKOS Core	yes	SPARQL	available	available	2016

---

[9] https://jena.apache.org/documentation/serving_data/.

[10] Representational State Transfer that relies on a stateless, client-server, cacheable HTTP communications protocol.

6   Implementation of the Classification Server

The classification server has been implemented using Skosmos as a frontend for handling SKOS vocabularies, and Jena Fuseki as a SPARQL RDF store containing SKOS vocabulary data (see Fig. 1). The input of the system and the possible connections to the users or directly to Phaidra will be discussed in the sections below.

Alternatively, instead of Fuseki, we could use other SPARQL 1.1 compliant RDF stores, but the performance of other tools did not seem to be sufficient with large vocabularies since there is no text index support in generic SPARQL 1.1.

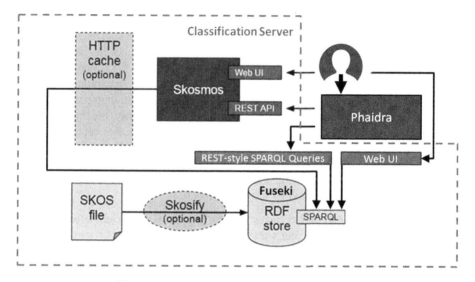

Fig. 1. System Architecture (Original source: [8])

6.1   Installation of Skosmos and Jena Fuseki

Skosmos and Fuseki require Apache and PHP running on the server. We have installed them on a Windows 7 environment (Professional 64 bit, Service Pack 1, Intel Core i7-56000 CPU, 2.6 GHz, 16 GB RAM) using Java 1.8 (jre1.8.0_40), with XAMPP (xampp-win32-1-8-3-4-VC11), as well as on a CENTOS 6.5 and on Ubuntu 16.04 virtual machine (Intel Xeon CPU E5-2670 0 @ 2.60 GHz).

A detailed installation guide can be found on GitHub[11] for the Linux version, but there are some deviations on the Windows version, as well as there are some important issues that are worth highlighting.

Jena Fuseki is a SPARQL server and triple store, which is the recommended backend for Skosmos. The Jena text extension can be used for faster text search. Simply download the latest Fuseki distribution and unpack the downloaded file to the intended folder of Fuseki.

---

[11] https://github.com/NatLibFi/Skosmos/wiki/Installation.

6.2 Configuration of Skosmos and Fuseki

Configuration of Skosmos. Skosmos can be configured basically in two files, `config.inc` for setting some general parameters, and `vocabularies.ttl` to configure the vocabularies shown in Skosmos.

In `config.inc` one can set the name of the vocabularies file, change the timeout settings, set interface languages, set the default SPARQL endpoint, and set the SPARQL dialect if Jena text index is needed.

Vocabularies are managed in the RDF store accessed by Skosmos via SPARQL. The available vocabularies are configured in the `vocabularies.ttl` file that is an RDF file in Turtle syntax.

Each vocabulary is expressed as a `skosmos:Vocabulary` instance (subclass of `void:Dataset`). The local name of the instance determines the vocabulary identifier used within Skosmos (e.g. as part of URLs). The vocabulary instance has the following properties: title of vocabulary (in different languages), the URI namespace for vocabulary objects, language(s) and the default language that the vocabulary supports, URI of the SPARQL endpoint containing the vocabulary, and the name of the graph within the SPARQL endpoint containing the data of the individual vocabulary.

In addition to vocabularies, the `vocabularies.ttl` file also contains a classification for the vocabularies expressed as SKOS. The categorization is used to group the vocabularies shown in the front page of Skosmos. You can also set the content of the About page in `about.inc`, and add additional boxes to the left and to the right of the front page in `left.inc` and in `right.inc`.

Configuration of Fuseki. Fuseki stores data in files. It is also possible to configure Fuseki for in-memory use only, but with a large dataset, this requires a lot of memory. The in-memory use of Fuseki is usually faster.

The Jena text enabled configuration file specifies the directories where Fuseki stores its data. The default locations are `/tmp/tdb` and `/tmp/lucene`. To flush the data from Fuseki, it is simply required to clear or to remove these directories.

The Jena text extension can be used for faster text search, and Skosmos needs to have a text index to work with vocabularies of medium to large size. The limit is a few thousand concepts, depending on the performance of the SPARQL endpoint and on how much latency is acceptable to the users.

If Fuseki is started in the TDB with `./fuseki-server –config config.ttl` it will run using text index. To use Fuseki in TDB, the TDB location for Jena text index should be set, and the Lucene text directory in `config.ttl`. If Fuseki is started in the memory with `./fuseki-server –update –mem /ds`, then there is no text indexing by default.

It is also possible to use in-memory TDB and text index, but it requires a Fuseki configuration file (`config.ttl`) with special "file names" that are actually in-memory (for TDB: `tdb:location` "–mem–"; and for Jena text: `text:directory` "mem";).

Timeout settings. If there is more data than Skosmos is able to handle, some queries can take very long time. The slow queries are usually the statistical queries (number of concepts per type, number of labels per language) as well as the alphabetical index.

Short execution timeout for PHP scripts can trigger Runtime IO Exception. To change the timeout values, check PHP and Apache's time out settings (e.g. in `php.ini` the `max_execution_time`). It is highly recommended to find this setting and change it to a higher value (say to 5 or 10 min).

Skosmos also has a `HTTP_TIMEOUT` setting in `config.inc`, that should only be used for external URI requests, not for regular SPARQL queries, but there might be unknown side-effects. The EasyRdf HTTP client has a default timeout of 10 s. It is also recommended to change this value.

It is also suggested to change the timeout value of the browsers – if possible – from where it is planned to access Skosmos.

6.3   Getting and Setting Vocabularies

The basic usage of our Classification Server is to store the classifications locally (if its access time is acceptable), and we also provide the links to the remote SPARQL endpoints of the classifications, if they are available.

If certain vocabularies are planned to use locally, they have to be in SKOS format, and should be uploaded to the local SPARQL server, that is to Jena Fuseki.

Downloading and converting vocabularies. Vocabularies can be downloaded from the original dataset provider (e.g. from Getty, COAR, Statistics Austria, etc.), or in case of a small dataset, they can be created manually. The vocabularies need to be expressed using SKOS Core representation in order to publish it via Skosmos directly, but SKOS-XL representation or even files in Excel can be also easily converted to SKOS Core. For the SKOS-XL to SKOS Core conversion we have used the owlart converter[12]. SKOS-XL labels can be converted to SKOS Core labels by executing SPARQL Update queries, as well. If the classification is available in Excel or CVS, then VBA macros can convert it to SKOS Core structures. The format of the file that is accepted by Fuseki can be `rdf/xml` (`.rdf` or `.xml`), turtle (`.ttl`) or N-Triples (`.nt`).

When SKOS files are coming from external resources or they have been converted from other formats, it is recommended to pre-process the vocabularies using a SKOS proofing tool, like Skosify[13]. This ensures, e.g., that the broader/narrower relations work in both directions, and that related relationships are symmetric. Skosify reports and tries to correct a lot of potential problems in SKOS vocabularies. It can also be used to convert non-SKOS RDF data into SKOS. Online version of the Skosify tool is also available, where the default options can be used after selecting the vocabulary to be checked.

Uploading files to Fuseki. If Skosmos is used for accessing classifications in the local SPARQL triple store, then the datasets have to be uploaded to Fuseki. First, it has to be considered if Fuseki will run either in memory or in a predefined folder, usually called TDB. If Fuseki runs in the memory, then all uploads and updates (if it is allowed) will

---

[12] https://bitbucket.org/art-uniroma2/owlart/downloads.

[13] https://code.google.com/p/skosify/.

be temporary. If Fuseki runs in the TDB, then the uploads and updates will remain there even if we exit from Fuseki and restart it.

In a SPARQL triple store there is always a default (unnamed) graph, and there can also be multiple named graphs. In other words, there is only one default graph (with no name), but there can be any number of named graphs in a SPARQL endpoint/dataset. The URI namespaces can be used as graph names. E.g. http://vocab.getty.edu/tgn/ would store Getty's TGN data.

The datasets can be uploaded to Fuseki online, when Fuseki is running, or offline, when Fuseki is not running. To upload the dataset online the control panel of the web interface of Fuseki or command line instructions can be used. For offline upload the datasets can be directly loaded to the TDB.

When uploading datasets online to Fuseki through its control panel, the Graph can be set to "default" or a graph name should be provided. If a graph name is used, it should be the same graph name of its dataset in `skosmos:sparqlGraph` (e.g. http://vocab.getty.edu/tgn/) in `vocabularies.ttl`.

The Fuseki file upload handling is not very good at processing large files. It loads the dataset first into memory, only then to the on-disk TDB database (and also the Lucene/Jena text index). It can run out of memory on the first step ("Out-OfMemoryError: java heap space" is a typical error message when this happens). If we give several GB memory to Fuseki (for example Setting JVM heap to 8 GB: `export JVM_ARGS=-Xmx8000M`) it should be able to upload large (several hundreds of MB) files, though it might take a while and it is recommended to restart Fuseki afterwards to free some memory.

6.4   Some Examples and Problems of Adding Individual Vocabularies

In this section we are going to describe some examples for individual vocabularies that we are using in our Classification Server, that show typical problems and solutions.

Getty vocabularies[14] contain structured terminology for art and other cultural, archival and bibliographic materials. They provide authoritative information for cataloguers and researchers, and can be used to enhance access to databases and web sites.

Getty has its own SPARQL endpoint, but it is not responding in the right way to our Classification Server. There seems to be some incompatibility between Skosmos (in practice, the EasyRdf library which is used to perform SPARQL queries) and the Getty SPARQL endpoint.

Even if we could access the Getty SPARQL endpoint, it would most likely be extremely slow to use it with Skosmos, since it doesn't have a text index that Skosmos could use. The lack of text index prevents any actual use of Skosmos with the Getty endpoint.

Therefore, we have tried to upload Getty vocabularies to our own local Fuseki SPARQL endpoint with the Jena text index. But unfortunately Getty vocabularies do not work well in Skosmos due to their very large size.

---

[14] http://vocab.getty.edu/.

There are two sets of each Getty vocabulary, the "explicit" set and the "full" set (Total Exports). With the "explicit" set, which is smaller, we had to configure Fuseki to use inference so that the data store can infer the missing triples. With the full set this is not needed, but the data set is much larger so we had difficulties loading it. We could finally upload the full set of Getty's vocabularies using the `tdbloader` utility of Jena Fuseki.

The downloaded export file of the full set includes all statements (explicit and inferred) of all independent entities. It's a concatenation of the Per-Entity Exports in N-Triples format. Because it includes all required Inference, it can be loaded to any repository (even one without RDFS reasoning).

We had to download the External Ontologies (SKOS, SKOS-XL, ISO 25964), from http://vocab.getty.edu/doc/#External_Ontologies to get descriptions of properties, associative relations, etc. We have downloaded the GVP Ontology from http://vocab.getty.edu/ontology.rdf.

Finally we have loaded the full.zip export files (aat, tgn and ulan) form http://vocab.getty.edu/dataset/.

In this way we have made some Getty vocabularies available in our Classification Server, but due to their huge size, they are rather slow.

COAR Resource Type Vocabulary[15] defines concepts to identify the genre of a resource. Such resources, like publications, research data, audio and video objects, are typically deposited in institutional and thematic repositories or published in journals.

The main problem with COAR is that it only represents labels using SKOS XL properties. Skosmos doesn't support SKOS XL currently. Unfortunately, the remote endpoint of COAR[16] cannot be used either, because the COAR endpoint data currently is not SKOS Core, but SKOS-XL. Since we wanted to use COAR data in our Classification Server, we have converted to SKOS Core labels using *owlart* (see Downloading and converting vocabularies).

ÖFOS[17] is the Austrian version of the Field of Science and Technology Classification (FOS 2007), maintained by Statistics Austria. The Austrian classification scheme for branches of science (1-character and 2-character) is a further development modified for Austrian data.

ÖFOS can be downloaded in PDF and CSV format, but neither in SKOS structure (in xml/rdf, turtle or N-Triples), nor Linked Open Data through a SPARQL Endpoint is available.

Since we have received it directly from Statistics Austria in Excel format, the simplest way of converting it to SKOS was using VBA macros. These macros simply read the content of the Excel file, extend them with the appropriate RDF and SKOS labels, and write it to the desired xml/rdf or turtle format.

---

[15] https://www.coar-repositories.org/news-media/release-of-coar-resource-type-vocabulary-for-open-access-repositories/.

[16] http://vocabularies.coar-repositories.org/sparql/repositories/coar.

[17] http://unstats.un.org/unsd/cr/ctryreg/ctrydetail.asp?id=1017.

7   Available Services and Usage of the Classification Server

Currently from our Classification Server four general on-line classifications from external triple stores (AGROVOC, Eurovoc, STW, UNESCO), some other general local classifications (e.g. Getty, GND, ÖFOS, COAR Resource Type Vocabulary, etc.) and two local, Phaidra specific classifications are available. The local classifications have been uploaded to our local triple store in order to make them accessible from Skosmos.

The operation of the Classification Server is quite simple: from the opening page of Skosmos (See Fig. 2) we simply have to click on one of the classifications, and then the selected classification will be opened. We can see its vocabulary information, and we can select basically between alphabetical and hierarchical view. Depending on the configuration we can see the change history of the vocabulary or the group of concepts. We can also search for specific contents directly in our entire triple store server, or simply in the selected classification.

Fig. 2. Opening page of the Classification Server

8   Connecting Phaidra to the Classification Server

Phaidra requires the Classification Server when the user ingests new items and wants to add metadata to this from a controlled vocabulary. Another scenario when the user searches for some documents classified with some metadata from a controlled vocabulary and wants to display or resolve them.

The connection between Phaidra and the Classification Server has been realized using the REST API of Skosmos and/or with the REST-style SPARQL Queries of Jena Fuseki (See Fig. 1). These are read-only interfaces over HTTP to the data stored in the Classification Server, where requests can be built in the URL. The returned data is in UTF-8 encoded JSON-LD format.

Skosmos provides a REST-style API and Linked Data access to the underlying vocabulary data. The REST URLs must begin with the `/rest/v1` prefix. Most of the methods return the data as UTF-8 encoded JSON-LD, served using the `application/json` MIME type. The data consists of a single JSON object which includes JSON-LD context information (in the `@context` field) and one or more fields which contain the actual data.

Jena Fuseki provides REST-style SPARQL HTTP Update, SPARQL Query, and SPARQL Update using the SPARQL protocol over HTTP. Fuseki implements W3C's SPARQL 1.1 Query, Update, Protocol and Graph Store HTTP Protocol.

9 Conclusions

In the described research we have successfully completed our research objectives that were to collect some available methods and tools for classification, with which we could implement a Classification Server. The selected tools (i.e. Skosmos with Jena Fuseki) seemed to be a good choice, however we had some difficulties during the implementation, and with the access and upload of certain classifications.

The Classification Server has fulfilled the general and technical requirements according to their priorities that we have set up. The current stable version contains at the moment 14 internal and 4 external classifications.

The direct usage of the Classification Server from Phaidra is under development.

Acknowledgements. We would like to express our special thanks to Osma Suominen, the main developer of Skosmos at the National Library of Finland, who were very helpful by answering any Skosmos or Jena Fuseki related question.

References

1. Digital Preservation Coalition: Introduction: Definitions and Concepts. Digital Preservation Handbook. York (2008)
2. Day, M.: The Long-Term Preservation of Web Content. Web Archiving, pp. 177–199. Springer, Heidelberg (2006)
3. W3C Working Group: SKOS Simple Knowledge Organization System Primer, 18 August 2009. https://www.w3.org/TR/skos-primer/
4. W3C Semantic Web: Introduction to SKOS. https://www.w3.org/2004/02/skos/intro
5. Zeng, M.L., Chan, L.M.: Semantic interoperability. In: Encyclopedia of Library and Information Sciences, 4th edn., p. 8 (2015)
6. DuCharme, B.: Learning SPARQL: Querying and Updating with SPARQL 1.1., 2nd edn. O'Reilly Media, Sebastopol (2013)
7. Suominen, O., Ylikotila, H., Pessala, S., Lappalainen, M., Frosterus, M., Tuominen, J., Baker, T., Caracciolo, C., Retterath, A.: Publishing SKOS vocabularies with Skosmos. Manuscript submitted for review (2015)
8. Suominen, O.: Publishing SKOS concept schemes with Skosmos. AIMS Webinar, 6 April 2016, Slide 25 (2016)

Enriching Scientific Publications from LOD Repositories Through Word Embeddings Approach

Arben Hajra[1(✉)] and Klaus Tochtermann[2]

[1] South East European University (SEEU), Tetovo/Skopje, Republic of Macedonia
a.hajra@seeu.edu.mk
[2] Leibniz Information Centre for Economics (ZBW), Kiel/Hamburg, Germany
k.tochtermann@zbw.eu

Abstract. The era of digitalization is increasingly emphasizing the role of Digital Libraries (DL), by increasing requirements and expectations of services provided by them. The interoperability among repositories and other resources continues to be a subject of research in the field. Retrieving publications related to a particular topic from different DLs, especially from diverse domains, require several clicks and online visits of many different points of access. However, achieving interoperability by cross-linking publications, authors and other related data would facilitate the scholarly communication in general. Starting from a single point, a scholar would be able to find resources i.e., publications and authors, previously enriched with several other information from different repositories. Repositories available as semantic web content, such as bibliographic Linked Open Data (LOD) datasets are the focus of this study. Primarily, we consider existing alignments among concepts between repositories. Improvements regarding the semantic measurements of relatedness of different resources are possible by the application of text-mining techniques. The paper introduces preliminary experiments conducted by vector space models through the application of TF-IDF and Cosine Similarity (CS). Additionally, the paper discusses experiments of applying a word embedding approach, with which we are focusing mainly on the context by distributed word representations, instead of word frequency, weighting and string matching. We apply the contemporary Word2Vec model as a similar deep learning approach to model semantic word representations.

Keywords: Digital Libraries · Linked Open Data · Semantic web · Word embeddings · Data mining · Recommended systems

1 Introduction

Traditionally, libraries provide the basic information infrastructures for scholarly communication. The era of digitalization emphasized their role in this process, but at the same time, requirements and expectations of services provided by them increased. In this situation, Digital Libraries (DL) successfully managed to adapt to these challenges by improving the utilization of resources [14]. Nonetheless, there is still a gap

© Springer International Publishing AG 2016
E. Garoufallou et al. (Eds.): MTSR 2016, CCIS 672, pp. 278–290, 2016.
DOI: 10.1007/978-3-319-49157-8_24

between the demand and offer. Interoperability among resources continues to be the subject in this field even today [15–18].

Often, scientific digital libraries are specialized in specific domains such as: economics, social sciences, computer sciences, agronomics, etc. Retrieving similar publications within the same DL is a common practice in most of DLs. However, recommending semantically similar publications from two or more different repositories is still an open field of research. Today, retrieving publications related to a particular topic, from different DLs and especially from different domains, is still very heuristic, and often require step-wise or as far as possible simultaneous navigations through the affected DLs. The current practice of Google Scholar gives an idea for such recommendations, however there are much more resources which are not made visible by services like this. To this end, achieving the interoperability among DLs by cross linking publications, authors and other related data would facilitate the scholarly communication in general. The idea is as follows: Starting from a single point of access, a scholar would be able to find resources i.e., publications and authors, previously enriched with several other information from different repositories. And when a scholar fetches a publication in a DL, the system will offer the scholar a list of semantically related publications from other repositories, an extended list of co-authors, and other related data corresponding to that publication.

Repositories available as semantic web content, such as bibliographic Linked Open Data (LOD) repositories [15, 29], are in the focus of this study. Primarily, we consider the existing alignments among concepts between repositories, by exploring best practices for consuming them. After that, we investigate the role of thesauri, including descriptors with the corresponding narrowed, broadened and extended concepts through Simple Knowledge Organization System Reference - SKOS[1] vocabulary. Improvements regarding the semantic measurements between resources are achieved by evaluating several text-mining techniques. In this study, we present preliminary experiments conducted by vector space models through the application of TF-IDF and Cosine Similarity (CS). Additionally, we extend the experiments by applying a word embedding approach, in which we are focusing mainly on the context by distributed word representations, instead of words frequency, weighting and string matching. The contemporary Word2Vec[2] model is applied as a similar deep learning approach to model semantic word representations.

The main intention of our work is to find a novel and automatic approach for cross-linking scientific publications from different repositories. In our view, the implementation of deep learning approach for language processing is proposed as the most comprehensive approach for this purpose. To this end, we show how we can automatically determine the semantic similarity between publications, even if only a small set of metadata is available.

---

[1] https://www.w3.org/TR/swbp-skos-core-spec.
[2] https://code.google.com/p/word2vec/.

2   Motivation and Problem Statement

Recommender systems are applied in several fields, therefore it is inevitable to explore their application in scholarly communication, particularly in digital libraries [11–13]. However, the common implementation of recommending systems in DLs is mainly a practice used within the same repository. Recommending and interlinking publications by cross-linking relevant information from several repositories still remains a challenge [19, 20]. At the moment, repositories are considered as isolated silos, which make it difficult to process matching similar resources by using the same query string in different repositories. Cross-linking resources, i.e., scientific publications with assured degree of semantic similarity, certainly presents a complex process of lexical or string matching, mostly due the diversity of ontologies and metadata vocabularies used for describing resources.

3   Proposed Approach and Related Work

Recommender systems for scientific publications are generally grounded on content analysis, user profiles and collaborative filtering with incontestable role of social data [21–24]. However, in this work we are following a different strategy for initiating and retrieving the list of recommended relevant resources. In essence, the user triggers the search and selects a paper from a DL that best fits his or her requirements. In a next step, the selected publication is enriched with closely related publications, authors and similar information found in other repositories.

The interoperability is initiated from one repository by considering all existing metadata for a single publication, such as: title, authors, abstract and keywords. Using this information, we are connecting to other external repositories to search for possible semantically related publications and other related information (e.g. author details) to the initial publication (Fig. 1).

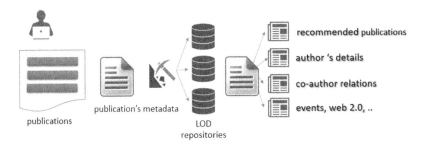

publications   publication's metadata   LOD
                                                  repositories

recommended publications

author 's details

co-author relations

events, web 2.0, ..

Fig. 1.   Enriching a scientific publication with information from other repositories.

In order to achieve this, we leverage already available contents on the semantic web, such as Linked Open Data (LOD) repositories, as one of the most promising data sources [30]. As such, the existing alignments among concepts between repositories are considered with the corresponding narrowed, broadened and extended concepts through the

SKOS vocabulary. At the same time, the deployment of several data mining techniques is crucial in this process [21]. In our work we apply two approaches the vector space model and word embedding approach.

This work was evaluated with the content of the EconStor[3] repository, which is a leading Open Access repository in Germany. Through EconStor, the German National Library of Economics - Leibniz Information Centre for Economics (ZBW) offers a platform for Open Access publishing to researchers in economics. ZBW also maintains the Standard Thesaurus Wirtschaft (STW)[4], which is the Thesaurus for Economics used for description and indexing purposes.

3.1 Aligned Concept Between Repositories and Thesauruses

Most of LOD repositories as part of LOD cloud[5], offer a number of incoming/ongoing links to other datasets for mapping several resources or concepts that have the same meaning. EconStor, through the STW thesaurus, has numerous mappings to other thesauri and vocabularies. For instance, for Agrovoc (Multilingual Agricultural Thesaurus)[6] 1,027 *skos:exactMatch* alignments exist, while for TheSoz[7] (Thesaurus Social Sciences) 3,022 *skos:exactMatch* and 1,397 *skos:narrowMatch* are available. According to this, the initial experiments are done between EconStor and OpenAgris[8] based on structural similarity between these two repositories. Both of them offer an open catalog as part of LOD cloud with available SPARQL endpoints and RDF dump files, as well as thesauri on both sides, STW and Agrovoc respectively.

Based on our previous evaluation conducted using 112 publications, the list of retrieved publications according to the aligned concepts between repositories was extremely wide [8]. For example, in order to deliver more details, the concept "biofuel" from EconStor is aligned to Agrovoc as "biofuels", and is used for describing 7083 documents in OpenAgris catalog. By including all the existing aligned concepts describing a paper, the list will be even broader. Hierarchical navigation between concepts with the use of knowledge organization systems by broadening and narrowing the concepts, e.g., the notion of Germany broadened to Europe and narrowed to Berlin, helps to reduce complexity by narrowing down the number of results. However, the outcome is not satisfactory for measuring similarity among publications and offering a shorter list of recommended publications (Fig. 2).

Therefore, we also use alignments between repositories/thesauruses for retrieving an initial set of publications, especially for reformulating a search query from one

---

[3] http://www.econstor.eu/.

[4] http://zbw.eu/stw.

[5] http://linkeddata.org/.

[6] http://aims.fao.org/agrovoc.

[7] http://www.gesis.org/en/services/research/thesauri-und-klassifikationen/social-science-thesaurus/.

[8] http://aims.fao.org/openagris.

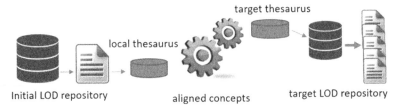

Fig. 2. Retrieving scientific publications from LOD repositories based on concepts' alignments

vocabulary to another [8]. The presence of thesauri in the primary and targeting repository can be useful for extending the corpus of metadata concepts, which, as we will show later, is very significant for further analyses.

3.2 Publications Metadata and Vector Space Model

In such a situation, the involvement of other metadata, such as title, abstract and keywords is mandatory. By including these elements in the implementation of data mining approaches among the set of metadata and thesauri concepts, the similarity between publications is calculated and used for ordering purposes.

We use the vector space model, in which we weight each concept from the selected metadata by applying the TF-IDF algorithm. The similarity among publications, i.e., vectors of concepts, is measured as deviation of angles between each document vector, by using the CS. Thus, iteratively we measure the similarity between metadata of our initial publication with the metadata of publications from the target repository.

Fig. 3. Combination of metadata components from a scientific paper for retrieving recommended publications from other repositories.

For this purpose, we conducted heuristic evaluations when analyzing the impact of each element. As shown in Fig. 3, the developed prototype makes it possible to adjust the relevance of each metadata set by weighting the title, abstract, keywords, and considering all the aligned concepts (including narrowed, broadened and related terms). The example in Fig. 3 shows that for the selected publication with current adjustment of the metadata, the words "food" and "price" become crucial. This results in retrieving also publications semantically distant with the initial publication, which are related to *food* and *agriculture* rather than *economy*.

The combination among the metadata is crucial for determining the semantic relativeness among the initial and retrieved publications. Different combinations among these parameters would result in different list of retrieved publications from the targeted repository. The impact can also be seen in the generated results. In this case, the first ranked publication is semantically very near to the initial publication.

Concerning this, in our previous work we have achieved very significant results by enriching author profiles with additional information from different digital libraries [25]. In another study [8], considering different cases, different combinations of these metadata also led to good results. Based on the evaluations done with 112 publications, the count-based approach with TF-IDF and Cosine Similarity [8], repeatedly shows that irrelevant terms are highly ranked. This results in compromising outcomes, i.e., recommending semantically distant publications to a particular publication. Therefore, the right combination of metadata terms for this purpose is very experimental. The above mentioned data mining techniques, TF-IDF and CS do not offer much to achieve a completely automated process [27].

4   Deep Learning Approach

Determining the semantic similarity between two texts represent a complex and challenging process. In general, there are several approaches introduced based on lexical matching, handcrafted patterns, term-weighting and syntactic parse trees [9, 10]. Indeed, lexical features, like string matching and frequency of words in a text, do not capture semantic similarity in a satisfied level [9, 27]. Hence, the deep learning approach for language processing based on neural network language models outperforms traditional count-based distributing models on word similarity [2]. Current trends for determining word similarities, i.e., semantic similarities among texts, rely on vector representations of words by using neural networks, known as *word embeddings* or word representations [1–7, 9, 26–28].

4.1   Word Embeddings

In deep learning, word embeddings currently represent the most outstanding field. It is the main discussed subject in almost every publication regarding the semantic representation of words in a low-dimensional vector [1–7, 9, 26–28]. Their presence is evident in many areas, such as in Natural Language Processing (NLP), Information Retrieval (IR) and generating search query strings. Word embeddings insert the complete vocabulary into a low-dimensional linear space. The embedded word-vectors are trained over large collections of text corpuses through neural networking models. Thus, words are embedded in a continuous vector space where semantically similar words are mapped to vectors. Learning the word embeddings is totally unsupervised method computed on a predefined text corpus.

Word embeddings currently have two well-known models of implementation: the Word2Vec algorithms proposed by Mikalov et al. for Google [7] and GloVe model from Pennington et al. at Stanford [28]. Our experiments and evaluations are based on Word2vec due to the performance and computational cost.

Word2Vec Embeddings. As noted before, Word2Vec is a novel word embeddings approach, which learns a vector representation for each word using neural network language model [7]. Two implementations of Word2Vec can be found, continuous bag-of-words (CBOW) and Skip-gram. CBOW predicts a word from the context of input text (surrounding words), while Skip-gram predicts the input words from the target context (surrounding words are predicted from one input word).

Word2Vec uses the hierarchical softmax training algorithm, which best fits for infrequent words while negative sampling better for frequent words and low dimensional vectors. Based on the previous analyses in [7, 26, 27], the skip-gram model with the use of hierarchical softmax algorithm is particularly efficient regarding the computational cost and performance. CBOW is recommended as more suitable for larger datasets. As such, the model can be trained on conventional personal machines with billions of words, achieving the ability to learn complex word relationships [7, 26].

Currently, there are several implementations of Word2Vec in different environments. The native proposed code is optimized in the C programming language. However, Deeplearning4j[9] implements a distributed form of Word2Vec for Java and Scala, while Gensim[10] and TensorFlow[11] offer a python implementation of Word2Vec.

4.2 Training and Building the Model

The experiments in this work are based on the Gensim package, which is a python implementation of Word2Vec model. Gensim provides very significant optimization regarding the computational speed, which overpasses even the native C implementation. Currently, there are several pre-trained models on different datasets, such as Google News, DBpedia and Freebase. However, considering the specificity of the domain, we prefer to train our own word vectors for deploying the experiments.

The model is trained in a text corpus for generating a set of vectors, which are word representations of words in that corpus. Thus, through a SPARQL query we retrieve all the titles, abstracts and keywords of 37,917 publications from EconStor. Since Gensim's Word2Vec expects a sequence of sentences as input, several preprocessing steps are performed at the corpus, such as conversion to utf8 unicode, lowercasing, removing

---

[9] http://deeplearning4j.org/word2vec.html.
[10] https://radimrehurek.com/gensim/models/word2vec.html.
[11] https://tensorflow.org/versions/r0.8/tutorials/word2vec/index.html.

numbers and punctuations. Finally, the model is trained in corpus of 12,329,307 raw words and 683,937 sentences. Before training the process, several parameters are determined that affect the training speed and performance. Based on our dataset size, only words that appear more than two times in the corpus are considered. The dimensionality space of the words inside a vector is set to 100, which means that each word is represented with 100 most similar words in that vector. More words in a vector means better quality, although bigger dataset must be used. The model has been trained in the hierarchical skip-gram architecture in a laptop with i5 CPU 1.7 GHz, 8 GB RAM memory. Surprisingly, the time it took was very fast, 109.5 s, and thus far beyond our expectations.

4.3   Analyzing the Model

This section presents the investigation of the learned model. We performed several analyses on top of trained model in Sect. 4.2. One of the most interesting analyses regarding the word representation approach is about finding the set of closest words based on a particular entered word. For instance, regarding the economic domain of the trained corpus, we are interested to see what the model learned about the economic concept "money" and a more general one, "food". Table 1, lists ten nearest terms that Word2Vec has calculated for these words.

Table 1.   Top ten most similar words based on the words "money" and "food", generated through word2vec from our text corpus.

a. for the word *"money"*		b. for the word *"food"*	
Word	**Similarity**	**Word**	**Similarity**
liquidity	.764	energy	.789
credit	.723	agricultural	.786
loan	.709	water	.767
debt	.654	land	.756
lending	.644	crop	.701
borrowing	.643	fuel	.694
asset	.642	transport	.694
short-term	.634	agriculture	.691
bank	.633	electricity	.690
bond	.632	milk	.684

The generated results are very impressive. For example, the word "liquidity" is ranked as the most similar to "money" with a degree of similarity .764 out of 1, and all others are intuitively very close to it. Moreover, a word is represented in a relationship to hundred words like this, as defined at the training parameters. To our knowledge it is almost impossible to generate such a result through dictionaries or thesauruses. Thus, if we are referring to the STW thesaurus described in Sect. 3, the concept "money" is not represented with many meaningful terms, regarding the SKOS vocabulary. Even the usage of other external resources, such as WordNet synonyms, does not offer such an impressive set of related terms.

The trained model can be used for several other semantic language processing. Accordingly, there is a possibility to retrieve a list of most similar words by subtracting words from a given set of words. Thus, from a set of metadata we have the possibility to include and exclude several concepts. For example, from the set of metadata concepts defined for a publication, we want to consider the terms "bank", "oil" and "price" by excluding the term "food". Therefore, based on this formula [(bank + oil + price) − (food)], the trained model offers the term *currency* with .764 similarity, *liquidity* with . 734 and *spreads* with .695. This means that the retrieved publications according to this expression are semantically related to these terms.

5 Results and Discussions

This study presents several approaches with regard to the initial purpose to enrich scientific publications of a DL with other relevant information from other repositories. However, the main challenge is the determination of semantic relatedness between the initial and retrieved publications.

As emphasized in Sect. 3, the implementation of count-based approach through TF-IDF and Cosine Similarity, requires a large set of metadata from the publications, to measure the similarity degree. Moreover, the right combination of metadata elements is crucial. Hence, in several cases the presence of a more general concept used in these metadata had negative impact on the result. For example, regarding the publication titled *"The long run impact of biofuels on food prices"*, the word *"food"* has been determinant in the similarity measurements when only the title has been considered for the calculation. Thus, the retrieved publications have been related to *"agriculture"* and *"food diets"*, which semantically are not that close to the initial publication. By including the abstract and keywords, improvements were evident. However, this applies heuristic involvements in the evaluation of results. Moreover, the count-based approach shows significant weakness in recognizing relationships among terms, even in the cases when the presence of thesauri is evident.

Based on the developed prototype, we have evaluated randomly 37 publications from EconStor. For each selected publication, the prototype retrieves and orders the most semantically similar publications from OpenAgris. The process is the same as in Fig. 3, however the similarity is calculated through Word2Vec instead of TF-IDF and CS. The top ten retrieved publications are manually analyzed in order to determine the semantic relevance with the initial publication. In a situation like this, the implementation of word embeddings approach shows outstanding results, even with smaller amount of metadata and combinations among them. In 100 % of the cases, the Word2Vec embedding approach overcome TF-IDF with Cosine Similarity.

Table 2 depicts the results of one from the 37 evaluated publications, by comparing the results generated in both approaches with two different sets of metadata. Firstly, the similarity degree between publications A and B is calculated only on titles (T), such as $sim(T_a, T_b)$. As such, for the first retrieved publication on that list, Word2Vec has generated **.804** similarities with the EconStor publication titled *"The long run impact of biofuels on food prices"*. The count-based implementation of Cosine Similarly gives **.5103** similarities between the same titles. In the same example, analyses are extended by including other metadata terms in the similarity calculations. Hence, from the EconStor publications the **title(T_a)**, **abstract(A_a)**, **keywords(K_a)** and **descriptors(D_a)** are considered, while from the OpenAgris publications the **title(T_b)**, **abstract(A_b)** and **descripts(D_b)**. The last two columns of Table 2 show the similarity among these metadata comparatively, $sim(T_aA_aK_aD_a, T_bA_bD_b)$. By considering the first publication from Table 2, TF-IDF with CS generates **.428** similarity degree among them, while Word2Vec gives **.962**. The row number ten emphasizes even more the discrepancy between the generated results. In that case, we realized that the retrieved publication is closely related to the EconStor publication, however the first method has generated only .2757 similarity degree compared to .9343 from Word2Vec.

Table 2. The similarity degree between a particular EconStor publication with OpenAgris publications, calculated with TF-IDF & CS versus Word2Vec.

	Title	$Sim(T_a, T_b)$		$sim(T_aA_aK_aD_a, T_bA_bD_b)$	
		TF-IDF with CS	Word2Vec	TF-IDF with CS	Word2Vec
1	Biofuels versus food production: Does biofuels production increase food prices?	.5103	**.8040**	.4280	**.9620**
2	The "not-so-modern" consumer – considerations on food prices, food security, new technologies and market distortions	.2970	**.7780**	.4275	**.8904**
3	High food commodity prices: will they stay? who will pay?	.2357	**.7740**	.4241	**.9204**
4	Consumers' perceptions regarding tradeoffs between food and fuel expenditures: A case study of U.S. and Belgian fuel users	.0871	**.6521**	.4163	**.9368**
5	Impact of biofuel production and other supply and demand factors on food price increases in 2008	.1925	**.8660**	.4159	**.9594**
6	Biofuels and food security: Micro-evidence from Ethiopia	.3086	**.6592**	.4023	**.8903**
7	Food Versus Biofuels: Environmental and Economic Costs	.3087	**.6723**	.3991	**.8043**
8	Rising food prices intensify food insecurity in developing countries	.3693	**.7710**	.3647	**.9461**
9	How much hope should we have for biofuels?	.1443	**.4420**	.3631	**.9318**
10	Oil price, biofuels and food supply	.3086	**.8320**	.2757	**.9343**

The word embeddings approach evidently overcome the count-based and text-matching approach. The results generated here are significantly better even with smaller amount of concepts included in similarity calculation. The similarity calculated by Word2Vec shows outstanding performance, even when only titles are compared. The presence of other metadata, such as the abstract and keywords, improves the calculation of semantic similarity between publications. By considering the performed evaluations, the word embeddings approach evidently contribute for enriching a scientific publication with semantically related information, such as other publications from different repositories.

6 Summary

The main intention of this work was to emphasize the advantages resulting from an improved interoperability among different Digital Libraries and to investigate different algorithms to achieve this interoperability. Thus, by cross-linking data from different places, a particular resource would be enriched with several other information. This results in a significant enhancement of scholarly communication in general, regarding time consuming and quality of the required information. The idea is to perform a single query in a single place (e.g. their favorite DL) and still to offer scholars information from different repositories, based upon this single query. Ultimately, a selected publication in a DL, will be enriched with a list of recommended publications from other DLs, such as, additional information about authors, conferences, etc.

In order to achieve this, we needed to find this information and then determine its relevance i.e., semantic similarity between two different resources. For this purpose, bibliographic Linked Open Data repositories are considered by investigating the alignments among them. We applied several data mining techniques, such as TF-IDF and Cosine Similarity, among the publications metadata. The generated results, showed that the traditional count-based and text-matching approach require a heuristic way to determine a satisfactory level of semantic similarity among publications. Given this, we also followed the deep learning approach to model semantic word representations. The implementation of a contemporary Word2Vec model results in an outstanding outcome. This is achieved by simplifying the combination process between the metadata, and even more, by performing it on a smaller set of metadata, such as title's concepts only. However, significant improvements are evident by extending the set of metadata with concepts from the abstract and keywords. More detailed analysis with these sets of metadata, and the expansion of the evaluations range will be investigated in our future work.

References

1. Lebret, R., Collobert, R.: Rehabilitation of count-based models for word vector representations. In: Gelbukh, A. (ed.) CICLing 2015. LNCS, vol. 9041, pp. 417–429. Springer, Heidelberg (2015). doi:10.1007/978-3-319-18111-0_31
2. Levy, O., Goldberg, Y., Dagan, I.: Improving distributional similarity with lessons learned from word embeddings. Trans. Assoc. Comput. Linguist. 3, 211–225 (2015)
3. Bengio, Y., Schwenk, H., Senécal, J.S., Morin, F., Gauvain, J.L.: Neural probabilistic language models. In: Holmes, D.E., Jain, L.C. (eds.) Innovations in Machine Learning, vol. 194, pp. 137–187. Springer, Heidelberg (2006). doi:10.1007/3-540-33486-6_6
4. Collobert, R., Weston, J.: A unified architecture for natural language processing: deep neural networks with multitask learning. In: Proceedings of the 25th International Conference on Machine Learning, pp. 160–167. ACM (2008)
5. Mnih, A., Hinton, G.E.: A scalable hierarchical distributed language model. In: Advances in Neural Information Processing Systems, pp. 1081–1088 (2009)
6. Turian, J., Ratinov, L., Bengio, Y.: Word representations: a simple and general method for semi-supervised learning. In: Proceedings of the 48th Annual Meeting of the Association for Computational Linguistics, pp. 384–394. Association for Computational Linguistics (2010)

7. Mikolov, T., Chen, K., Corrado, G., Dean, J.: Efficient estimation of word representations in vector space. arXiv preprint arXiv:1301.3781 (2013)

8. Hajra, A., Latif, A., Tochtermann, K.: Retrieving and ranking scientific publications from linked open data repositories. In: Proceedings of the 14th International Conference on Knowledge Technologies and Data-driven Business, p. 29. ACM (2014)

9. Kenter, T., de Rijke, M.: Short text similarity with word embeddings. In: Proceedings of the 24th ACM International on Conference on Information and Knowledge Management, pp. 1411–1420. ACM (2015)

10. Robertson, S., Zaragoza, H.: The probabilistic relevance framework: BM25 and beyond. Inf. Retr. **3**(4), 333–389 (2009)

11. Mooney, R.J., Roy, L.: Content-based book recommending using learning for text categorization. In: Proceedings of the Fifth ACM Conference on Digital Libraries, pp. 195–204. ACM (2000)

12. Huang, Z., Chung, W., Ong, T.H., Chen, H.: A graph-based recommender system for digital library. In: Proceedings of the 2nd ACM/IEEE-CS Joint Conference on Digital Libraries, pp. 65–73. ACM (2002)

13. Smeaton, A.F., Callan, J.: Personalisation and recommender systems in digital libraries. Int. J. Digit. Lib. **5**(4), 299–308 (2005)

14. Kling, R., McKim, G.: Scholarly communication and the continuum of electronic publishing. arXiv preprint arXiv:cs/9903015 (1999)

15. Paepcke, A., Chang, C.C.K., Winograd, T., García-Molina, H.: Interoperability for digital libraries worldwide. Commun. ACM **41**(4), 33–42 (1998)

16. Borgman, C.L.: Challenges in building digital libraries for the 21st Century. In: Lim, E.-P., Foo, S., Khoo, C., Chen, H., Fox, E., Urs, S., Costantino, T. (eds.) ICADL 2002. LNCS, vol. 2555, pp. 1–13. Springer, Heidelberg (2002). doi:10.1007/3-540-36227-4_1

17. Besser, H.: The next stage: moving from isolated digital collections to interoperable digital libraries. First Monday **7**(6) (2002). doi:10.5210/fm.v7i6.958

18. Sheth, A.P.: Changing focus on interoperability in information systems: from system, syntax, structure to semantics. In: Goodchild, M., Egenhofer, M., Fegeas, R., Kottman, C. (eds.) Interoperating Geographic Information Systems, vol. 495, pp. 5–29. Springer, Heidelberg (1999). doi:10.1007/978-1-4615-5189-8_2

19. Dietze, S., Sanchez-Alonso, S., Ebner, H., Qing, Y.H., Giordano, D., Marenzi, I., Pereira, N.B.: Interlinking educational resources and the web of data: a survey of challenges and approaches. Program **47**(1), 60–91 (2013)

20. Horava, T.: Challenges and possibilities for collection management in a digital age. Lib. Resour. Tech. Serv. **54**(3), 142–152 (2011)

21. Park, D.H., Kim, H.K., Choi, I.Y., Kim, J.K.: A literature review and classification of recommender systems research. Expert Syst. Appl. **39**(11), 10059–10072 (2012)

22. Bobadilla, J., Ortega, F., Hernando, A., Gutiérrez, A.: Recommender systems survey. Knowl. Based Syst. **46**, 109–132 (2013)

23. Lops, P., De Gemmis, M., Semeraro, G.: Content-based recommender systems: state of the art and trends. In: Ricci, F., Rokach, L., Shapira, B., Kantor, P.B. (eds.) Recommender Systems Handbook, pp. 73–105. Springer, US (2011)

24. Sugiyama, K., Kan, M.Y.: Scholarly paper recommendation via user's recent research interests. In: Proceedings of the 10th Annual Joint Conference on Digital Libraries, pp. 29–38. ACM (2010)

25. Hajra, A., Radevski, V., Tochtermann, K.: Author profile enrichment for cross-linking digital libraries. In: Kapidakis, S., Mazurek, C., Werla, M. (eds.) TPDL 2015. LNCS, vol. 9316, pp. 124–136. Springer, Heidelberg (2015). doi:10.1007/978-3-319-24592-8_10

26. Kusner, M., Sun, Y., Kolkin, N., Weinberger, K.Q.: From word embeddings to document distances. In: Proceedings of the 32nd International Conference on Machine Learning (ICML-15), pp. 957–966 (2015)
27. Baroni, M., Dinu, G., Kruszewski, G.: Don't count, predict! A systematic comparison of context-counting vs. context-predicting semantic vectors. In: ACL (1), pp. 238–247 (2014)
28. Pennington, J., Socher, R., Manning, C.D.: Glove: global vectors for word representation. EMNLP. **14**, 1532–1543 (2014)
29. Latif, A., Scherp, A., Tochtermann, K.: LOD for library science: benefits of applying linked open data in the digital library setting. KI-Künstliche Intelligenz **30**(2), 149–157 (2016)
30. Berners- Lee, T., Hendler, J., Lassila, O.: The semantic web. Sci. Am. **284**(5), 28–37 (2001)

MusicWeb: Music Discovery with Open Linked Semantic Metadata

Mariano Mora-Mcginity[✉], Alo Allik, György Fazekas, and Mark Sandler

Queen Mary University, London, UK
{m.mora-mcginity,a.allik,g.fazekas,mark.sandler}@qmul.ac.uk

Abstract. This paper presents MusicWeb, a novel platform for music discovery by linking music artists within a web-based application. MusicWeb provides a browsing experience using connections that are either extra-musical or tangential to music, such as the artists' political affiliation or social influence, or intra-musical, such as the artists' main instrument or most favoured musical key. The platform integrates open linked semantic metadata from various Semantic Web, music recommendation and social media data sources. Artists are linked by various commonalities such as style, geographical location, instrumentation, record label as well as more obscure categories, for instance, artists who have received the same award, have shared the same fate, or belonged to the same organisation. These connections are further enhanced by thematic analysis of journal articles, blog posts and content-based similarity measures focussing on high level musical categories.

Keywords: Semantic Web · Linked open data · Music metadata · Semantic audio analysis · Music information retrieval

1 Introduction

In recent years we have witnessed an explosion of information, a consequence of millions of users producing and consuming web resources. Researchers and industry have recognised the potential of this data, and have endeavoured to develop methods to handle such a vast amount of information. There are two main approaches to music recommendation [1]: the first is known as *collaborative filtering* [2], which recommends music items based on the choices of similar users. The second model is based on audio content analysis, or *music information retrieval*. The task here is to extract low to high-level audio features such as tempo, key, metric structure, melodic and harmonic sequences, instrument recognition and song segmentation, which are then used to measure music similarity.

There are, however, limitations in both approaches to music recommendation. Most users participating in collaborative filtering listen to a very small percentage of the music available, the so called "short-tail", whereas the much larger "long-tail" remains mainly unknown [3]. Many music listeners follow artists because of their style and would be interested in music from similar artists.

© Springer International Publishing AG 2016
E. Garoufallou et al. (Eds.): MTSR 2016, CCIS 672, pp. 291–296, 2016.
DOI: 10.1007/978-3-319-49157-8_25

There are many different ways in which people are attracted to new artists: word of mouth, their network of friends, music magazines or blogs, songs heard in a movie or a T.V. commercial, they might be interested in a musician who has played with another artist or been mentioned as an influence, etc. The route from listening to one artist and discovering a new one would sometimes seem very disconcerting were it to be drawn on paper. A listener is not so much following a map as exploring new territory, with many possible forks and shortcuts. Music discovery systems generally disregard this kind of information, often because it is very nuanced and difficult to parse and interpret.

MusicWeb is a music discovery platform which offers users the possibility of exploring editorial, cultural and musical links between artists. It gathers, extracts and manages metadata from many different sources, including DBpedia.org, Sameas.org, MusicBrainz, the Music Ontology, Last.FM and Youtube as well as editorial and content-derived information. The connections between artists are based on YAGO categories [4], which successfully extracts categories from each wikipedia entry after contrasting them with WordNet. These are categories such as style, geographical location, instrumentation, record label, but also more obscure links, for instance, artists who have received the same award, have shared the same fate, or belonged to the same organisation or religion. These connections are further enhanced by thematic analysis of journal articles, blog posts and content-based similarity measures focusing on high level musical categories.

2   MusicWeb Architecture

MusicWeb provides a browsing experience using connections that are either extra-musical or tangential to music, such as the artists' political affiliation or social influence, or intra-musical, such as the artists' main instrument or most favoured musical keys. It does this by pulling data from several different web knowledge content resources and presenting them for the user to navigate in a faceted manner [5]. The listener can begin his journey by choosing or searching an artist. The application offers Youtube videos, audio streams, photographs and album covers, as well as the artist's biography (see example in Fig. 1). The page also includes many box widgets with links to artists who are related to the current artist in different, and sometimes unexpected ways.

MusicWeb was originally conceived as a platform for collating metadata about music artists using already available online linked data resources. The core functionality of the platform relies on available SPARQL endpoints as well as various commercial and community-run application programming interfaces (APIs).

The MusicWeb API uses a number of linked open data (LOD) resources and Semantic Web ontologies to process and aggregate information about artists:

Musicbrainz[1] is an online, open, crowd-sourced music encyclopedia, that provides reliable and unambiguous identifiers for entities in music publishing metadata, including artists, releases, recordings, performances, etc.

---

[1] http://musicbrainz.org.

Fig. 1. Example of a MusicWeb artist page.

DBPedia[2] is a crowd-sourced community effort to extract structured information from Wikipedia and make it available on the Web.

Sameas.org[3] manages Universal Resource Identifier (URI) co-references on Web of Data.

Youtube API is used to query associated video content for the artist panel.

Last.fm[4] is an online music social network and recommender system that collects information about users listening habits and makes available crowd-sourced tagging data through an API.

YAGO is a semantic knowledge base that collates information and structure from Wikipedia, WordNet and GeoNames.

The Music Ontology [6] provides main concepts and properties for describing musical entities, including artists, albums, tracks, performances, compositions, etc., on the Semantic Web

The global MusicBrainz identifiers enable convenient and concise means to disambiguate between potential duplicates or irregularities in metadata across resources, a problem which is all too common in systems relying on named entities. Besides identifiers, the MusicBrainz infrastructure is also used for the search functionality of MusicWeb. However, in order to query any information in DBpedia, the MusicBrainz identifiers need to be associated with a DBpedia resource,

---

2 http://dbpedia.org.
3 http://sameas.org.
4 http://last.fm.

which is a different kind of identifier. This mapping is achieved by querying the Sameas.org co-reference service to retrieve the corresponding DBpedia URIs. The caveat in this process is that Sameas does not actually keep track of MusicBrainz artist URIs, however, by substituting the domain for the same artist's URI in the BBC domain[5], MusicWeb can get around this obstacle. Once the DBpedia artist identity is determined, the service proceeds to construct the majority of the profile, including the biography and most of the linking categories to other artists. The standard categories available include associated artists and artists from the same hometown, while music group membership and artist collaboration links are queried from MusicBrainz. The core of the Semantic Web linking functionality is provided by categories from YAGO. The Last.fm API provides with information on artists it deems similar.

3   Artist Similarity

There are many ways in which artists can be considered related: similarity may be based on a particular style or genre, but it may also mean that artists are followed by people from similar social backgrounds, political inclinations, or age groups. Artists can also be associated because they have collaborated, participated in the same event, or their lyrics touch upon similar themes. Linked data facilitates faceted searching and displaying of information [7]: an artist may be similar to many other artists in one of the ways just mentioned, and to a completely different plethora of artists in other senses, all of which might contribute to music discovery. Semantic Web technologies can help us gather different facets of data and shape them into representations of knowledge. MusicWeb does this by searching similarities in three different domains: socio-cultural, research and journalistic literature and content-based information retrieval.

Socio-cultural connections between artists in MusicWeb are primarily derived from YAGO categories that are incorporated into entities in DBpedia. Many categories, in particular those that can be considered extra-musical or tangential to music, stem from the particular methodology used to derive YAGO information from Wikipedia. While DBpedia extracts knowledge from the same source, YAGO leverages Wikipedia category pages to link entities without adapting the Wikipedia taxonomy of these categories [4]. The hierarchy is created by adapting the Wikipedia categories to the WordNet concept structure. This enables linking each artist to other similar artists by various commonalities such as style, geographical location, instrumentation, record label as well as more obscure categories, for example, artists who have received the same award, have shared the same fate, or belonged to the same organisation or religion. YAGO categories can reveal connections between artists that traditional isolated music datasets would not be able to establish.

Literature-based linking is achieved by data-mining research articles and online publications using natural language processing. MusicWeb uses

---

[5] http://www.bbc.co.uk/music/artists/.

Mendeley[6] and Elsevier[7] databases for accessing research articles that are curated and categorised by keywords, authors and disciplines. Online publications, such as newspapers, music magazines and blogs focused on music, on the other hand, constitute non-curated data. Relevant information in this case must be extracted from the body of the text. The data is collated by crawling websites by keywords or tags in the title and by following external links contained in pages. Many texts contain references to an artist name without actually being relevant to MusicWeb. A search for Madonna, for example, can yield many results from the fields of sculpture, art history or religion studies. The first step is to model the relevance of the text, and discard texts which are of no interest to music discovery. Texts and abstracts are then subjected to semantic analysis. The text as a bag of words is used to query the Alchemy[8] language analysis service for named entity recognition and keyword extraction. The entity recogniser provides a list of names that appear mentioned in the text together with a measure of relevance. MusicWeb identifies musical artists using its internal artist database as well as DBpedia, MusicBrainz and Freebase. Keyword extraction is used for non-curated sources and involves checking keywords against WordNet for hypernyms. Artists that share keywords or hypernyms are considered to be relevant to the same topic in the literature. MusicWeb also offers links between artists who appear in different articles by the same author, as well as in the same journal.

Content-based linking involves methodology of Music Information Retrieval (MIR) [8] which facilitate applications that rely on perceptual, statistical, semantic or musical features derived from audio using digital signal processing and machine learning methods. These features may include statistical aggregates computed from time-frequency representations extracted over short time windows. Higher-level musical features include keys, chords, tempo, rhythm, as well as semantic features like genre or mood, with specific algorithms to extract this information from audio. High-level stylistic descriptors can correlate with lower level features such as the average tempo of a track, the frequency of note onsets, the most commonly occurring keys or chords or the overall spectral envelope that characterises instrumentation. To exploit different types of similarity, we model each artist using three main categories of audio descriptors: rhythmic, harmonic and timbral. We compute the joint distribution of several low-level features in each category over a large collection of tracks from each artist. We then link artists exhibiting similar distributions of these features. The features are obtained from the AcousticBrainz[9] Web service which provides descriptors in each category of interest. Tracks are indexed by MusicBrainz identifiers enabling unambiguous linking to artists and other relevant metadata. For each artist in our database, we retrieve features for a large collection of their tracks in the above categories, including beats-per-minute and onset rate (rhythmic), chord histograms (harmonic) and Mel-Frequency Cepstral Coefficients (timbral) features.

---

[6] http://dev.mendeley.com/.
[7] http://dev.elsevier.com/.
[8] AlchemyAPI is used under license from IBM Watson.
[9] https://acousticbrainz.org/.

4   Conclusions

MusicWeb is an emerging application to explore the possibilities of linked data-based music discovery. The methods of linking artists employed in the system are intended to overcome issues such as infrequent access of lesser known artists in large music catalogues (the "long tail" problem) or the difficulty of recommending artists without user ratings in systems that employ collaborative filtering ("cold start" problem) [3]. This facilitates users to engage in interesting discovery paths through the space of music artists. Although similar to recommendation, this is in contrast with most recommender systems which operate on the level of individual music items. We aim at creating links between artists based on stylistic elements of their music derived from a collection of recordings and complement the social and cultural links. Future work will address investigating various different approaches to music discovery and how they can benefit from linked music metadata. The next steps are directed towards evaluating the potential acceptance of MusicWeb by end users to find out which linking methods listeners find appealing or interesting, and which they would use most.

References

1. Song, Y., Dixon, S., Pearce, M.: A survey of music recommendation systems and future perspectives. In: 9th International Symposium on Computer Music Modeling and Retrieval (2012)
2. Sneha, S., Jayalakshmi, D.S., Shruthi, J., Shetty, U.R.: Recommending music by combining content-based and collaborative filtering with user preferences. In: Sridhar, V., Sheshadri, H.S., Padma, M.C. (eds.) ICERECT 2012. LNCS, vol. 248, pp. 507–515. Springer, Heidelberg (2014)
3. Celma, Ò.: Music Recommendation and Discovery: The Long Tail, Long Fail, and Long Play in the Digital Music Space. Springer, Heidelberg (2010)
4. Fabian, M.S., Gjergji, K., Gerhard, W.: Yago: a core of semantic knowledge unifying wordnet and wikipedia. In: 16th International World Wide Web Conference, WWW, pp. 697–706 (2007)
5. Marchionini, G.: Exploratory search: from finding to understanding. Commun. ACM 49(9), 41–46 (2006)
6. Raimond, Y., Abdallah, S.A., Sandler, M.B., Giasson, F.: The music ontology. In: ISMIR, pp. 417–422. Citeseer (2007)
7. Rodríguez-García, M., Colombo-Mendoza, L.O., Valencia-García, R., Lopez-Lorca, A.A., Beydoun, G.: Ontology-based music recommender system. In: Omatu, S., Malluhi, Q.M., Gonzalez, S.R., Bocewicz, G., Bucciarelli, E., Giulioni, G., Iqba, F. (eds.) Distributed Computing and Artificial Intelligence, 12th International Conference, vol. 373, pp. 39–46. Springer, Heidelberg (2015)
8. Casey, M.A., Veltkamp, R., Goto, M., Leman, M., Rhodes, C., Slaney, M.: Content-based music information retrieval: current directions and future challenges. IEEE Proc. 96(4), 668–696 (2008)

Highlighting Timely Information in Libraries Through Social and Semantic Web Technologies

Ioannis Papadakis[1], Konstantinos Kyprianos[1(✉)], and Apostolos Karalis[2]

[1] Department of Archives, Library Science and Museum Studies, Ionian University,
Ioannou Theotoki 72, 49100 Corfu, Greece
papadakis@ionio.gr, k.kyprianos@gmail.com
[2] Department of Informatics, University of Piraeus, Karaoli & Dimitriou 80,
18534 Piraeus, Greece
akaralis@hotmail.com

Abstract. Until now, timeliness of information in libraries is commonly used to underpin collection development and is directly related to quality in terms of realizing whether information is sufficiently up-to-date and available for use.

In this paper, it is argued that timely information could be exploited from libraries in another context. More specifically, it would be meaningful for the users of a library to explore resources based on their relevancy to popular events that occasionally occur to the society. Along these lines, a prototype digital library service is implemented and accordingly deployed, based on popular crowd-sourcing services (i.e. Twitter) that are integrated with semantic web technologies. The proposed service is evaluated in terms of its ability to provide accurate and timely information.

Keywords: Information timeliness · Social web · Twitter · Hashtags · DBpedia · Semantic web · Linked data

1 Introduction

Timeliness of information in libraries is commonly used to underpin collection development and is directly related to quality in terms of realizing whether information is sufficiently up-to-date and available for use [1]. Traditionally, libraries identify timely information through the employment of statistical methods such as the measurement of the resources' circulation [10, 11] and/or through log file analysis capturing the searches users perform against the Online Public Access Catalog – OPAC and other services [12].

Such methods focus on the analysis of user transactions over a certain period of time and therefore provide insight into the past. This way, the decision making process within a library does not take into account current trends, since they are not yet reflected to user transactions. Additionally, it is difficult to identify the occasional popularity of specific resources, having in mind that resources related to trending topics have a bigger demand in specific periods of time [9].

In this paper, it is argued that libraries need to come up with new methods that distinguish and accordingly promote timely information. The advent of the social Web

© Springer International Publishing AG 2016
E. Garoufallou et al. (Eds.): MTSR 2016, CCIS 672, pp. 297–308, 2016.
DOI: 10.1007/978-3-319-49157-8_26

points towards this direction, since the corresponding technologies provide the opportunity to focus on users and their demands [3] in a more direct fashion [14–17].

Along these lines, a methodology is proposed that facilitates the identification and utilization of trending topics captured by Twitter's hashtags. Hashtags are further expanded through the employment of adequate semantic web technologies and are matched against the subjects of a library's catalog. The proposed methodology is realized as a digital library service that is accordingly evaluated.

The remainder of the paper is structured as follows. In the following section, the importance of timeliness of information in libraries is highlighted and the techniques that libraries traditionally employ to study user needs in relation to the problem of information timeliness are discussed. Then, a short synopsis of the employment of social Web technologies in libraries and their potentials is presented. Particular attention is given to social tagging and hashtags. After that, as a proof of concept, a service capable of suggesting timely resources to library users is presented and various implementation issues are discussed. Next, the proposed service is accordingly assessed and the corresponding results are shown. The final section concludes the paper and points directions for future work.

2   Information Timeliness and Libraries

Information has a life cycle that depends on how quickly new information arrives to replace old one and whether such information is delivered in a timely fashion [2]. Timeliness of information in the library domain is mainly utilized in collection development and is traditionally measured through the employment of the following two factors: a) to what extent information provided by the library is up-to-date and b) the delivery speed of information; how long it takes to process and accordingly deliver new information to the user [13].

2.1   Timeliness for Collection Development and Beyond

Crawley-Low [20] proposes interlibrary loan and circulation as services that facilitate libraries in gathering information about the collection usage for specific periods of time in the past. In a similar fashion, the Finnish Collection Map Consortium [21] proposes a collection-mapping method to evaluate libraries' collections. Among others, the method utilizes the average age and languages of the collection, usage statistics (e.g. circulation, interlibrary loan, in-house use, turn-over rate etc.) and shelf-scanning. In another approach, Agee [10] and Borin & Yi [22] consider usage as one of the most important factors for collection development. They also suggest various methods to study usage, namely surveys, discussion groups, interviews and experimental settings.

The aforementioned collection development strategies are based on information that comes from the past. Such information varies from implicit feedback originating from user transactions that have been recorded (e.g. interlibrary loan, circulation, etc.) to explicit feedback from users (e.g. surveys, interviews, etc.). Both types of strategies aim in providing insight about the age of the underlying resources and the delivery speed of information.

In this paper, it is argued that timely information should be exploited from libraries in another context. Thus, a library should be able to suggest resources based on their relevancy to popular events that occasionally occur to the society.

The impact of trending topics to the society has been studied for many years in the fields of sociology, communication, marketing and political science and it seems to start gaining attention in the field of social networking as well [25].

To sum up, libraries need to pay close attention to what actually happens to the society in a timely fashion and accordingly highlight the corresponding resources to their users. Social Web tools point towards this direction, since they can aid libraries in studying their users in a more efficient, quick and direct way. In the following section, the social Web is examined as a communication tool between libraries and their users.

3   Social Web and Libraries

During the past few years, social Web has emerged as a communication channel capable of facilitating personalized information sharing and collaboration in a timely manner. It refers to *"a collection of technologies such as blogs, wikis, RSS feeds, social networks etc. where users are able to add, share and edit the content, creating a socially networked web environment"* [4].

One social Web activity that is starting to gain attention from libraries during the past few years is social tagging. Tagging enables users to add keywords to shared content [5]. A tag is a non-hierarchical keyword or term assigned to a piece of information (e.g. a website) [6]. Additionally, tagging allows libraries to partner with their users in an effort to provide enhanced subject access [4].

Twitter is a social Web tool that allows users to broadcast brief text updates about things that are happening to their life. Users refer to Twitter when they want to find information about breaking news, real-time events, people information and topical information [18, 19]. Such features establish Twitter as a tool that can provide timely information quicker than any other mass media (e.g. television, radio etc.).

Hashtags are employed by Twitter to classify messages, propagate ideas and also promote specific topics and people. More specifically, hashtags can classify a tweet according to its meaning. *"By simply adding a hash symbol (#) before a string of letters, numerical digits or underscore signs (_), it is possible to tag a message, helping other users to find tweets that have a common topic. Hashtags allow users to create communities of people interested in the same topic by making it easier for them to find and share information related to it"* [7].

Along these lines, this paper examines Twitter as a third-party tool that can provide libraries with timely information about the present. More specifically, hashtags within tweets are exploited to identify trending topics in a specific domain. The corresponding hashtags may then be mapped to the library's authority indices (i.e. subject headings, titles, etc.) that, in turn, point to relevant resources. This process aids libraries in highlighting interesting and timely resources to their users.

3.1 Vocabulary Alignment Through the Employment of Semantic Web Technologies

Integrating information from Twitter within a library is not a straightforward process. Implications arise due to the fact that Twitter and libraries represent two fundamentally different ecosystems. Crowdsourcing tools such as Twitter allow their users to express their information needs in any way they see fit. Minimum rules are employed in an effort to attract as many users as possible. The lack of authority control results in the production of information that is very difficult to regulate.

On the other hand, the ultimate goal of a library is to provide information that is well organized. Along these lines, libraries traditionally employ controlled vocabularies in an effort to apply authority control to their collections. Thus, it is apparent that information originating from Twitter needs to be further processed before it finds its way into the library.

In this paper, semantic web technologies are utilized as an intermediate layer that facilitates the alignment of Twitter hashtags and controlled vocabularies. During the past few years, semantic web technologies and linked data in particular are continuously gaining ground within libraries in an effort to overcome various issues that prevent libraries from interacting with other information services. Thus, there are many examples of national libraries around the world that have created semantic web services to facilitate topical data interoperability beyond the library community [23].

Perhaps the most widely used information service of the semantic web is DBpedia [24]. DBpedia has emerged as an effort to extract structured data from Wikipedia and make this information available as linked data. Nowadays, DBpedia acts as an intermediate service that interlinks information from various datasets on the Web. As it will be shown later in this paper, DBpedia underpins a mapping mechanism between trending topics originating from Twitter and controlled vocabularies that are commonly employed within libraries.

The next section introduces a service capable of highlighting timely information about a certain discipline in libraries through the employment of social and semantic web technologies.

4   Proof of Concept

In this section, a service is introduced that maps popular hashtags about informatics to their corresponding resources within a digital library. More specifically, the proposed service is based on the following workflow; initially, the service discovers the most popular hashtags of a certain discipline (i.e. informatics). Then, such hashtags are filtered through the employment of Bing's spelling suggestions service in an effort to transform them to readable keywords. The resulting keywords are tunneled to DBpedia, a third-party semantic web service capable of expanding the coming keywords with semantically related terms to the initial hashtags. Finally, the resulting keywords appear as query suggestions to the digital library's homepage.

At this point, it should be mentioned that hashtags have particular information value, since they act as concentrated representations of groups of tweets that talk about the same thing.

As shown in Fig. 1, the service's architecture is comprised of three layers: *client*, *server* and *backend*. The *client layer* consists of the interaction module that is user accessible. The *server layer* is based on the retrieval module acting as a mediator between the following services: Twitter search API[1], Bing spelling suggestions service[2], DBpedia and DSpace[3]. Additionally, it facilitates the communication between the *client layer* and the *backend layer*. Finally, the *backend layer* keeps the service's data realized as a Mongo Database[4].

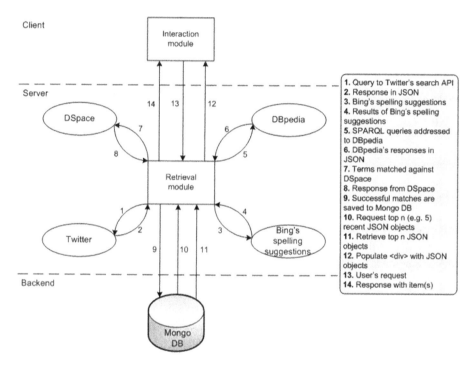

Fig. 1. The overall architecture of the proposed service

4.1   Information Model

The functionality of the proposed service is realized through an information model that facilitates communication between the various modules of the aforementioned architecture. The information flow within the service is outlined in the following steps:

---

[1] Twitter search API, avail. at: https://dev.twitter.com/rest/public/search [accessed: 12/06/2016].

[2] Bing API, avail. at: http://datamarket.azure.com/dataset/bing/search [accessed: 12/06/2016].

[3] DSpace, avail. at: http://dspace.org/ [accessed: 12/06/2016].

[4] MongoDB, avail. at: https://www.mongodb.org/ [accessed: 12/06/2016].

Step 1: Initially, the *Retrieval module* addresses a query to Twitter's Search API about informatics.

Step 2: The corresponding tweets are delivered to the *Retrieval module* as a JSON object. The module isolates the hashtags from the surrounding text within each tweet.

Step 3-4: Twitter hashtags are single words that are quite often logically comprised of two or more words without a space between them. Consequently, hashtags need to split in proper phrases. For this purpose, Bing's spelling suggestions service [8] is employed. For example, "CloudComputing" will be transformed to "Cloud Computing".

Step 5-6: Since the crowdsourcing nature of Twitter implies that hashtags most likely belong to an informal vocabulary, the proposed service aligns the corresponding hashtags to a more formal vocabulary through the employment of DBpedia's SPARQL endpoint[5]. Thus, for each hashtag/spelling suggestion, a SPARQL query[6] is addressed to DBpedia's endpoint in an effort to semantically expand the hashtags with their possible categories and/or related terms. For example, a SPARQL query about the hashtag "data science" results in the discovery of the related term "Information science"[7]. The corresponding results are ultimately received by the *Retrieval module* in JSON format.

Step 7-8: The trending terms as well as the broad categories they belong to together with possible related terms are matched against the subject index of DSpace. Such a process is implemented through the employment of DSpace's REST API[8].

Step 9: In case of a positive response (i.e. resources related to the specific term actually exist), a JSON object is stored at the Mongo DB, which represents the related bibliographic records corresponding to the term. Each JSON object consists of the following fields:

(a) Keyword: the term (i.e. trending term, category, related term) that matches against the digital library's index.

(b) URL: the URL of the corresponding search query. Upon request, the resource(s) that match against the digital library's index are returned.

(c) Hashtag: the initial hashtag that triggered the entire process.

(d) Timestamp.

Step 10-11: The n (e.g. 5) most recent JSON objects are requested from Mongo DB and accordingly retrieved from the *Retrieval module*.

Step 12: The retrieved JSON objects are visualized as a division HTML element (i.e. $<$ div $>$) inside the homepage of DSpace.

Step 13-14: Upon a user interaction with the service's division HTML element, a corresponding search request is addressed to DSpace and the resulting response is presented as a list of suggested resources within the digital library's collection.

---

[5] DBpedia's SPARQL endpoint, avail. at: http://dbpedia.org/sparql [accessed: 12/06/2016].

[6] SPARQL Query Language, avail at: www.w3.org/TR/rdf-sparql-query/ [accessed: 12/06/2016].

[7] The corresponding SPARQL query would be:
SELECT ?related WHERE{ <http://dbpedia.org/resource/Data_science> <http://purl.org/dc/terms/subject> ?related}.

[8] DSpace REST API, avail. at: https://wiki.duraspace.org/display/DSDOC5x/REST+API [accessed: 12/06/2016].

In the following section, the deployment of the proposed methodology to a specific DSpace installation is demonstrated.

4.2 Deployment of the Proposed Service

The proposed service is offered by Dione, the DSpace-based, institutional repository of the University of Piraeus[9]. Dione contains resources related to the four schools of the University[10]. The proposed service focuses on the resources related to the School of Information and Communication Technologies.

As shown in Fig. 2, the proposed service is visualized as a division HTML element (i.e. <div>) containing trending topics at the top right of the digital library's homepage.

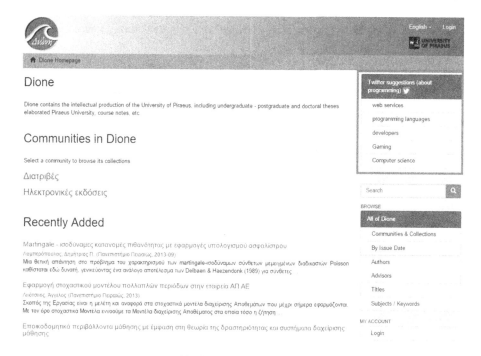

Fig. 2. Dione's homepage

4.3 Implementation Issues

At the beginning, the goal was to work with all the most popular hashtags on Twitter, regardless of discipline to which they belong. However, it was practically impossible to map generic hashtags to the digital library's subject index (too much noise was detected), so it was decided to limit the scope of the proposed service to a certain discipline. Since the University hosts the School of Information and Communication Technologies, the service looked for tweets about informatics through the employment of a search query

---

[9] Dione, avail. at: http://dione.lib.unipi.gr/ [accessed: 12/06/2016].

[10] The four schools of University are presented in Fig. 4.

containing the keyword "programming". In the following section, the evaluation of the proposed service is presented.

5   Evaluation

The proposed service has been deployed to Dione, the institutional repository of the University of Piraeus in Greece since the end of January, 2016 and the corresponding evaluation is primarily based on a log file analysis of user interactions that have been recorded during February, 2016. The main goal of the study is to measure the popularity of the proposed service as compared to the traditional navigational interactions that are commonly provided to the users of Dione. Moreover, an effort is made to assess the performance of the core modules that constitute the proposed service[11]. According to the log file analysis, 27,168 interactions to Dione were recorded during February, 2016.

5.1   Popularity

Nearly one third (i.e. 8,905 of 27,168) of the interactions to Dione are navigational interactions that have been recorded through specific components of the system's user interface. More specifically, Dione consists of 11 navigational interactions, grouped as follows:

– Twitter suggestions about programming (i.e. the proposed service),
– 6 browse indices (i.e. communities and collections, dateissued, author, advisor, title and subject) and
– 4 discovery filters (i.e. type, dateissued, departments and graduate studies).

As shown in Fig. 3, the most popular navigational interaction is against the subject index (43.01 %), followed by the author index (31.67 %). The advisor index (15.44 %) ranks third and Twitter suggestions appear in the fourth place (2.15 %). The rest of the interactions are not very popular since they cover below 2 % of the total sum.

At a first glance, it seems that the impact of the proposed service is not very strong among the visitors of Dione. However, it must be taken into account the fact that the proposed suggestions focus on a School (i.e. the School of Information and Communication Technologies) that represents a particular subset of the overall user community The other three Schools are related to largely different disciplines. Since it is not possible to isolate the logs originating from one particular School, it would be meaningful to substitute the percentages appearing in Fig. 3 with their weighted versions, normalized to the number of interactions that correspond to the user community of the School of Information and Communication Technologies (see Fig. 4).

Along these lines, Fig. 5 presents the popularity of each navigational interaction provided by Dione, projected to the population of the School of Information and Communication Technologies (e.g. the 50 "dateissued" interactions that refer to the whole university population, are normalized to 9 interactions that correspond to the School of Information and Communication Technologies). Same as in Fig. 3,

---

[11] The corresponding log files as well as access to Mongo DB's tables can be found at: http://aimashup.org/timeliness.

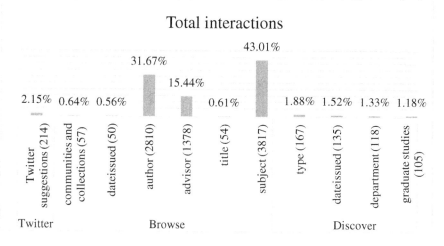

Fig. 3. Total interactions

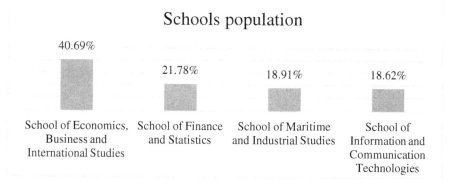

Fig. 4. Schools population

interactions against the proposed service rank to the fourth place. However, this time, the distance from the third place is considerably shorter (i.e. 11.68 %–14.00 %).

We have also measured how many of the interactions against the proposed service have led to actual downloads of the corresponding resources in an effort to identify interactions that have led to satisfied information needs. Thus, 166 (77.57 %) interactions have led to actual downloads, whereas 48 (22.43 %) did not lead to any downloads at all. The number of suggestions that lead to actual downloads can be safely considered high, particularly if we compare it against the corresponding number of downloads of another interaction, e.g. the discovery type 'filter'; 107 (64 %) interactions against the 'filter' type have led to actual downloads, whereas 60 (36 %) did not lead to any downloads at all.

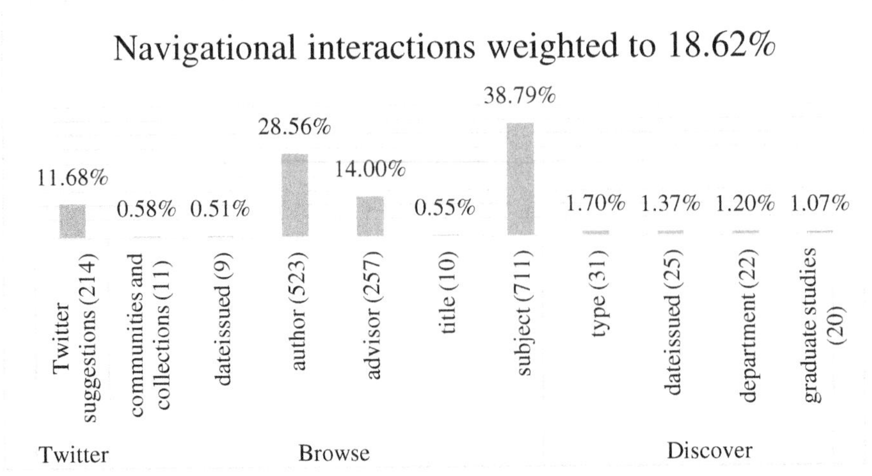

Navigational interactions weighted to 18.62%

Fig. 5. Total interactions weighted to 18.62 % of the total population

5.2   Under the Hood: Performance of the Underlying Modules

In the previous section, log file analysis was employed to assess the popularity of the proposed service among the members of Dione's user community. In this section, we analyse and accordingly assess the various interactions between the core modules of the service and the remote online services that have been employed (namely: Twitter, Bing and DBpedia).

During the evaluation period, 5,000 trending hashtags have been compared against the subject index of Dione and accordingly 1,153 successful matches have been identified[12]. The 1,153 successful matches correspond to 45 distinct hashtags. At this point, it should be mentioned that according to the service's architecture, each hashtag has been processed by two modules, namely Bing's spelling suggestions service and DBpedia. Thus, each hashtag appears as Bing's recommendation (since it is more readable than the original hashtag) in Dione's homepage and corresponds to a query containing all the semantically related terms (originating from Bing and DBpedia). For example, the hashtag "#datascience" appears as "data science" in Dione's homepage and is accordingly addressed to Dione's information retrieval module as "Information science" OR "Data science".

Dione's user community visited 26 out of the 45 hashtags offered by the proposed service. More specifically, from the 26 visited hashtags, only 5 originated directly from Twitter without any intervention from Bing or DBpedia. From the remaining 21 hashtags, 9 originated from Bing's spelling suggestions, 5 originated from DBpedia and 7 hashtags originated both from DBpedia and Bing (see Fig. 6). Thus, it is apparent that Bing's spelling suggestion service had a positive impact to the proposed one and more than half of the visited hashtags offered by the system were affected by DBpedia.

---

[12]   The Bing's spelling suggestions service allows 5,000 remote requests per month.

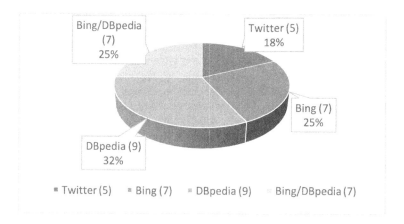

Fig. 6. Performance of Bing and DBpedia modules

6   Conclusions – Future Work

In this paper, a novel approach is proposed that takes advantage of information existing outside the library environment in favor of value-added services that promote resources about trending topics. As a proof of concept, an online service is introduced that maps popular hashtags to their corresponding resources within a digital library and suggests them to the users. The proposed service is integrated to the institutional repository of an academic library and has been accordingly evaluated.

The assessment process was based on quantitative methods. A log file analysis revealed that the proposed service drew the attention of a considerable number of library users. It is also encouraging the fact that most of the interactions with the service were followed by downloads of the corresponding resources. Moreover, further analysis on the core modules of the service showed that query expansion through DBpedia improved both the quantity and the quality of the provided suggestions.

The assessment of the proposed service clearly indicates that libraries could benefit a lot from the employment of crowdsourcing services such as Twitter. It is also apparent that semantic web technologies could play a crucial role in enhancing the overall performance of services based on information outside the library. Future work focuses on minimizing the inherent "noise" of hashtags and on finding ways of applying the proposed approach to the entire digital library collection.

References

1. Goncalves, M.A., Moreira, B.L., Fox, E.A., Watson, L.T.: What is a good digital library? – a quality model for digital libraries. Inf. Process. Manage. **43**, 1416–1437 (2007)
2. Miller, H.: The multiple dimensions of information quality. Inf. Syst. Manage. **13**(2), 79–82 (1996)
3. Anderson, P.: What is web? 2.0 ideas, technologies and implications for education. JISC Technol. Stand. Watch **1**(1), 1–64 (2007). http://www.jisc.ac.uk/media/documents/techwatch/tsw0701b.pdf. Accessed 09 Jun 2016

4. Lwoga, E.T.: Measuring the success of library 2.0 technologies in the african context: the suitability of the DeLone and McLean's model. Campus-Wide Inf. Syst. **30**(4), 288–307 (2013)
5. Lawson, K.G.: Mining social tagging data for enhanced subject access for readers and researchers. J. Acad. Librariansh. **35**(6), 574–582 (2009)
6. Spitieri, L.: User-generated metadata: boon or bust for indexing and controlled vocabularies? (2014). http://www.slideshare.net/cleese6/usergenerated-metadata-boon-or-bust-for-indexing-and-controlled-vocabularies. Accessed 11 Jun 2016
7. Cunha, E., Magno, G., Comarcel, G., Almeida, V., Goncalves, M.A., Benevenuto, F.: Analyzing the dynamic evolution of hashtags on twitter: a language-based approach. In: Proceedings of the Workshop on Language in Social Media (LSM 2011), pp. 58–65 (2011)
8. Movahedian, H., Khayyambashi, M.R.: Folksonomy-based User interest and disinterest profiling for improved recommendations: an ontological approach. J. Inf. Sci. **40**(5), 594–610 (2014)
9. Smith, I.M.: What do we know about public library use? Aslib Proc. **51**(9), 302–314 (1999)
10. Agee, J.: Collection evaluation: a foundation for collection development. Collect. Build. **24**(3), 92–95 (2005)
11. Knievel, J.E., Wicht, H., Connaway, S.L.: Use of circulation statistics and interlibrary loan data in collection management. Coll. Res. Libr. **67**(1), 35–49 (2006)
12. Covey, D.T.: Usage and Usability Assessment: Library Practices and Concerns. Digital Library Federation, CLIR, Washington (2002)
13. Weaver-Meyers, P., Stolt, W.: Delivery speed, timeliness and satisfaction. J. Libr. Adm. **23**(1–2), 23–42 (1997)
14. Gerolimos, M., Konsta, R.: Services for academic libraries in New Era. D-Lib Mag. **17**(7/8), 1 (2011)
15. Walia, P.K., Gupta, M.: Application of web 2.0 tools by national libraries. Webology **9**(2), 21–30 (2012)
16. Linh, N.C.: A survey of the application of Web 2.0 in Australian university libraries. Libr. Hi Tech **26**(4), 630–653 (2008)
17. Mahmood, K., Richardson Jr., J.V.: Impact of web 2.0 technologies on academic libraries: a survey of ARL libraries. Electron. Libr. **31**(4), 508–520 (2013)
18. Teevan, J., Ramage, D., Morris, M.R.: #TwitterSearch: a comparison of microblog search and web search. In: Proceedings of the Fourth ACM International Conference on Web Search and Data Mining (WSDM 2011), pp. 35–44 (2011)
19. Lau, C.H., Li, Y., Tjondronegoro, D.: Microblog retrieval using topical features and query expansion. In: Text REtrieval Conference (2011)
20. Crawley-Low, J.V.: Collection analysis techniques used to evaluate a graduate-level toxicology collection. J. Med. Libr. Assoc. **90**(3), 310–316 (2002)
21. Hyödynmaa, M., Ahlholm-Kannisto, A., Nurminen, H.: How to evaluate library collections: a case study of collection mapping. Collect. Build. **29**(2), 43–49 (2010)
22. Borin, J., Yi, H.: Indicators for collection evaluation: a new dimensional framework. Collect. Build. **27**(4), 136–143 (2008)
23. Papadakis, I., Kyprianos, K., Stefanidakis, M.: Linked data URIs and libraries: the story so far. D-Lib Mag. **21**(5/6), 5 (2015). doi:10.1045/may2015-papadakis
24. Auer, S., Bizer, C., Kobilarov, G., Lehmann, J., Cyganiak, R., Ives, Z.: DBpedia: a nucleus for a web of open data. In: Aberer, K., et al. (eds.) ASWC/ISWC -2007. LNCS, vol. 4825, pp. 722–735. Springer, Heidelberg (2007). doi:10.1007/978-3-540-76298-0_52
25. Cha, M., Haddadi, H., Benevenuto, F., Gummadi, P.K.: Measuring user influence in twitter: the million follower fallacy. In: Proceedings of the 4th International AAAI conference on Weblogs and Social Media, ICWSM, pp. 10–17 (2010)

Track on Open Repositories, Research Information Systems and Data Infrastructures

GACS Core: Creation of a Global Agricultural Concept Scheme

Thomas Baker[1](✉), Caterina Caracciolo[1](✉), Anton Doroszenko[2], and Osma Suominen[3]

[1] Food and Agriculture Organization of the United Nations, Rome, Italy
tom@tombaker.org, caterina.caracciolo@fao.org
[2] CAB International, Wallingford, UK
[3] National Library of Finland, Helsinki, Finland

Abstract. The most frequently used concepts from AGROVOC, CABT, and NALT – three major thesauri in the area of food and agriculture – have been merged into a Global Agricultural Concept Scheme, with 15,000 concepts and over 350,000 terms in 28 languages in its beta release of May 2016. This set of core concepts ("GACS Core") is seen as the first step towards a more comprehensive Global Agricultural Concept Scheme. In the context of a new Agrisemantics initiative, GACS is intended to serve as hub linking user-oriented thesauri with semantically more precise and specialized domain ontologies linked, in turn, to quantitative datasets. The goal is to improve the discoverability and semantic interoperability of agricultural information and data for the benefit of researchers, policy-makers, and farmers in support of innovative responses to the challenges of food security under conditions of climate change.

1 A Shared Concept Scheme

The Food and Agriculture Organization of the United Nations (FAO), CAB (Centre for Agriculture and Biosciences) International (CABI), and the National Agricultural Library of the USDA (NAL) maintain separate thesauri about agriculture, food, and nutrition for indexing bibliographic databases. The AGROVOC Concept Scheme (created 1982)[1], CAB Thesaurus (1983)[2], and NAL Thesaurus (1990s)[3] are used to index, respectively, AGRIS (8 million records), CAB Abstracts (11.5 million), and Agricola (5.2 million).

Having collaborated in the 1990s on mappings and common classifications, the three organizations joined forces again in 2013 to explore the feasibility of creating a shared Global Agricultural Concept Scheme (GACS).[4] The project aimed at facilitating search across databases, at improving the semantic reach of their databases by supporting queries that freely draw on terms from any mapped thesaurus, and at achieving efficiencies of scale from collaborative maintenance.

---

[1] http://aims.fao.org/agrovoc.
[2] http://www.cabi.org/cabthesaurus/.
[3] http://agclass.nal.usda.gov/.
[4] http://agrisemantics.org/gacs.

© FAO, 2016
E. Garoufallou et al. (Eds.): MTSR 2016, CCIS 672, pp. 311–316, 2016.
DOI: 10.1007/978-3-319-49157-8_27

2   Creating GACS Core

The process began in March 2014 with the formation of a joint GACS Working Group. After a preliminary analysis found that some 98 % of the indexing fields in AGRIS used just 10,000 out of the 32,000–plus concepts in AGROVOC, mapping began with three selections of 10,000 most frequently used concepts. These were algorithmically mapped to each other, pairwise, by adapting the AgreementMakeLight ontology matching system[5]; mappings were verified by hand; a second algorithm checked for clusters of inconsistent mappings ("lumps"); the lumps were discussed online or in meetings; as a result of decisions taken, the mappings were corrected by hand (to remove mappings or to change their meaning); and the corrected mappings were used to generate new concepts algorithmically. Concepts in the new concept scheme were given URIs in a new namespace[6] and represented in RDF using the W3C standard, Simple Knowledge Organization Scheme (SKOS).[7] This initial set of core concepts is called GACS Core in the expectation that GACS will become more comprehensive in scope and less centralized in its maintenance.

Figure 1 shows a lump detected by algorithmic analysis of the manually verified mappings, the meanings of which are spelled out in Table 1. In this case, the working group determined that *energy intake* had to do with organisms and that *energy consumption*, along with the narrower *fuel consumption*, had to do with natural resources. By deleting the mapping NA550, redefining CN6768 as narrow-to-broad, and letting the concept-creating algorithm pick the most popular labels, three new GACS concepts were created, with mappings back to their sources (see Fig. 2).

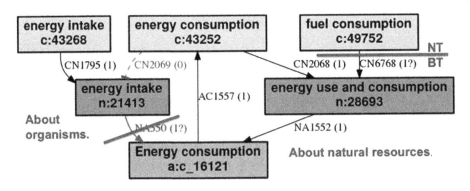

Fig. 1. Cluster of mappings between AGROVOC (a:), CAB Thesaurus (c:), and NAL Thesaurus (n:) flagged as a "lump"

---

[5] https://github.com/AgreementMakerLight/AML-Jar.

[6] http://id.agrisemantics.org/gacs/.

[7] https://www.w3.org/TR/skos-reference/.

Table 1. Set of manually verified mappings (before correction)

ID	Source concept	Mapping	Target concept
AC1557	agro:c_16121	fully equivalent to	cabt:43252
CN2069	cabt:43252	not related to	nalt:21413
NA5507	nalt:21413	probably equivalent to	agro:c_16121
CN1795	cabt:43268	fully equivalent to	nalt:21413
CN2068	cabt:43252	fully equivalent to	nalt:28693
NA1552	nalt:28693	fully equivalent to	agro:c_16121
CN6768	cabt:49752	probably equivalent to	nalt:28693

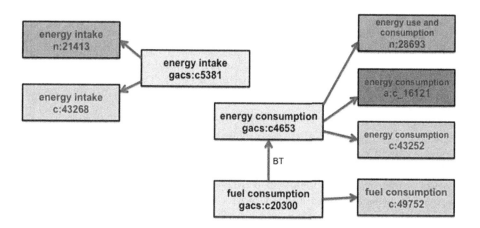

Fig. 2. Corrected mappings form concepts in GACS Core

3 Correcting GACS Core

GACS Core Beta 3.1, soft-launched in May 2016, provides 15,000 concepts labeled with 350,000 terms, some in more than twenty-five languages. This set of concepts is considered stable, with URIs that are not expected to change. The reconciliation of diverse source concepts into common GACS Core concepts, illustrated above, is largely complete. Some problems resulting from the integration process, such as overlapping labels, have been substantially fixed, though much detailed quality control remains to be done. During this test phase, implementers are encouraged to use GACS Core on an experimental basis and provide feedback.

The evolving editorial policies for GACS Core follow best practices of modern thesaurus design as per ISO 25964, "Thesauri and interoperability with other vocabularies": concepts, described with natural-language labels, clarified with definitions and scope notes, mapped to other concepts with associative and hierarchical relations, and organized into thematic groups. For Version 1.0, GACS Core must be cleaned and corrected with respect to the following:

Thematic groups. Many thesauri, including NAL Thesaurus, General Finnish Ontology (YSO), UNESCO, and STW Thesaurus for Economics (Germany) provide a thematic division of concepts into clusters orthogonal to the hierarchy of broader and narrower concepts. To provide this, the GACS working group revived an existing classification scheme that had been jointly developed by their predecessors in the 1990s, incorporated into the 1999 release of CAB Thesaurus, then set aside. Thematic group information gleaned from the 1999 CAB Classified Thesaurus (a separate thesaurus, soon to be re-released in electronic form) quite evenly covers 82 % of the concepts in GACS Core, leaving circa 2,750 unassigned.

Custom relations. AGROVOC and CAB Thesaurus each use a few properties to specify the nature of a relation between two concepts beyond the generic thesaurus relations of broader, narrower, and related. Previous efforts to "ontologize" thesauri with such additional relations revealed practical obstacles to ensuring that the properties would be applied consistently, comprehensively, and maintainably. For GACS Core, the working group decided that custom relations must meet use cases salient enough to justify the effort. Two properties qualified: `hasProduct`, and `productOf`, for relating *fish (product)* to *fish)*, the organism.

Hierarchical relations. When concepts from the three sources were merged into GACS Core concepts, their hierarchical and associative relations were also merged. GACS Core has some 600 "top concepts," or concepts with no broader concept. Top concepts are typically meant to facilitate faceted browsing or the creation of microthesauri. Ideally, top concepts should fit on just a page or two. Likewise as a result of mapping, almost one third of the concepts in GACS Core ended up with more than one broader concept. While a certain amount of polyhierarchy may be inevitable, even desirable, best practice is to keep the hierarchy as simple and pyramid-like as possible. The working group will examine how similar thesauri define their top concepts and evaluate the use cases for top concepts in light of the thematic groups. Once a set of top concepts is agreed, along with a set of principles for assigning hierarchical relations, existing relationships will be carefully vetted, pruned, and adjusted.

Semantic types. Some thesauri differentiate concepts by type, such as *organisms* or *places*. Thesauri can use the hierarchies under top-level concepts to roughly group concepts of a given type (as with AGROVOC), though hierarchies may not follow the principle of general-to-specific (hyponymy) strictly enough to ensure that an "isa" ("type of") relationship would always hold; hierarchies may also contain "part of" (meronym) relationships. Type can be assigned to concepts using subject categories (as with CAB Thesaurus) or other type systems, such as the UMLS Semantic Network (as with NAL Thesaurus). While recognizing that semantic types could usefully clarify the meaning of concepts, provide transitive "isa" relationships, and pull together concepts from across the hierarchy, the GACS working group opted to explore the benefits of committing to types by starting with simple set of *Chemical, Geographical, Organism, Product,* and the generic *Topic*.

4   GACS Extensions and Modules

Almost one third of the concepts in AGROVOC (11,000) are now tightly coupled to GACS Core, leaving a long tail of circa 21,000 concepts that are both unmapped and less frequently used. Continuing to maintain this long tail under the AGROVOC brand is possible but poses problems if GACS Core is to be officially preferred. Users would face one set of GACS Core concepts and a larger set of AGROVOC concepts, with different URIs and browsable separately. As one possible solution, the unmapped concepts of AGROVOC could be assigned GACS URIs but not marked as being in GACS Core, creating an AGROVOC-based extension to GACS. GACS URIs would be promoted while AGROVOC URIs remained mapped to their GACS equivalents and thus usable in perpetuity.

Generalizing from this case, the notion of a GACS Extension could be defined as a set of concepts within the general scope of GACS but with no overlap to GACS Core. GACS Extensions would not be subject to the constraints of shared maintenance. GACS Extensions would provide a home for concepts pruned from GACS Core. Ideally, they would be searchable through a single interface with GACS Core simply by selecting from a menu.

Concepts of well-defined types, such as organisms, geographical names, or chemicals, could in principle be defined as GACS Modules, the maintenance of which could in principle be delegated entirely to other, more expert communities. Exploration of this option will begin with vocabularies for soil data [1].

5   Towards an Agrisemantics Ecosystem

GACS Core is intended to serve as a hub within Agrisemantics, an emerging community network of semantic assets relevant to agriculture and food security.[8] The Agrisemantics idea was explored in a July 2015 workshop, with support from the Gates Foundation[9], and elaborated in the Chania Declaration of May 2016[10], which looks towards an "ecosystem of linked data repositories, data management services and virtual collaboration environments to increase the pace of knowledge production for agricultural innovation" (see Footnote 10). This goal is currently being pursued by a new Agrisemantics Working Group of the Research Data Alliance (RDA).

Like other thesauri, GACS Core provides topics for tagging information resources from bibliographic abstracts, journal articles, and grey literature, to Web resources such as videos, podcasts, and courseware. Its topical concepts, such as *farmers' attitudes* and *family relations*, are fuzzy enough to accommodate the perspectives of a broad diversity of information seekers. In contrast, datasets for quantitative analyses, such as sensor readings and crop yields, are composed of data elements defined with precision and at a fine level of granularity. Datasets

---

[8] http://agrisemantics.org.

[9] http://aims.fao.org/sites/default/files/Report_workshop_Agrisemantics.pdf.

[10] http://blog.agroknow.com/?p=5067.

are typically defined in the context of a particular software application and serialized in formats specific to that application. Interoperability across datasets is limited by the sheer effort required to determine equivalences among differently named elements, then to extract those elements from a diversity of formats.

5.1 Semantic Authority Control of Quantitative Data Elements

Between thesauri such as GACS Core and quantitative datasets lie ontologies—focused sets of concepts with precise definitions, global identifiers, and strongly typed semantic relationships. The Agrisemantics initiative proposes to test the idea that ontologies can provide a bridge between general-purpose thesauri and application-specific datasets. Ontologies can provide stable, global identity to concepts found under a diversity of local names and embedded in a diversity of software applications, in effect functioning as authorities for data elements, analogously to the library science notion of "authority control."

Semantic authority control for data elements could improve food security by supporting, for example, an analysis of the yield gap in sub-Saharan Africa. Such an analysis would need to draw both on crop-related datasets and on relevant research and multimedia resources indexed in bibliographic databases. A wheat data element, labeled 'GW' in a phenotype dataset, could be mapped to the concept 'grain weight' as defined and globally identified in the CGIAR Crop Ontology [2][11]. In turn, the Crop Ontology concept could be mapped to the broader concept 'Grain' in GACS Core. Searches could return not only datasets about grain weight, but references to published papers where the weight of the grain was studied.

In the context of Agrisemantics, GACS can serve as a hub for a richly linked network of thesauri and domain-specific ontologies, linked to innumerable quantitative datasets. By facilitating the integration of data and research results from many sources, such a semantic platform can support innovation in agriculture and contribute to the creation and management of sustainable food systems.

Acknowledgement. The views expressed in this information product are those of the author(s) and do not necessarily reflect the views or policies of FAO.

References

1. L'Abate, G., Caracciolo, C., Pesce, V., Geser, G., Protonotarios, V., Costantini, E.A.: Exposing vocabularies for soil as linked open data. Inf. Process. Agric. **2**(3–4), 208–216 (2015). http://dx.doi.org/10.1016/j.inpa.2015.10.002
2. Shrestha, R., Matteis, L., Skofic, M., Portugal, A., McLaren, G., Hyman, G., Arnaud, E.: Bridging the phenotypic and genetic data useful for integrated breeding through a data annotation using the crop ontology developed by the crop communities of practice. Front. Physiol. **3** (2012). http://dx.doi.org/10.3389/fphys.2012.00326

---

[11] http://www.cropontology.org.

The Metadata Ecosystem of DataID

Markus Freudenberg[1]([✉]), Martin Brümmer[2], Jessika Rücknagel[3],
Robert Ulrich[4], Thomas Eckart[5], Dimitris Kontokostas[1],
and Sebastian Hellmann[1]

[1] Institut für Angewandte Informatik (InfAI), AKSW/KILT,
Universität Leipzig, Leipzig, Germany
{freudenberg,kontokostas,hellmann}@informatik.uni-leipzig.de
[2] eccenca GmbH, Hainstr. 8, 04109 Leipzig, Germany
martin.bruemmer@eccenca.com
[3] re3data, Göttingen, Germany
ruecknagel@sub.uni-goettingen.de
[4] re3data, Karlsruhe, Germany
robert.ulrich@kit.edu
[5] Abteilung Automatische Sprachverarbeitung, Universität Leipzig, Leipzig, Germany
teckart@informatik.uni-leipzig.de
http://aksw.org/Groups/KILT, http://eccenca.com,
http://www.re3data.org, http://asv.informatik.uni-leipzig.de/en

Abstract. The rapid increase of data produced in a data-centric economy emphasises the need for rich metadata descriptions of datasets, covering many domains and scenarios. While there are multiple metadata formats, describing datasets for specific purposes, exchanging metadata between them is often a difficult endeavour. More general approaches for domain-independent descriptions, often lack the precision needed in many domain-specific use cases. This paper introduces the multilayer ontology of DataID, providing semantically rich metadata for complex datasets. In particular, we focus on the extensibility of its core model and the interoperability with foreign ontologies and other metadata formats. As a proof of concept, we will present a way to describe *Data Management Plans (DMP)* of research projects alongside the metadata of its datasets, repositories and involved agents.

1 Introduction

In 2006, Clive Humby coined the phrase "the new oil" for (digital) data[1], heralding the ever-expanding realm of what is now summarised as: Big Data. Attributed with the same transformative and wealth-producing abilities, once connected to crude oil bursting out of the earth, data has become a cornerstone of economical and societal visions. In fact, the amount of data generated around the world has increased dramatically over the last years, begging the question if those visions have already come to pass.

The steep increase in data produced can be ascribed to multiple factors. To name just a few: (a) The growth in content and reach of the World Wide Web.

---

[1] https://www.theguardian.com/technology/2013/aug/23/tech-giants-data.

© Springer International Publishing AG 2016
E. Garoufallou et al. (Eds.): MTSR 2016, CCIS 672, pp. 317–332, 2016.
DOI: 10.1007/978-3-319-49157-8_28

(b) The digitalising of former analogue data. (c) The realisation of what is called the Internet of Things (IoT)[2]. (d) The shift of classic fields of research and industry to computer-aided processes and digital resource management (e.g. digital humanities, industry 4.0). (e) Huge data collections about protein sequences or human disease taxonomies are established in the life sciences. (f) Research areas like natural language processing or machine learning are generating and refining data. (g) In addition, open data initiatives like the Open Knowledge Foundation are following the call for 'Raw data, Now!'[3] of Tim Berners-Lee, demanding open data from governments and organisations.

As a new discipline, data engineering is dealing with the fallout of this trend, namely with issues of how to extract, aggregate, store, refine, combine and distribute data of different sources in ways which give equal consideration to the four V's of Big Data: Volume, Velocity, Variety and Veracity[4]. Instrumental to all of this, is providing rich metadata descriptions for datasets, thereby enabling users to discover, understand and process the data it holds, as well as providing provenance on how a dataset came into existence. This metadata is often created, maintained and stored in diverse data repositories featuring disparate data models that are often unable to provide the metadata necessary to automatically process the datasets described. In addition, many use cases for dataset metadata call for more specific information depending on the circumstances. Extending existing metadata models to fit these scenarios is a cumbersome process resulting often in non-reusable solutions.

In this paper we will present the improved metadata model of DataID (cf. Sects. 4 and 5), a multi-layered metadata system, which, in its core, describes datasets and their different manifestations, as well as relations to agents like persons or organisations, in regard to their rights and responsibilities. In a previous version of DataID [1] we already provided a solution for an accessible, compatible and granular best-practice of dataset descriptions for Linked Open Data (LOD).

We want to build on this foundation, presenting improvements in regard to PROVENANCE, LICENSING and ACCESS. In particular, we want to address the aspects EXTENSIBILITY and INTEROPERABILITY of dataset metadata, demonstrating the universal applicability of DataID in any domain or scenario. As a proof of concept for its EXTENSIBILITY we will show how to provide extensive metadata for Data Management Plans (DMP) of research projects (cf. Sect. 6) by extending the DataID model with properties specific to this scenario. The INTEROPERABILITY with other metadata models is exemplified by the mapping of common CMDI (CLARIN) profiles to DataID in Sect. 7.

2   Related Work

The Data Catalog Vocabulary (DCAT) is a W3C Recommendation [2] and serves as a foundation for many available dataset vocabularies and application profiles.

---

[2] http://siliconangle.com/blog/2015/10/28/page/3#post-254300.

[3] http://www.wired.co.uk/news/archive/2012-11/09/raw-data.

[4] http://www.ibmbigdatahub.com/infographic/four-vs-big-data.

In [3] the authors introduce a standardised interchange format for machine-readable representations of government data catalogues. The DCAT vocabulary includes the special class Distribution for the representation of the available materialisations of a dataset (e.g. CSV file, an API or RSS feed). These distributions cannot be described further within DCAT (e.g. the type of data, or access procedures). Applications which utilise the DCAT vocabulary (e.g. datahub.io[5]) provide no standardised means for describing more complex datasets either. Yet, the basic class structure of DCAT (Catalog, CatalogRecord, Dataset, Distribution) has prevailed. Range definitions of properties provided for these classes are general enough to make this vocabulary easy to extend.

DCAT, as opposed to PROV-O, expresses provenance in a limited way using a few basic properties such as `dct:source` or `dct:creator`, thus it does not relate semantically to persons or organisations involved in the publishing, maintenance etc. of the dataset. There is no support or incentive to describe source datasets or conversion activities of transformations responsible for the dataset at hand. This lack is crucial, especially in a scientific contexts, as it omits the processes necessary to replicate a specific dataset, a feature easily obtainable by the use of PROV-O.

Metadata models vary and most of them do not offer enough granularity to sufficiently describe complex datasets in a semantically rich way. For example, CKAN[6] (Comprehensive Knowledge Archive Network), which is used as a metadata schema in data portals like datahub.io, partially implements the DCAT vocabulary, but only describes resources associated with a dataset superficially. Additional properties are simple key-value pairs which themselves are linked by `dct:relation` properties. This data model is semantically poor and inadequate for most use cases wanting to automatically consume the data of a dataset.

While not implementing the DCAT vocabulary, META-SHARE [4] does provide an almost complete mapping to DCAT, providing an extensive description of language resources, based on a XSD schema. In addition it offers an exemplary way of describing licenses and terms of reuse. Yet, META-SHARE is specialised on language resources, thus lacking generality and extensibility for other use cases.

Likewise the Asset Description Metadata Schema[7] (ADMS) is a profile of DCAT, which only describes a specialised class of datasets: so-called Semantic Assets. Highly reusable metadata (e.g. code lists, XML schemata, taxonomies, vocabularies etc.), which is comprised of relatively small text files.

DCAT-AP (DCAT Application Profile for data portals in Europe[8]) is a profile, extending DCAT with some ADMS properties. It has been endorsed by the ISA Committee in January of 2016[9]. Due to the stringent cardinality restrictions,

---

[5] http://datahub.io/.

[6] http://ckan.org/.

[7] https://www.w3.org/TR/vocab-adms/.

[8] https://joinup.ec.europa.eu/asset/dcat_application_profile/asset_release/dcat-ap-v11.

[9] https://joinup.ec.europa.eu/community/semic/news/dcat-ap-v11-endorsed-isa-committee.

extending DCAT-AP to serve more elaborate purposes will prove difficult. As remarked in Sect. 7 the representation of different agent roles is lacking in the current version of DCAT-AP. Neither DCAT-AP nor ADMS give any consideration to defining responsibilities of agents, extending provenance or providing thorough machine-readable licensing information.

Similar problems afflicted the previous version of the DataID ontology [1]. Rooted in the Linked Open Data world, it neglected important information or provided properties (e.g. `dataid:graphName`) which are orphans outside this domain. While already importing the PROV-O ontology, it was lacking a specific management of rights and responsibilities.

3   Motivation

In 2011, the European Commission published its *Open Data Strategy* defining the following six barriers[10] for "open public data":

1. a lack of information that certain data actually exists and is available,
2. a lack of clarity of which public authority holds the data,
3. a lack of clarity about the terms of re-use,
4. data made available in formats that are difficult or expensive to use,
5. complicated licensing procedures or prohibitive fees,
6. exclusive re-use agreements with one commercial actor or re-use restricted to a government-owned company.

Taking these as a starting point, enriched by requirements of multiple use cases (e.g. Sect. 6) and considering the existing and missing features of related vocabularies described in the previous section, we contrived the following short list of important aspects of dataset metadata:

(A1) PROVENANCE: a crucial aspect of data, required to assess correctness and completeness of data conversion, as well as the basis for trustworthiness of the data source (no trust without provenance).

(A2) LICENSING: machine-readable licensing information provides the possibility to automatically publish, distribute and consume only data that explicitly allows these actions.

(A3) ACCESS: publishing and maintaining this kind of metadata together with the data itself serves as documentation benefiting the potential user of the data as well as the creator by making it discoverable and crawlable.

(A4) EXTENSIBILITY: extending a given core metadata model in an easy and reusable way, while leaving the original model uncompromised expands its application possibilities fitting many different use cases.

(A5) INTEROPERABILITY: the interoperability with other metadata models is a hallmark for a widely usable and reusable dataset metadata model.

When regarding aspects **(A4)** and **(A5)**, taking into account the intricate requirements of many use cases (as we will see in Sect. 6), EXTENSIBILITY and

---

[10] http://europa.eu/rapid/press-release_MEMO-11-891_en.htm.

INTEROPERABILITY seem contradictory when leaving the more general levels of a domain description. A vocabulary capable of interacting with other metadata vocabularies might be too general to fit certain scenarios of use. Restrictive extensions to a vocabulary might encroach on its ability to translate into other useful metadata formats. This notion is corroborated by this document [5]. Note: We (the authors) do not differentiate between EVOLVABILITY and EXTENSIBILITY in the context of this paper. The discrepancies with INTEROPERABILITY are true for both concepts.

We conclude, not only is there a gap between existing dataset metadata vocabularies and requirements thereof, but it seems unlikely that we are able to solve all these diverse problems with just one, monolithic ontology.

4   The Multi-layer Ontology of DataID

While trying to solve the different aspects, which we discussed in the previous section, and tending to the needs of different usage scenarios, the DataID ontology grew in size and complexity. In order not to jeopardise EXTENSIBILITY and INTEROPERABILITY, we modularised DataID in a core ontology and multiple extensions. The onion-like layer model (cf. Fig. 1) illustrates the import restrictions of different ontologies. An ontology of a certain layer shall only import DataID ontologies from layers below their own. The mid-layer (or common extensions) of this model is comprised of highly reusable ontologies, extending DataID core to cover additional aspects of dataset metadata. While non of them are a mandatory import for use case specific extensions, as opposed to DataID core, in many cases some or all of them will be useful contributions.

DataID core provides the basic description of a dataset (cf. Sect. 5) and serves as foundation for all extensions to DataID.

Linked Data[11] extends DataID core with the VOID vocabulary [6] and some additional properties specific to LOD datasets. Many VOID and Linked Data references from the previous version of DataID were outsourced into this ontology.

Activities & Plans[12] provides provenance information of activities which generated, changed or used datasets. The goal is to record all activities needed to replicate a dataset as described by a DataID. Plans can describe which steps (activities, precautionary measures) are put in place to reach a certain goal. This extension relies heavily on the PROV-O ontology [7].

Statistics will provide the necessary measures to publish multi-dimensional data, such as statistics about datasets, based on the Data Cube Vocabulary [8].

Ontologies under the DataID multilayer concept do not offer cardinality restrictions, making them easy to extend and adhere to OWL profiles. An application profile for the DataID service (cf. Sect. 8) was declared using SHACL[13].

---

[11] https://github.com/dbpedia/DataId-Ontology/tree/master/ld.

[12] https://github.com/dbpedia/DataId-Ontology/tree/DataManagementPlan Extension/acp.

[13] http://w3c.github.io/data-shapes/shacl/.

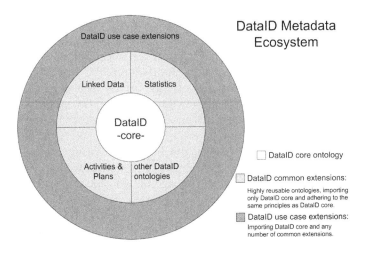

Fig. 1. The metadata ecosystem of DataID

Extending this ecosystem of dataset metadata with domain-specific OWL ontologies adds further opportunities for applications clustered around datasets, as we will showcase in Sect. 6.

5   DataID Core

This section provides a concise overview of the DataID-core ontology, highlighting important features and improvements to the previously presented version in 2014 [1]. The current version (2.0.0) adheres to the OWL profile OWL2-RL[14]. Figure 2 supplies a depiction of this ontology. DCTERMS is used for most general metadata of any concept.

DataID is founded on two pillars: the DCAT and PROV-O ontologies. The class `dataid:DataId` subsumes `dcat:CatalogRecord`, which describes a dataset entry in a `dcat:Catalog`. It does not represent a dataset, but provenance information about dataset entries in a catalog. It is the root entity in any DataID description.

In addition the VOID vocabulary plays a central role, as the dataset concept of both the DCAT and VOID were merged into `dataid:Dataset`, providing useful properties about the content of a dataset from both ontologies. In particular, the property `void:subset` allows for the creation of dataset hierarchies, while `dcat:distribution` points out the distributions of a dataset.

The class `dcat:Distribution` is the technical description of the data itself, as well as documentation of how to access the data described (`dcat:accessURL` / `dcat:downloadURL`). This concept is crucial to be able to automatically retrieve and use the data described in the DataID, simplifying, for example, data analysis. We introduced additional subclasses (e.g. `dataid:ServiceEndpoint`), to further distinguish how the data is available on the web.

---

[14] https://www.w3.org/TR/owl2-profiles/.

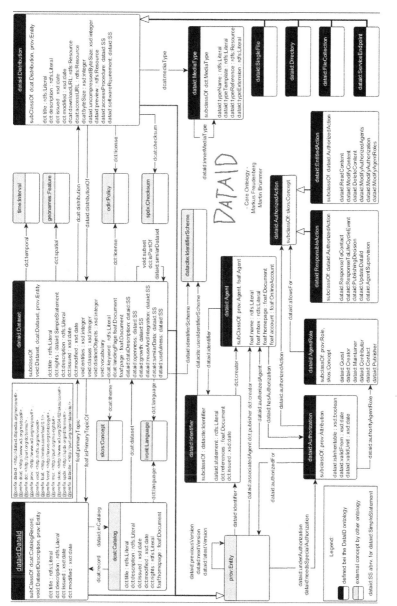

Fig. 2. DataID core

An exact description of all classes and properties can be found under the DataID namespace uri `http://dataid.dbpedia.org/ns/core` including this depiction. The ontology RDF document is also available there: `http://dataid.dbpedia.org/ns/core.ttl` (.owl)

DCAT does not offer an intrinsic way of specifying the exact format of the content described by a distribution. While the property `dcat:mediaType` does exist, its expected range `dct:MediaTypeOrExtend` is an empty class (without any further definitions). Therefore, we created `dataid:MediaType` to remedy this matter. With the property `dataid:innerMediaType` we can even describe nested formats (e.g. .xml.bz2), useful in pipeline processing.

The most important change to the previous version of DataID is the possible expression of which role an agent can take in regard to metadata entities (e.g. the whole DataID and all datasets, a single distribution etc.). This is achieved by the class `dataid:Authorization`, which is a subclass of `prov:Attribution`, a qualification of the property `prov:wasAttributedTo`. Basically it states, which role(s) (`dataid:authorityAgentRole`) an agent (`dataid:authorizedAgent`) has regarding a certain collection of entities (`dataid:authorizedFor`). This mediator is further qualified by an optional period of time for which it is valid and authoritative restrictions by the entities themselves, allowing only specific instances of `dataid:Authorization` to exert influence over them (`dataid:needsSpecialAuthorization`).

The role an agent can take (`dataid:AgentRole`) has only one property, pointing out actions it entails. A `dataid:AuthorizedAction` shall either be a `dataid:EntitledAction`, representing all actions an agent could take, as well as the actions an agent has to take (`dataid:ResponsibleAction`). Actions and roles defined in this ontology (e.g. `dataid:Publisher`) are only examples of possible implementations and can be replaced to fit a use case. Hierarchical structures of agent roles or actions can provide additional semantics.

6   Data Management Plans

Over the last years Data Management Plans (DMP) have become a requirement for project proposals within most major research funding institutions. It states what types of data and metadata are employed, The use case described here will introduce an extension to the DataID ontology to extensively describe a Data Management Plan for digital data in a universal way, laying the foundation for tools helping researchers and funders with the drafting and implementing DMPs. Based on multiple requirements, raised from different DMP guidelines, we will showcase the creation of a DataID extension. We incorporated the re3data ontology to describe repositories and institutions, exemplifying the use of external ontologies.

Requirements of Data Management Plans. The following requirements were distilled from an extensive list of DMP guidelines of different research funding bodies, covering most of the non-functional demands raised pertaining to digital datasets. A complete list of funding organisations and their DMP guidelines involved in this analysis is available on the web[15].

---

[15] http://wiki.dbpedia.org/use-cases/data-management-plan-extension-dataid# Organisation.

1. Describe how data will be shared (incl. repositories and access procedures).
2. Describe the procedures put in place for long-term preservation of the data.
3. Describe the types of data and metadata, as well as identifiers used.
4. Provisioning of copyright and license information, including other possible limitations to the reusability of the data.
5. Outline the rights and obligations of all parties as to their roles and responsibilities in the management and retention of research data.
6. Provision for changes in the hierarchy of involved agents and responsibilities (e.g. a Primary Investigator (PI) leaving the project).
7. Include provenance information on how datasets were used, collected or generated in the course of the project. Reference standards and methods applied.
8. Include statements on the usefulness of data for the wider public needs or possible exploitations for the likely purposes of certain parties.
9. Provide assistance for dissemination purposes of (open) data, making it easy to discover it on the web.
10. Is the metadata interoperable allowing data exchange between different meta data formats, researchers and organisations?
11. Project costs associated with implementing the DMP during and after the project. Justify the prognosticated costs.
12. Support the data management life cycle for all data produced.

To implement these demands in an ontology we can already make the following observations: 1.making further use of PROV-O is necessary to deal with the extensive demands for provenance, 2. a clear specification of involved agents and their responsibilities is needed and, 3. an extensive description of repositories retaining the described data is inescapable.

Our goal is to provide aid for researchers in drafting a DMP and implementing it with all requirements in mind: during the proposal phase, while the project is ongoing and the long term implementation of the DMP.

Registry of Research Data Repositories - re3data. The re3data[16] registry currently lists over 1.600 research repositories, making it the largest and most comprehensive registry of data repositories available on the web. By providing a detailed metadata description of repositories, the registry helps researchers, funding bodies, publishers and research organisations to find an appropriate data repository for different purposes [9]. Initiated by multiple German research organisations, funded by the German Research Foundation[17] from 2012 until 2015, re3data is now a service of DataCite[18]. In 2014 re3data merged with the DataBib registry for research data repositories into one service[19].

One central goal of re3data is to enhance the visibility of existing research data repositories and to enable all those who are interested in finding a repository

---

[16] http://www.re3data.org/.
[17] http://www.dfg.de/.
[18] https://www.datacite.org/.
[19] http://www.re3data.org/tag/databib/.

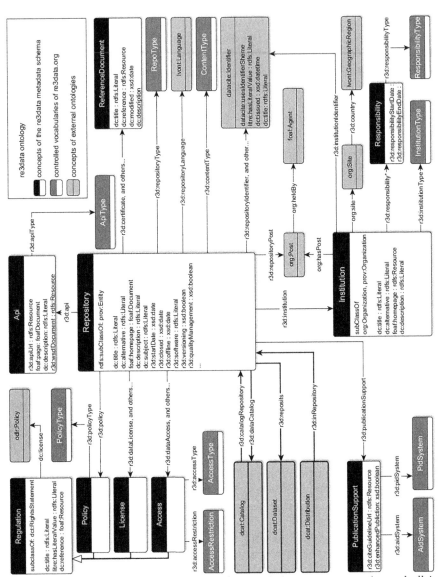

Fig. 3. re3data ontology

Note: This is a reduced version of the ontology omitting some properties and all instances of controlled vocabularies (white font on grey boxes). The re3data ontology has not been finalised by the time of submission. Some minor changes are still being discussed with re3data. The current version can be accessed here:

https://github.com/re3data/ontology/blob/master/r3dOntology.ttl.

to assess a respective information service. This is achieved by an extensive and quality approved metadata description of the listed research data repositories. The basis for this description is the "Metadata Schema for the Description of Research Data Repositories", having 42 properties in the current version 3.0 [10]. Considering the increasing number of funding bodies demanding a research data management plan as an integral part of a grant proposal, information regarding research data repositories is of great importance. The re3data schema does provide a thorough description of repositories and the unique opportunity to incorporate an existing, up-to-date collection of research repositories in future DataID-based applications. To accomplish the integration into the DMP ontology extension, we transformed the current XML-based schema into an OWL-ontology, using established vocabularies like PROV-O and ORG. The schema as well as the data provided by re3data will be available as Linked Data (e.g. via re3data ReSTful-API), thus making it discoverable and more easily accessible for services and applications, reaching a larger circle of users (Fig. 3).

Alongside the repository-concept, a rudimentary description of institutions which are hosting or funding a repository is needed to ensure long-term sustainability and availability of a repository. The derived re3data ontology supplements `r3d:Repository` and `r3d:Institution` with fitting PROV-O subclasses making them subject to provenance descriptions. The ORG ontology is used to further extend the Institution class, providing organisational descriptions.

Access regulations to the repository and the research data must be clarified, as well as the terms of use. The re3data ontology unifies all license and policy objects under the class `r3d:Regulation`, using the property `dct:license` to point out `odrl:Policy` descriptions of licenses, as used in the DataID ontology.

By linking to `dcat:Catalog` via `r3d:dataCatalog` and `dcat:Dataset` with `r3d:reposits`, we introduced the necessary means to relate descriptions of data stored inside a repository. By providing this interface with the DCAT vocabulary, DataIDs can be used for the description of data in the re3data context.

Implementation. The DataID core ontology, the Activities & Plans extension (cf. Sect. 4) and the re3data ontology are the foundational components of the DMP extension (depiction: Fig. 4). On top of which we added additional semantics, solving the requirements listed in Sect. 6.

Extensive use of the PROV-O ontology and the concepts and properties introduced by the Activities & Plans extension is key to DMP, providing the means for describing sources and origin activities of datasets **(R7)**.

In the same vein, using the `dataid:Authorization` concept, augmented with a DMP specific set of `dataid:AgentRole` and `dataid:AuthorizedAction`, adds necessary provenance and satisfies requirement **(R5)** and **(R6)**.

A description of repositories involved in a DMP is provided by the concept `r3d:Repository`, including exact documentation of APIs and access procedures **(R1)**. More detailed information on the type of data or additional software necessary to access the data, was introduced with `dataid:Distribution`.

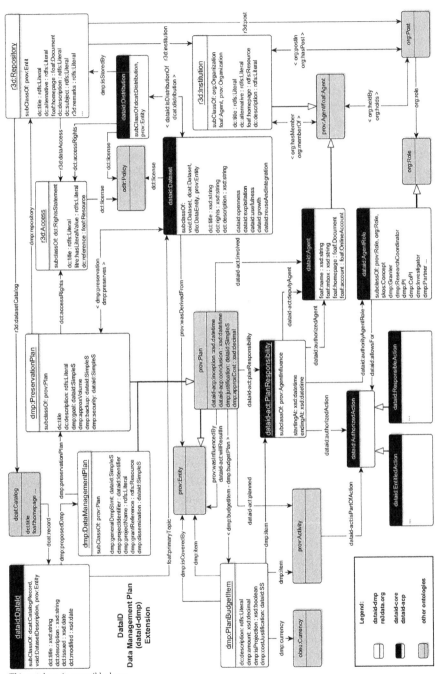

Fig. 4. Data management ontology

This ontology is accessible here:
https://github.com/dbpedia/DataId-Ontology/blob/DataManagementPlanExtension/dmp/dataManagementPlanExt.ttl

As in DataID core, information about licenses and other limitations are provided via dct:license and dct:rights (**R4**), or the complementary properties of the re3data ontology concerning access and other policies. Helpful information on usefulness, reusability and other subjects for possible users of the portrayed datasets are added to the dataid:Dataset concept: dataid:usefulness, dataid:reuseAndIntegration, dataid:exploitation etc. (**R8**).

Requirement (**R3**) is intrinsic to DataID and needs no further representation, while (**R10**) is exemplified by the next section.

Several functional requirements raised by the guidelines of research funding bodies (which are not included in the requirements of this section) will be covered by the DataID service (cf. Sect. 8). It will provide a versioning system for DataIDs (based on properties like dataid:nextVersion), enabling features like tracking changes to a DataID over time. Thereby, the full data management life cycle of datasets is supported (**R12**), which spans all phases of a Data Management Plan, but this is outside of the scope of this document.

The heart of the DMP extension are two subclasses of prov:Plan: The dmp:DataManagementPlan provides the most general level of textual statements about the DMP itself or the planned dissemination process (**R9**), as well as the necessary references to pertaining projects. While dmp:PreservationPlan entities can describe different approaches for preservation of different datasets (**R2**) or provide temporal scaling (e.g. regarding embargo periods). Besides textual statements about general goals and provisions for security and backup, using the dataid-acp:planned property to point out specific tasks, put in place to preserve data long term, is one of the more notable provenance information.

The concept dmp:BudgetItem is an optional tool to list costs pertaining to activities, responsibilities (consequently costs of agents) and any entity involved in a plan like dmp:PreservationPlan. Together with dmp:approxCost and dmp:justification it satisfies requirement (**R11**).

As a summary; we created 3 classes and 17 properties, which, together with the concepts and properties introduced by the re3data ontology, can describe Data Management Plans as demanded by the requirements of Sect. 6. An example of a DataID with DMP extension has been created by the ALIGNED H2020 project (e.g. the English DBpedia dataset[20]).

7   CMDI – Component MetaData Infrastructure

The Component MetaData Infrastructure (CMDI) is a component-based framework for the creation and utilisation of metadata schemata [11]. It allows the distributed development of metadata components (defined as sets of related elements) and their combination to profiles in any level of detail, forming the basis for the creation of resource-specific XML Schemata and around one million publicly available metadata files. CMDI is a flexible metadata framework, which can be applied to resources from any scientific field of interest. It is especially relevant

---

[20] http://downloads.dbpedia.org/2015-10/core-i18n/en/2015-10_dataid_en.ttl.

Table 1. Most popular CMD profiles and their completeness regarding DataID classes

CMD profile	CMD instances (in % of all)	Supported properties of dataid:Dataset	Supported dataid:AgentRoles
OLAC-DcmiTerms	156.210 (17,4 %)	13	3
Song	155.403 (17,3 %)	9	1
imdi-session	100.423 (11,2 %)	9	2
teiHeader	87.533 (9,7 %)	10	2

in the context of the European research infrastructure CLARIN [12] where it is used to describe resources with a focus on the humanities and social sciences.

The very flexible and open approach of the CMDI which allows for its wide applicability, may lead in parts to problems regarding consistency and INTEROPERABILITY. Despite being rich in descriptive metadata, some CMD profiles lack consistent information of the kind stated in Sect. 6. This includes the explicit specification of involved persons, descriptions of authoritative structures as well as technical details and actual download locations. Earlier work on the conversion of CMD profiles into RDF/RDFS [13] reflects the complete bandwidth of CMDI-based metadata, but also some idiosyncrasies that may constrain its usage in other contexts. It is expected that a transformation of relevant data to a uniform, DataID-based vocabulary will enhance visibility and exploitation of CMDI resources in new communities. We created explicit mappings for CMD profiles, accountable for 56 % of all publicly available metadata files, matching the appropriate DataID classes and applied them on all respective instance files via XSPARQL[21]. An overview of created mappings can be found on Github[22].

The creation and further adaptation of these mappings showed that the support of data considered essential in DataID differs between all profiles. The summary Table 1 demonstrates this effect for primary properties of `dataid:Dataset` and the support of different agent roles specified in `dataid:Agent`. Apparently there is a varying degree of conformance of both approaches, indicating possible shortcomings in specific CMD profiles. An example for such a potential deficit is the fine-grained modelling of involved persons or organisations via DataID's Agent concept that is only partially supported in most profiles.

8   Lessons Learned and Future Work

We modularised the DataID ontology into a multilayer composition arranged around a single core ontology. This was necessary to preserve EXTENSIBILITY and INTEROPERABILITY, as the vocabulary was growing due to a plethora of requirements of different use cases. An example of multiple DataIDs already

---

[21] https://www.w3.org/Submission/xsparql-language-specification/.
[22] https://github.com/dbpedia/Cmdi-DataID-mappings.

in use can be found with the latest version of DBpedia (2015-10), we stored alongside the datasets (e.g. for the English DBpedia[23]).

We have shown that by extending DataID core with existing addendums and even external ontologies, we could satisfy complex metadata requirements like those of Data Management Plans, while keeping the ability to inter-operate with other metadata vocabularies (like CMDI) in turn. In the wake of this process we incorporated the re3data XML schema into our metadata system, resulting in homogenised metadata. This holds not only for merging external repositories, but also for the identification of potential shortcomings within the same repository as has been shown by converting CMD profiles. The conversion process especially helps to uncover data quality issues and schema gaps.

We are in the process of implementing a DataID service and website to simplify and automate the creation, validation and dissemination of DataIDs, supporting humans in creating DataIDs manually, as well as automation tasks with a service endpoint. Additional work has to be done with DataID extensions, to offer additional dataset description options. Integrating DataID fully into the processes and tools defined by the ALIGNED project is another outstanding task. DataID core is planned to be published as a W3C member submission.

Acknowledgements. This research has received funding by grants from the H2020 EU projects ALIGNED (GA 644055) and FREME (GA-644771) as well as the Smart Data Web (GA-01MD15010B).

References

1. Brüummer, M., et al.: DataID: towards semantically rich metadata for complex datasets. In: Proceedings of the 10th International Conference on Semantic Systems, SEM 2014, pp. 84–91. ACM, Leipzig (2014)
2. Deri, F.M., Galway, N.: Data Catalog Vocabulary (DCAT). W3C Recommendation. https://www.w3.org/TR/vocab-dcat/
3. Maali, F., Cyganiak, R., Peristeras, V.: Enabling interoperability of government data catalogues. In: Wimmer, M.A., Chappelet, J.-L., Janssen, M., Scholl, H.J. (eds.) EGOV 2010. LNCS, vol. 6228, pp. 339–350. Springer, Heidelberg (2010)
4. McCrae, J.P., Labropoulou, P., Gracia, J., Villegas, M., Rodríguez-Doncel, V., Cimiano, P.: One ontology to bind them all: the meta-share owl ontology for the interoperability of linguistic datasets on the web. In: Gandon, F., Guéret, C., Villata, S., Breslin, J., Faron-Zucker, C., Zimmermann, A. (eds.) ESWC 2015. LNCS, vol. 9341, pp. 271–282. Springer, Heidelberg (2015)
5. Nielsen, H.F.: Interoperability and evolvability. https://www.w3.org/Protocols/Design/Interevol.html
6. Alexander, K., et al.: Describing linked datasets with the VoID vocabulary. W3C Interest Group Note. https://www.w3.org/TR/void/
7. McGuinness, D., Lebo, T., Sahoo, S.: The PROV ontology. W3C Recommendation. http://www.w3.org/TR/prov-o/
8. Cyganiak, R., et al.: The RDF data cube vocabulary. W3C Recommendation. https://www.w3.org/TR/vocab-data-cube/

---

[23] http://downloads.dbpedia.org/2015-10/core-i18n/en/2015-10_dataid_en.ttl.

9. Pampel, H., et al.: Making research data repositories visible: the re3data.org registry. PLoS ONE **8**(11), e78080 (2013)
10. Rücknagel, J., et al.: Metadata schema for the description of research data repositories. In: GFZ Germans Research Center for Geosciences
11. Broeder, D., et al.: A data category registry- and component-based metadata framework. In: Proceedings of LREC. European Language Resources Association (2010). ISBN: 2-9517408-6-7
12. Hinrichs, E., Krauwer, S.: The CLARIN research infrastructure: resources and tools for e-Humanities scholars. In: Proceedings of LREC 2014. European Language Resources Association (ELRA) (2014)
13. Durco, M., Windhouwer, M.: From CLARIN component metadata to linked open data. In: LDL 2014, LREC Workshop (2014)

DSCrank: A Method for Selection and Ranking of Datasets

Yasmmin Cortes Martins[1,3(✉)], Fábio Faria da Mota[2],
and Maria Cláudia Cavalcanti[1]

[1] Military Institute of Engineering, Rio de Janeiro, Brazil
`yoko@ime.eb.br`
[2] IOC/FIOCRUZ, Rio de Janeiro, Brazil
`fabio@ioc.fiocruz.br`
[3] National Laboratory of Scientific Computing, Petrópolis, Brazil
`yasmmin@lncc.br`

Abstract. Considerable efforts have been made to build the Web of Data. One of the main challenges has to do with how to identify the most related datasets to connect to. Another challenge is to publish a local dataset into the Web of Data, following the Linked Data principles. The present work is based on the idea that a set of activities should guide the user on the publication of a new dataset into the Web of Data. It presents the specification and implementation of two initial activities, which correspond to the crawling and ranking of a selected set of existing published datasets. The proposed implementation is based on the focused crawling approach, adapting it to address the Linked Data principles. Moreover, the dataset ranking is based on a quick glimpse into the content of the selected datasets. Additionally, the paper presents a case study in the Biomedical area to validate the implemented approach, and it shows promising results with respect to scalability and performance.

1 Introduction

The Semantic web is an extension of the traditional Web. In order to build it, most efforts and initiatives involve the introduction of semantic annotations to describe resources, such as data and texts, using the Resource Description Framework (RDF[1]).

These resources should be identified by accessible URIs (Uniform Resource Identifiers). The idea is to build a global data space containing billions of described data - the Web of Data [1], through which it is possible to navigate. The LOD[2](Linking Open Data) initiative is one of the main efforts that has been contributing to the growth of the Web of Data. It provides a set of best practices that should be adopted by data publishers to facilitate the linking of

---

[1] http://www.w3.org/rdf.
[2] http://www.w3.org/wiki/SweoIG/TaskForces/CommunityProjects/
LinkingOpenData.

© Springer International Publishing AG 2016
E. Garoufallou et al. (Eds.): MTSR 2016, CCIS 672, pp. 333–344, 2016.
DOI: 10.1007/978-3-319-49157-8_29

data. One of these practices is to use uniform vocabularies to form RDF assertions. Ontologies may provide even richer assertions, as it is a formal, explicit specification of a shared conceptualization [11]. The use of ontologies to describe data, enables machine reasoning and new assertions emerge from inferences over these data.

The Biomedical area is one of the areas that published several datasets according to the LOD principles, using ontologies and vocabularies. Nowadays, there are more than 600 ontologies in BioPortal [10] and in the Open Biological and Biomedical Ontologies (OBO)[3].

One of the main LOD principles is to publish already interlinked data. This is very important because it amplifies knowledge discovery. But before publishing a new dataset according to these best practices, it is important to know which is the best dataset to link to [3]. And, when there are many distributed alternatives, it is important to have some method to investigate which are the most relevant datasets. The main idea is to reduce the number of datasets, by ranking and selecting them, and consequently, to reduce the costs of the subsequent dataset mapping process. Nevertheless, it is out of the scope of this work to effectively map items of a dataset to other datasets' items.

This paper proposes the DSCrank method, which was built to address this issue. Taking into account the new dataset content, it starts with a simple list of URLs, which point to dataset catalogues (listings of datasets), and it ends with a list of the best ranked datasets. In order to do that, it uses a combination of SPARQL[4] queries and an adaptation of the focused crawling strategy [2]. In addition, it performs a relevance analysis, ranking the datasets according to their relevance with respect to the new dataset, taking into account the frequency and coverage of its terms inside the target datasets.

Experiments were performed in the context of a project named BIOKNOWL-OGY[5], in the biomedical area, and showed interesting results. The MetaResistomeDB database was used as the dataset to be published. Two experiments were run, using two differemt datasets catalogues, each one with a different HTML structure. The results are given in terms of the time taken to run each part of the DSCrank implementation (crawling and ranking), the amount of data noise filtered from the dataset catalogues, and the scores of the best ranked datasets. A specialist was invited to validate the method by confirming the relevance of each dataset according to the description presented.

2   DSCrank Method

DSCrank is a method to find and select semantic data sources whose domain are close to a dataset that needs to be published (publishing dataset) to be linked to others. The method is composed of two main components: Navigation & Filtering and Analysis & Ranking. The first component is in fact a focused

---

[3] http://www.obofoundry.org/.

[4] https://www.w3.org/TR/sparql11-overview/.

[5] http://bioknowlogy.biowebdb.org.

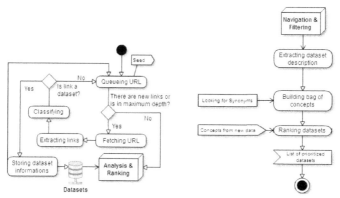

(a) Navigation & Filtering compo-   (b) Analysis & Ranking com-
nent   ponent

Fig. 1. Components of DSCrank method.

crawler, which was adapted to find and extract information from catalogues of datasets and identify the semantic links that provide access to these datasets. The second component comprises a set of steps to calculate the relevancy of the datasets found with respect to the publishing dataset.

2.1 Navigation and Filtering

The crawler component functionality is based on a generic focused crawler [2], and extends it with some new features. The focused crawler is a crawler that have seeds to guide the navigation's start. It also has a classification module to decide which URL to visit or not. Figure 1a shows the internal steps of the crawler. The new features added include datasets filtering based on the information about them in the catalogues pages (where they are listed), given as seeds.

The *Queueing* module takes as input a set of seeds (URLs of datasets catalogues). It is able to deal with each URL and manages the queue of URLs to visit. Then, the *Fetching URL* module gets each URL (which contains a list of items - potential datasets) from the queue, and gets each item information from the list of datasets: description text, name and URLs to detailed pages. This extraction is based on HTML page templates. Three templates were used: two for the catalogues whose designs were based on the CKAN[6] structures (one is for the listing page and another for the detailed dataset page), and one for the Bio2RDF catalogue page[7]. The new URLs found are delivered to the *Extracting Links* module, which searches them for external URLs, filtering out images and style or javascript files.

The new list of filtered URLs (links) proceed to verification by the *Classifying* module. This module evaluates all the links found to select those who point

---

[6] http://ckan.org/about/.
[7] http://download.bio2rdf.org/release/3/release.html.

to RDF files or SPARQL endpoints. This evaluation takes into account the *availability* dimension of a data quality metric presented in [13], which measures the portion of the dataset that is present, obtainable and ready for use. This module tests the URL by submitting a SPARQL query to it and evaluating the type of data returned. It also tests if they return a valid RDF file. These tests correspond to the following availability dimension metrics: A1 (if the server responds to a SPARQL query), A2 (if an RDF dump is provided and can be downloaded) and A3 (detection of dereferenceability of URIs). If one of the URLs tested returns a semantic content, the data found about each potential target dataset is saved in the Sesame[8] Triple Store.

Each queued URL, when fetched is removed from the queue. When the queue is empty, the *Crawler* component activates the *Relevance analysis* module. So, at the end of this component, those datasets which have no description or semantic content associated are removed. This filtering process decreases the number of datasets to be analysed by the next component.

2.2 Analysis and Ranking

This component does the Relevancy analysis to rank the datasets involving tasks which calculate the affinity between data items from the source and target datasets. Algorithm 1 is used to describe the process of this analysis. Its input is the list of potential datasets found before, and the path where the publishing dataset can be found.

In the *Extracting dataset description* step, the datasets information saved previously are retrieved forming a list to be analysed case by case (this list is used as input to Algorithm 1). Each list item contains the name, description and the URLs of each target dataset. The list is delivered to the *Building bag of concepts* step. This step is responsible for building two bags of terms, one for the publishing dataset, and the other for each target dataset.

First, to build the bag (represented in Algorithm 1, line 3) related to the publishing dataset (named L_0), the *Building bag of concepts* module issues a generic SPARQL query over it, to retrieve distinct labels of *subjects* and *objects* from all triples. Then, NLP (natural language processing) tasks are applied, such as cleaning, for removing special characters, and stemming [4].

After, the same module processes the description texts of each potential target dataset (represented in Algorithm 1, line 5), stored in the *Datasets Database*. It applies NLP tasks, such as cleaning, tokenization [9], Part of Speech tagging [12], for selecting only nouns, adjectives and their variations, and Synonym enrichment using WordNet[9], for adding a cloud of similar words. At the end, each target dataset (D_k) has a list of terms (L_k) that represents it.

In the *Ranking datasets* step, the affinity between the publishing and each target dataset is defined according to an analysis of the target dataset contents, by querying their corresponding endpoints or RDF remote files. In the first case,

---

[8] http://rdf4j.org/about.docbook?view.
[9] http://wordnet.princeton.edu/wordnet/download.

Algorithm 1. Relevance analysis algorithm.

1: **Input:** *potential_datasets* //List of datasets with their informations; *path* // Path of the publishing dataset
2: **function** *analysis_for_ranking* (*potential_datasets, path*)
3:     *input_concepts* ← get_treat_concepts(*path*)
4:     **for** each *dataset* in *potential_datasets* **do**
5:         *bag_of_concepts* ← treat_description(*dataset['description']*)
6:         *dataset_score* ← 0
7:
8:         **for** each *url* in *dataset['urls']* **do**
9:             *type_url* ← verify_type_source(*url*)
10:
11:             **if** *type_url* = 'file' **then**
12:                 **if**   test_code_http(*url*)=200   and   test_size_limit(*url*)   and not(generic_namespace(*url*)) **then**
13:                     *query* ← prepare_query_graph(*url*)
14:                     *bag_of_concepts*   ←   *bag_of_concepts* + retrieve_remote_concepts(*query*)
15:                 **end if**
16:             **end if**
17:
18:             **if** *type_url* = 'endpoint' **then**
19:                 *url_test* ← *url_test* + *query_limited*
20:                 **if** test_code_http(*url_test*)=200) **then**
21:                     **for** each *concept* in *input_concepts* **do**
22:                         *query* ← prepare_query_count(*concept*)
23:                         *score* ← retrieve_count_matched_concepts(*query*)
24:                         save_score_concept(*score, concept*)
25:                     **end for**
26:                 **end if**
27:             **end if**
28:
29:         **end for**
30:
31:         calc_final_score(*potential_datasets, input_concepts, bag_of_concepts*)
32:     **end for**
33: **end function**

SPARQL queries are issued to count how many terms from the L_0 list are found in the target dataset. An example of this query for the term *"penicillin"* is: *select (count(?label_target) as ?cont) where { ?uri_target rdfs:label ?label_target . filter(regex(?label_target, 'penicillin', 'i')) . }*. The returned results are used to compose a list called LE_k, which maintains pairs (t_i, f_i), where t_i is the term from L_0 and f_i is the frequency with which t_i occurs in a D_k dataset. Before sending the SPARQL queries, it verifies if the URL is available (code HTTP equals to 200). This case procedure is described in Algorithm 1, from line 18 to 27.

Algorithm 2. Algorithm for the final dataset score calculus

1: **Input:** *potential_datasets* //List of datasets with their informations; *input_concepts* // List of treated concepts of source dataset; *final_bag_of_concepts* // Final list of target bags of concepts

2: **function** *calc_final_score* (*potential_datasets, input_concepts, final_bag_of_concepts*)

3:     **for** each *dataset* in *potential_datasets* **do**

4:         **for** each *concept* in *input_concepts* **do**

5:             *concept_add_score* ← grep_search_hits(*concept, final_bag_of_concepts*)

6:             change_score_concept(*concept_add_score, concept*)

7:         **end for**

8:

9:         $\alpha \leftarrow 0.5$

10:        $\beta \leftarrow 0.5$

11:        *sum_tf_idf* ← calculate_sum_tf_idf()

12:        *coverage* ← calculate_ratio_present_terms()

13:        *dataset_score* ← $(\alpha * sum\_tf\_idf) + (\beta * coverage)$

14:        save_dataset_score(*dataset_score, dataset*)

15:     **end for**

16: **end function**

The second case (remote RDF file) is described in Algorithm 1, from line 11 to 16. For this case, an aggregation query would be too expensive, once every counting query for each L_0 term would need a full scan of the remote file of a D_k database. Thus, a simple triple pattern query is issued to retrieve all labels (objects of triples whose predicates are *rdfs:label*) available at the RDF file. These labels are added to the original L_k list (generated at the *Extracting dataset description* step), enriching it. In this case, the availability dimension is also verified testing if the URLs has a valid content and size lower than 1 GB.

After that, the rest of the ranking calculus is described in Algorithm 2, from line 3 to 7. It takes as input the list of datasets, the list of terms of the publishing dataset (*input_concepts*) and the list of terms of the target dataset L_k (*final_bag_of_concepts*). It does the frequency counting for the second case, differently from the first case. LE_k is updated with the respective (t_i, f_i) pairs. Each frequency counting is made using the GREP strategy[10] for efficient keyword search in text files. In this case, the keyword is a term from L_0 and the text file is the list L_k.

Finally, the relevance of a dataset D_k is given by the $Score_k$ value, which is calculated according to the formula below. It sums and weights two parts, which correspond to relevance parameters. The first part is based on the TF/IDF calculus [6], which is used here as a way to calculate the relevance of a term t_i from D_k, verifying its relative frequency in all the target datasets. The other part calculates the coverage of terms from D_0 in D_k, which means how many

---

[10] http://info.ils.indiana.edu/~stevecox/unix/s603/man/grep.1.pdf.

terms from D_0 are present in D_k dataset.

$$Score_k = \alpha * (\sum_{i=1}^{|L_0|} calc\_TfIdf(t_i, f_i)) + \beta * (|LE'_k|/|L_0|)$$

Following the $Score_k$ formula, α and β are weights given to each part. They assume values in the interval [0,1], and are complementary ($\alpha + \beta = 1$). The function $calc\_TfIdf(t_i, f_i)$ calculates the TF-IDF value of each term t_i based on its frequency f_i and each result coming from this function is summed up for dataset D_k. LE'_k is a list derived from LE_k which contains the terms of LE_k whose frequency f_i is greater than zero (or $LE'_k = \{(t_i, f_i) \in LE_k | f_i > 0\}$). So the coverage is given by the ratio between the quantity of terms in L_0 that were found at least one time in D_k and the total of terms in L_0. This final score calculus is described in Algorithm 2, between lines 9 and 14. As a default, α and β weights may assume the same value (0.5).

The results obtained by the two steps are saved for future usage. Information about each target dataset, such as its score and the terms (of the source dataset) coverage and frequency, could be used to feed some mapping mechanism. The DSCrank method was implemented as a JAVA application, and could be combined/extended with a dataset mapping application.

3   Experiments and Evaluation

The DSCrank implementation was used to perform experiments, and evaluate if the proposed method is able to identify and rank datasets. These experiments were performed in the context of a real case in the Biomedical area. The idea was to verify if the method is efficient in terms of performance, scalability and precision.

Two experiments were made, each one using a different dataset catalogue as input. One of the catalogues chosen is the *Linked Data Catalogue* hosted by Mannheim University, which contains 1921 datasets listed. The second choice was the Bio2RDF Catalogue, which contains 35 datasets listed. Both experiments used as the source dataset, a recently published dataset, the MetaResistomeDB[11], which contains 2899 triples and that should be linked to external databases. It includes data about bacteria resistance type, protein targets and antibiotics, such as Cloxacillin, Penicillin, Cefoxitin.

For the first step (Crawling component), DSCrank found/selected 87 datasets using the Mannheim Catalogue, and 35 datasets using Bio2RDF. The resulting list of 122 selected datasets was evaluated by a specialist, who classified each dataset, with respect to their relevance to the MetaResistomeDB, assigning 5 classification levels: 4 - Strongly relevant; 3- Relevant; 2 - Weakly relevant; 1 - Neutral and 0 - Not relevant. The list of datasets and their respective grades are compiled and available publicly[12]. The specialist participation was important

---

[11] http://bioknowlogy.biowebdb.org/metaresistomedb/sparql-vt.php.
[12] http://ypublish.info/pdf-validation-table.pdf.

because he knows the publishing dataset domain and will evaluate precisely the described target datasets. Furthermore, the specialist opinion was used as a reference (baseline) for evaluating the DSCrank implementation performance.

The experiments were performed using the following computer configuration: Windows 10 as Operational System, processor core i5 third generation, 8 GB of RAM and 500 GB of Hard disk.

3.1   First Experiment (Mannheim Catalogue)

The Mannheim experiment started with a URL that embeds a filtering option. The linked data Mannheim catalogue, without filters, has a total of 1921 datasets. In order to reduce the search space, we simulated the choice of a researcher to filter a subset of all of these datasets. Thus, the expression *life-science* was used as a keyword to reduce the search space to a subset related to the publishing dataset domain. After this search, the quantity of datasets returned was smaller but in a sufficient number to measure the scalability.

A total of 161 of the 1921 datasets were given for the navigation part using as seed one single URL[13] which initiates the listing pages. These pages, using the CKAN structure, formed the set of 9 seeds (9 URLs) that were queued and further analyzed.

The Navigation part to find and filter potential datasets operates in levels. In the first level, to analyze the first identified initial seeds, it spent 27 min and 20 s, returning 87 datasets. After this level, 13 new links were found and added to the queue for future processing. These links are added when the algorithm can not be sure the dataset is a potential candidate looking at the links related section only. So, it decides to investigate one more level to test a source metadata link to search an external source. In the second level, there were no datasets found.

Many datasets were filtered out. This is important because it decreases the noise for the relevance analysis part and improves performance. At the end of the navigation part, a total of 74 datasets were discarded, either because they did not have a description, or because the associated links did not return any semantic content.

The Relevance analysis part spent 22 h 49 min and 5 s. The list of the ten best ranked datasets, and their corresponding score values, for this catalogue was: [1] *Bio2RDF::Clinicaltrials* (21,737.54); [2] *BioSamples RDF* (16,029.43); [3] *Allie Abbreviation And Long Form Database in Life Science* (9,795.88); [4] *CHEMBL RDF* (3,279.05); [5] *Bio2RDF::Drugbank* (2,607.58); [6] *Bio2RDF::Omim* (1,298.53); [7] *Bio2RDF:Ncbigene* (912.49); [8] *Bio2RDF::Ctd* (612.15); [9] *Bio2RDF::Irefindex* (398.60) e [10] *CHEMBL-RDF (@Uppsala University)* (303.25).

According to the specialist, the classification for the datasets in the list summed: 3 Strongly relevant; 1 Relevant; 4 Little Relevant; 1 Neutral, and 1 Not Relevant. So, there were eight in the specialist relevance group within the

---

[13] http://linkeddatacatalog.dws.informatik.uni-mannheim.de/dataset? q=lifescience&sort=score+desc,+metadata_modified+desc.

ten DSCrank selected top datasets, resulting in a precision[14] of 80 %. However, the specialist based his analysis on his knowledge and on the brief description of the dataset, and as it was not possible for the specialist to navigate into the datasets, a further analysis were made for the two datasets that were not relevant according to him.

The *BioSamples RDF* dataset (classified as not relevant) had a high value for the *TF-IDF* calculus, for its corresponding list of terms. This means that several terms of the publishing dataset list (L_0) had a high frequency, and they also occurred in the other datasets. With respect to the terms' coverage, the maximum number of terms from L_0 found in all target datasets was 32, in a total of 337 terms (0.095), got by the *CHEMBL RDF* dataset. The *BioSamples RDF* dataset covered 27 terms, which means that it gets a good coverage if compared to the other top datasets. Therefore, taking into account that the *BioSamples RDF* dataset may be considered as relevant, despite of the specialist opinion, DSCrank obtained a precision of 90 %.

In terms of recall results, 27 datasets were classified as relevant by the specialist, but only 33.3 % of these were retrieved.

3.2   Second Experiment (Bio2RDF Catalogue)

Different from the first catalogue, Bio2RDF catalogue is not open for publication of datasets. In other words, this type of catalogue is just for consuming and its publication is curated. In addition, it is focused in the biomedical area, and the latest release had 35 datasets listed available[15]. The navigation part was very quick (2s). Moreover, since all the datasets had a corresponding description and an available endpoint, all of them passed to the relevance analysis part.

Relevance analysis spent 13 h 58 min and 8 s. Similar to the previous experiment, the top ranked datasets were selected: [1] *Saccharomyces Genome Database* (52,156.76); [2] *Clinical Trials.gov [clinicaltrials]* (7,120.59); [3] *Comparative Toxicogenomics Database [ctd]* (6,770.40); [4] *Drugbank [drugbank]* (1,873.39); [5] *Kyoto Encyclopedia of Genes and Genomes [kegg]* (998.84); [6] *NCBI Gene [ncbigene]* (599.29).

According to the specialist, the classification for the datasets in the list summed: 3 Strongly relevant; 2 Little Relevant and 1 Not Relevant. So, the specialist selected 5 in the 6 datasets top ranked by DSCrank. A similar motivation, as in the other experiment, led us to further investigate the *Saccharomyces Genome Database* (not relevant according to the specialist). It obtained a high value for the TF-IDF calculus, showing some importance in relation to the terms found and the frequency against other datasets. But, in this case, compared to the other target datasets, the *Saccharomyces Genome Database* dataset had a very low coverage. This happened because the terms found in a great quantity had a common stem. In this case, the specialist opinion was confirmed. In summary, for this experiment, the precision was 83.3 %. With respect to the recall,

---

[14] Ratio between relevant datasets retrieved and the number of top ranked datasets.
[15] http://download.openbiocloud.org/release/3/release.html.

the total of datasets considered relevant by the specialist was 21 in 35. Since using DSCrank, just 5 relevant datasets were selected, the recall was 23.81 %.

3.3   Discussion

Both experiments had as goal evaluating if the method attended the following criteria: (i) filtering efficiency; (ii) scalability; (iii) precision.

With respect to (i), the experiments show that the Relevance analysis part is a time consuming task. Experiments 1 and 2 took almost 23 h and more than 14 h for this part, respectively. The idea was to filter the datasets that worth to be ranked, avoiding the waste of time while performing the Relevance analysis. For the first experiment, filtering reduced in 46 % (74 discarded in 161) the number of the initial catalogue items, while the second, reduced in 40 % (14 in 35). Without such filtering, one can figure out the increase on the time that would have been spent for the Relevance analysis. Moreover, it is worth to notice that for the Navigation and filtering part, the worst time was 27 min, meaning it is worth to invest on the filter idea.

Since the Navigation part is relatively fast, with respect to scalability (ii), it is worth to analyze the Relevance analysis part. While the second experiment took about 14 h to rank 35 datasets, the first experiment (Mannheim) experiment took almost 23 h to rank 87 datasets. It shows that it scales linearly with the number of datasets to be analyzed.

Finally, with respect to item (iii), taking into account both experiments, from the 16 relevant datasets found by the DSCrank method, 13 were relevant according to the specialist. After a close analysis of the not relevant, we found a specialist mistake, and ended up with 14 relevant out of the 16. Therefore, we can say that the DSCrank method obtained an average precision of 87.5 %.

On the other hand, in both experiments, recall did not show good results. This may be due to the small size of some relevant datasets, if compared to the top ranked ones. Besides, maybe a different balance on the score formula, weighting the β term over the α term, could impact positively on these results. New experiments are planned in order to verify that. However, the idea of selecting relevant datasets to connect to, does not require that all possible relevant datasets should be addressed. In this sense, a good precision result is more important than a good recall.

Different cut-off points were used in each experiment. This was due to the different number and size of the analyzed databases. Since in both cases the results graphic (score x dataset) was characterized by a long-tail distribution, its beginning point was used as the cut-off point. Therefore, the low scored datasets were discarded.

4   Related Work

In [5], a method for dataset ranking is proposed and it is based on the calculus of the linkage capacity of a dataset. The main idea is to recommend the best ranked datasets for additional external links within the publishing dataset. However, this method is limited to the number and variety of external links present in the publishing dataset. They do not focus on the datasets' content and its similarity to the publishing dataset.

Another method is proposed in [8]. The crawling step takes the external datasets and navigates through them to find other connected external datasets. Once a list of datasets is obtained, it takes into account each dataset number of external links (equivalence relations), to calculate its relevance.

A third relevant work was proposed in [7]. This method uses a search by keyword to find potential target datasets. Additionally, it uses ontology matching techniques to filter irrelevant results. Some concepts of the publishing dataset are used to search the web for other similar datasets. Then, it applies ontology matching techniques to rank the identified datasets This may result in a very time consuming task, and may compromise the ranking quality.

Differently from all three works, DSCrank does not need to count on a set of predefined links. Instead, it uses dataset catalogues. This is a better choice as it enables to analyse dataset description, and discard the irrelevant datasets before analysing and ranking them. Moreover, relevance is calculated only for the selected datasets, using the frequency of words, and their synonyms, that are present in the publishing dataset. In addition, DSCrank uses a quicker method to calculate relevancy. The intention is to filter and prioritize datasets, as a way to speed up the matching process, which is a very time consuming technique.

5   Conclusion and Future Work

This paper presented the DSCrank method. It was developed to help discovering the best target datasets, given a new local dataset whose concepts must be linked to external datasets, as recommended by the best practices of the Linked Data initiative. DSCrank has two main steps, the navigation part, which extracts information from the patterns of HTML code and find accesses to semantic datasets, and the relevance analysis part, which learns how much the concepts of the new dataset are close to the data items of the target datasets.

Experiments on a real case study showed promising results in terms of the quantity of datasets filtered from the catalogues, showing a reduction of 46 % of the search space (for the first experiment). They also showed scalability, as DSCrank processed catalogues with more than 160 datasets. With respect to the quality of the selected datasets, more than 50 % of the best ranked datasets were considered relevant by the domain specialist.

Future works include exploring other dataset quality dimensions, such as interoperability and interlinking as other relevance analysis parameters. It means that the more external links they have, the better they would be ranked.

Acknowledgements. This work was partially funded by CAPES scholarship, CNPq (proc. 307647/2012-9) and FAPERJ (Proc.E-26/111.147/2011).

References

1. Bizer, C., Heath, T., Berners-Lee, T.: Linked data - the story so far. Int. J. Semant. Web Inf. Syst. **5**(3), 1–22 (2009)
2. Caliskan, K., Ozcan, R.: Comparing classification methods for link context based focused crawlers. In: 2013 International Conference on Electronics, Computer and Computation (ICECCO), pp. 143–146, November 2013
3. Hausenblas, M.: Exploiting linked data to build web applications. IEEE Internet Comput. **13**(4), 68–73 (2009). Accessed 01 May 2016
4. Hull, D.A.: Stemming algorithms: a case study for detailed evaluation. J. Am. Soc. Inf. Sci. (JASIS) **47**(1), 70–84 (1996)
5. Leme, L.A.P.P., Lopes, G.R., Nunes, B.P., Casanova, M.A., Dietze, S.: Identifying candidate datasets for data interlinking. In: Daniel, F., Dolog, P., Li, Q. (eds.) ICWE 2013. LNCS, vol. 7977, pp. 354–366. Springer, Heidelberg (2013). doi:10.1007/978-3-642-39200-9_29
6. Manning, C.D., Raghavan, P., Schtze, H.: Introduction to Information Retrieval. Cambridge University Press, New York (2008)
7. Nikolov, A., d'Aquin, M., Motta, E.: What should i link to? identifying relevant sources and classes for data linking. In: Pan, J.Z., Chen, H., Kim, H.-G., Li, J., Horrocks, I., Mizoguchi, R., Wu, Z., Wu, Z. (eds.) JIST 2011. LNCS, vol. 7185, pp. 284–299. Springer, Heidelberg (2012). doi:10.1007/978-3-642-29923-0_19
8. de Oliveira, H.R., Tavares, A.T., Lóscio, B.F.: Feedback-based data set recommendation for building linked data applications. In: International Conference on Semantic Systems, I-SEMANTICS 2012, pp. 49–55. ACM, New York (2012)
9. Raman, S., Chaurasiya, V., Venkatesan, S.: Performance comparison of various information retrieval models used in search engines. In: International Conference on Communication, Information Computing Technology (ICCICT), pp. 1–4 (2012)
10. Salvadores, M., Alexander, P.R., Musen, M.A., Noy, N.F.: Bioportal as a dataset of linked biomedical ontologies and terminologies in RDF. Semant. Web **4**(3), 277–284 (2013)
11. Studer, R., Benjamins, V.R., Fensel, D.: Knowledge engineering: principles and methods. Data Knowl. Eng. **25**(1–2), 161–197 (1998)
12. Toutanova, K., Klein, D., Manning, C.D., Singer, Y.: Feature-rich part-of-speech tagging with a cyclic dependency network. In: Conference of the North American Chapter of the Association for Computational Linguistics on Human Language Technology, pp. 173–180. Association for Computational Linguistics, Stroudsburg (2003)
13. Zaveri, A., Rula, A., Maurino, A., Pietrobon, R., Lehmann, J., Auer, S.: Quality assessment for linked data: a survey. Semant. Web **7**(1), 63–93 (2016)

WDFed: Exploiting Cloud Databases Using Metadata and RESTful APIs

Xin Wang[✉], Thanassis Tiropanis, and Ramine Tinati

Web and Internet Science Group, Electronics and Computer Science,
University of Southampton, Southampton, UK
{xwang,tt2,R.Tinati}@soton.ac.uk

Abstract. As a result of the development of Big Data and cloud databases, a huge amount of data are available on the Web, not only as dump files but also in databases. Due to the volume and heterogeneity of these data, it is a challenging task to find and consume them. To reduce the barrier of data sharing and reuse on the Web, we propose a data cataloguing framework called WDFed that combines the strength of both Linked Data and REST. WDFed adopts the Data Catalog Vocabulary (DCAT) to harmonise database metadata, and develops a two-way mapping between DCAT and a RESTful API. This framework provides an interoperable middle layer that enables humans as well as applications to discover and consume Big Data on the Web in a semi-automatic manner. The framework is implemented in a data portal called Web Observatory and we present several use cases to evaluate the framework.

Keywords: Linked data · Metadata · REST · DCAT · OAuth 2.0 · Cloud database · Schema.org · Distributed system

1 Introduction

Due to the Big Data hype volumes of data that are available on the Web increase significantly. In addition the development of cloud databases makes it effortless to store or publish data in databases on the Web. In the meantime, finding and consuming these data remain challenging since many of them are scattered on the Web, isolate behind different data stores, and lack metadata. These issues make data acquisition and federation a laborious and repetitive process, and leave gaps between data publishers and consumers.

Some dataset catalogues have been developed to provide common places for both publishers and consumers. They provide tools to streamline publishing, finding and using data, and provide metadata (usually brief) in a unified form. A representative of such catalogues is CKAN[1], which has been used by many organisations especially governments. Publishers are not only able to publish metadata, but also to upload data to CKAN. However, all data at CKAN gathered at a central point and no support of cloud database is available. Given the

---

[1] http://ckan.org/.

© Springer International Publishing AG 2016
E. Garoufallou et al. (Eds.): MTSR 2016, CCIS 672, pp. 345–356, 2016.
DOI: 10.1007/978-3-319-49157-8_30

volume and velocity of data on the Web, it is unlikely that a centralised approach can address data sharing issues at a large scale.

Mitchell and Wilson [13] discussed the probability of using Linked Data as a middle layer to consolidate heterogeneous data at metadata level, but leave blank many details of how data can be consumed. Meanwhile, Page et al. remind us the similarities between Linked Data and REST, that are: referring resources by HTTP URIs, encoding information (metadata) using standards, and including links among resources [4,8]. Inspired by the previous work, we propose a complete framework called WDFed that takes advantage of both Linked Data and REST. It curates detailed metadata of datasets and enables consumers to remotely access datasets via interoperable REST APIs. WDFed consists of three main building blocks:

- Data Catalog Vocabulary (DCAT) [12] that provides metadata of arbitrary datasets. It not only gives information that helps identify a dataset, but also describes the interface of the dataset through which data can be retrieved.
- A mapping that converts DCAT documents into REST [8] APIs. The REST API enables operations on datasets, especially allow consumers to query them from where they are. Since the REST APIs are solely determined by DCAT documents, catalogues built with WDFed can be recursively federated by combining their DCAT documents. The mapping is reversible that one can recover the DCAT document by traversing the REST API of a catalogue.
- A Schema.org[2] vocabulary that represents metadata as microdata embedded in Web pages. Schema.org is recognised by many search engines, which further improves the discoverability of datasets.

All three components are based on open standards to maximise interoperability. Especially the mapping between DCAT and REST APIs follows as many as possible conventions described in Request for Comments (RFC) documents by the Internet Engineering Task Force (IETF[3]). The proposed framework is implemented in a data catalogue platform called Web Observatory (WO) [17,18] at the University of Southampton.

The remaining sections of this paper are organised as follows. We firstly introduce the DCAT model in Sect. 2. Then details of the mapping between DCAT and REST APIs, including the mechanism of federating instances of the framework are given in Sect. 3. We further describe several use cases of the framework in Sect. 4. Conclusion and future plan are given in Sect. 5.

2   Representing Datasets in DCAT

DCAT is a RDF vocabulary that aims to facilitate interoperability between data catalogues on the Web, and well-suited to representing a collection of datasets. DCAT defines three main classes as shown in Fig. 1: *dcat:Catalog* that represents

---

[2] http://schema.org/.
[3] https://www.ietf.org/.

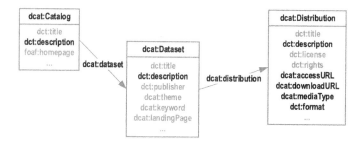

Fig. 1. A simplified diagram that demonstrates the three main classes of DCAT and their relationships. A complete diagram is given in [12].

the catalogue; *dcat:Dataset* that represents a dataset, and *dcat:Distribution* that represents an accessible form of a dataset. Each class has properties to give general information, such as *dct:title*, *dct:description*, *rdfs:label* etc. The dataset class in addition has finer information attached by properties like *dct:keyword*. The above mentioned general information is express in text and is mainly helpful for users (or applications such as recommendation systems) to identify potential datasets of interests. Besides, DCAT contains properties to provide machine understandable classification (*dcat:themeTaxonomy* for catalogue and *dcat:themes* for datasets), which, if carefully exploited, can be a promising way for automated datasets discovery, depending on the maturity of theme vocabularies.

In the proposed framework *dcat:Catalog* is used to refer to a collection of datasets and mainly serves as an entry point for exploring datasets in the system, by following *dcat:dataset* properties. Our main interest is on the other two classes, dataset and distribution. Beside general information described in the previous paragraph, a dataset can have one or more distributions, presented by the *dcat:distribution* property. This is the point from where we can extend the descriptions of datasets to services that allow operations on datasets, as described in the following section. Two key pieces of information are required to access a distribution of a dataset, the address of this distribution and its type definition (which in turn determines the procedure to access the distribution). The address is given by *dcat:accessURL* if the distribution can accessed via a service, or *dcat:downloadURL* in the case of a download file. The type definition is given by *dct:format*. In case the type is defined by IANA[4], *dct:mediaType*, which is a sub-property of *dct:format*, should be used to provide enhanced interoperability.

If the distribution is downloadable, given the address and media type of the distribution, it is sufficient for users and applications to access the distribution. However, it is not enough for users and applications to interact with distributions that are available as services, such as SPARQL [11] endpoints, SQL databases etc. The missing piece is the schema of the distribution which is necessary for constructing meaningful queries. Schemas can be included in descriptions of distributions

---

[4] http://www.iana.org/.

(by *dct:description*) for humans. For applications, however, specific vocabularies are required. For example, SPARQL endpoints can be described in detail by the Vocabulary of Interlinked Datasets (VoID) [1], which enables (semi-) automated query construction. Since RDF allows combining vocabularies without difficulty, the decision is between the benefit of automated datasets access and the efforts required to develop specific vocabularies for popular forms of datasets. The development of such vocabularies is beyond the scope of this paper.

3   Mapping DCAT Documents to REST APIs

Just like the source code of a program, DCAT carries enough information for accessing datasets, and can be "compiled" into operative REST APIs (mapping from static data to operative functions). Each resource in DCAT is mapped to a resource in the REST API. Applications can interact with REST resources via HTTP verbs that either get a representation of the resource (GET) or manipulate the resource (POST, PUT, PATCH, DELETE).

A crucial requirement of REST is Hypermedia as the Engine of Application State (HATEOAS). Simply put, it requires the representation of a resource to contain a "links" element specifying relationships of this resource to others. Figure 2 shows an catalogue and the its relationships to datasets in it. In the example each item with "links" has three fields: *href* gives the identifier of a resource; *rel* gives the relationships between the two resources of this link, and *method* gives supported methods on this resource. By sending a HTTP GET request to a dataset's identifier (e.g. https://api.example.com/catalogue/dataset_1) we should retrieve a representation of the dataset, including general information and another "links" element. Thus, HATEOAS enables users and applications to explore connected resources without referring to external documentations.

In the proposed framework a HTTP GET request to a resource retrieves a representation of the resource. For convenience we refer to the representation as a REST representation. In the remains of this section we follow the convention that values in DCAT documents are in *italic*, and values in REST representations are "double quoted".

General Mapping Rules. A top rule is that properties having resources as their values in a DCAT document are mapped to items in the "links" element of the REST representation, and properties having literal values go outside it. Key general information properties (e.g. title, description etc.) of all three classes (i.e. Catalog, Dataset and Distribution) are mapped to data fields in the REST representation by removing their namespaces.

$$\textbf{DCAT} \qquad \textbf{REST Rep.}$$
$$identifier \rightarrow identifier$$
$$dct:title \rightarrow title$$
$$dct:description \rightarrow description$$
$$dct:publisher \rightarrow publisher$$
$$dcat:keyword \rightarrow keyword$$
$$dcat:landingPage \rightarrow landingPage$$

```
GET https://api.example.com/cat

{
  "title" : "Example catalog",
  ...
  "links" : [
    {
      "href" : "/cat",
      "rel" : "self",
      "method" : "GET"
    },
    {
      "href" : "/dataset_1",
      "rel" : "item",
      "method" : "GET"
    },
    {
      "href" : "/dataset_2",
      "rel" : "item",
      "method" : "GET"
    },
    ...
  ]
  ...
}
```

Fig. 2. An HATEOAS example in JSON format shows a data catalogue and its link relations to datasets in it. Relative address are used to for simplicity.

It is worth mentioning that some properties listed above, such as *dct:publisher* and *dcat:landingPage*, have resources as recommended values and therefore should be mapped to items within "links". However it is common that the publisher only has a name, or the landing page is just a URL. In case such properties have resource values, there are two possible ways to do the mapping. One is to take the identifier of the resource value (e.g. the URI of a publisher) as the value of the corresponding field (e.g. the publisher field) in the REST representation, another is to extract a representative literal value from the resource value (e.g. the name or email of a publisher) as the value in the REST representation. We recommend the later approach since it gives immediate information and makes the mapping consistent (only literal values are presented outside "links"). In addition all properties having resource values are always (even they already occur in general information) mapped to relations in "links" to preserve the semantics of DCAT documents. Resources describing the current resource can also be mapped to "describedby" relations [3] in "links".

Mapping Rules of *dcat:Catalog*. A *dcat:Catalog* represents a collection of datasets and serves as the root (and should be the only entry) of mapped REST API. We describe the mapping rules by following an example shown in Fig. 3.

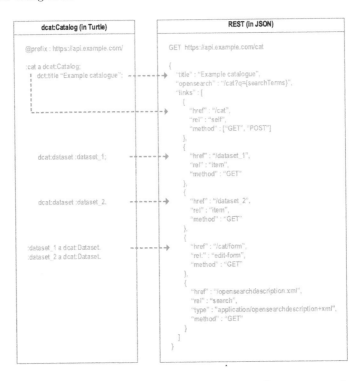

Fig. 3. Mapping rules of a *dcat:Catalog* to its REST representation.

General information is mapped according to rules described above (*dct:title* to "title"). As a catalogue it is likely that many datasets are contained and a searching function for filtering datasets would be helpful. We follow the OpenSearch 1.1 specification [5] to define the searching syntax. To emphasise the searching function the syntax is immediately available in a "opensearch" field in the REST representation. The *searchTerms* is a variable name defined by OpenSearch 1.1 and will be replaced by the keyword provided by users or application. When applicable, a more detailed OpenSearch description of the searching interface is referenced in the "search" field of "links", as described below.

The "links" element consists of following types of relationships:

– "self" [14] refers to the current catalogue.
– "item" [2] refers to a resource contained in the catalogue, i.e. a *dcat:Dataset*.
– "edit-form" [7] refers a form resource that can be used to add a new dataset.
– "search" [5] must refers to an OpenSearch description if available. The "type" must be "application/opensearchdescription+xml". Refer to the Open-Search 1.1 specification for details.
– "related" [15] refers to other related resources.

All these relations support a GET method that retrieves a representation of a resource. Besides, the "self" relation supports an extra POST method which is

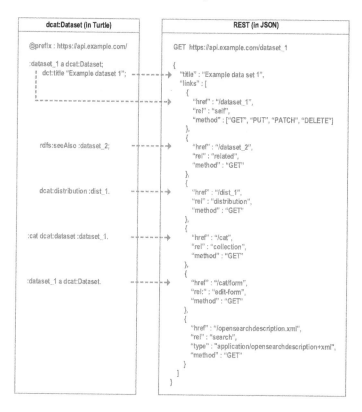

Fig. 4. Mapping rules of a *dcat:Dataset* to its REST representation.

used to create a new dataset by submitting (i.e. POSTing) a form given by the "edit-form" relation. The from is constructed according to the *dcat:Dataset* class. It is worth mentioning that there is a "create-form" relation which is specifically for creating new resources. In our case, however, creating a new dataset can be regarded as a special case of editing an empty dataset. Therefore only "edit-form" is given for simplicity.

Although it is not specified in DCAT, a catalogue (actually any resource) can have *rdfs:seeAlso* properties, and these are mapped to "related" relations. We assume that "related" relations refer to external resources which may not follow conventions in our system. They merely serve as references to other resources and do not carry the semantics of operable services. We exclude "related" relations in Fig. 3 to minimise vagueness of the semantics of the example.

Mapping Rules of *dcat:Dataset*. An exemplar of mapping rules is shown in Fig. 4. General information follows the same rules as for catalogues.

Things are changed slightly of the "links" element of datasets. We use the following relationships:

- "self" refers to the current dataset.
- "collection" [2] refers to a resource enclosing the dataset, i.e. a *dcat:Catalog*. This is the reverse relation of "item".
- "edit-form" refers a form resource that can be used to add a new dataset. This refers to the same form resource as the "edit-form" of a catalogue.
- "related" refers to other related resources.
- "search" refers to the same OpenSearch description as in catalogues if applicable.
- "distribution" refers to an access form of the dataset, i.e. *dcat:Distribution*. We propose this relation, since, to the best of our knowledge, no existing ones fitting our purpose.

The "self" relation supports four methods: GET, PUT, PATCH and DELETE. GET retrieves the representation of the dataset. By PUTing the form given in "edit-form" a new dataset is stored under the current identifier, i.e. the old dataset is replaced. PATCH allows partial update of the dataset, again using the form given by "edit-form". DELETE removes the dataset from the catalogue.

For the "related" relation, it is possible that the referred resources are datasets in the same catalogue, which in turn have well defined semantics and behaviours. As a result, unlike in catalogues, we include a "related" relation in the example, with the "method" field set to "GET". By following the "distribution" links applications can retrieve metadata that help interact with the distributions.

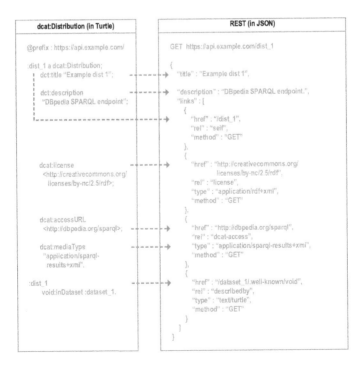

Fig. 5. Mapping rules of a *dcat:Distribution* to its REST representation.

Mapping Rules of *dcat:Distribution*. An exemplar of is shown in Fig. 5. We recommend to provide users information that is needed to explore the distribution in "description" since machine readable descriptions are not yet available for all types of distributions.

In "links" the following relations are used:

– "self" refers to the current dataset.
– "license" [16] refers to a license of the distribution. Depending on the format of the license a "type" specifying the media type should be given.
– Either "dcat-access" or "dcat-download", depending on whether the distribution is a service (e.g. SPARQL endpoint) or a downloadable file, gives the address to access this distribution. We propose these two relations with the prefix "dcat-" to minimise the probability of name conflicts.

If the distribution is a service that accepts queries (e.g. a SPARQL endpoint, a remote SQL database), OpenSearch syntax may be used to format the URL, in which {searchTerms} variable will be replaced by query strings. A preferred way is to provide an extra resource that describes this distribution by a "describedby" relation [3]. As shown in the example, a VoID document is given by a "describedby" relation, which helps construct basic queries automatically [9].

3.1 Protecting Proprietary Datasets Using OAuth 2.0

Many datasets are not open. To encourage data holders to share proprietary datasets mechanisms that can protect those datasets are necessary. The OAuth 2.0 Authorisation Framework [10] makes a good candidate since it has proved security and has been deployed in APIs of many companies. OAuth 2.0 guarantees that datasets are only accessed by authorised users and at the same time allows users to delegate applications without the needs of revealing their credentials.

4 Use Cases

The Southampton WO portal provides a catalogue of datasets, which have been developed using the proposed framework. It provides interfaces for both humans and applications to share, discover and access datasets in a secure way. The WO stores DCAT documents as JSON documents in MongoDB. Common operations on catalogued datasets, such as creating, reading, updating, deleting (CRUD) as well as searching can be straightforwardly implemented as native function calls in MongoDB.

4.1 Real Time Data Integration

The WO has been supporting analytics across several sectors of research including Web science, Data Science, Internet science, Open Innovation etc., with sites located in various international locations including, Europe, India, South Australia, Korea, Indonesia etc. A typical use case is sharing protected data from where they are without the hassle of copying them around or exposing credentials of databases.

A more elaborated use case that truly takes advantages of the framework is to enrich and annotate real time data streams using several remote heterogeneous datasets. As part of the data integration and processing in the Southampton Web Observatory, we publish real-time data from a collection of Web platforms such as Twitter, Wikipedia, and the popular citizen science platform Zooniverse. As part of our architecture, we transform and enrich these heterogeneous streams of data and unify them with a common WO Schema. These streams are then served via the Web Observatory API, enabling authentication and access to the streams.

In combination with the Southampton Web Observatory API, the real-time streams have been used to build several applications, which demonstrate the capabilities of the discoverability framework, and the ability to integrate multiple real-time streams[5]. Built as real-time analytical tools, which offer interactive visualisations of certain Web platform activity, the Web Observatory acts as a middle-layer to query relevant data and resources. Applications such as the Wikipedia activity dashboard makes use of the enriched Wikipedia stream to geographically map activity and communications within the platform, as well as interactions with other services, such as Twitter. Similarly, the Web Information Cascade application uses several of the real-time streams, along with historical datasets in order to provide an analytical representation of information passing within and between Web services. This may be the interaction of URLs, or simply shared messages, identified by common words or hashtags.

4.2   Populating Datasets Using Schema.org

Schema.org is a widely used general purpose vocabulary for embedding micro data in web pages. It is recognised by main search engines such as Google, Yahoo!, Bing etc. WO publishes a subset of metadata using Schema.org. These metadata will show in search engines and help populate datasets catalogued by WO.

Furthermore, full metadata are mapped to an extension of Schema.org vocabulary [6]. Although the extension is not yet fully supported by search engines, compatible applications can recover full DCAT metadata from these micro data. This mechanism provides a flexible way to obtain DCAT metadata of datasets. It also enables easy federation of WOs, as described below.

4.3   Catalogue Federation

The REST API of a catalogue is solely determined by DCAT documents, and therefore a bigger catalogue can be constructed by (1) combining DCAT documents of smaller catalogues, and (2) mapping the combined DCAT documents to a REST API. This feature makes the proposed framework to scale up very easily.

---

[5] http://app-001.ecs.soton.ac.uk/ramine.

There are two ways to obtain DCAT documents from WO. First, the mapping from DCAT to the REST API is bijective, and thus the DCAT of WO can be achieved by traversing its REST API. Since all resources are linked, the only knowledge required is the root URL of the WO.

The other way is via the Schema.org micro data published by WO. As stated before, WO has been deployed in several institutes in different countries. There is a demand of searching from one WO datasets in other WOs. To this end we developed an application[6] that keeps tracking all active WOs and crawls Schema.org micro data. The application is itself an aggregated WO without storing metadata in a database. Users can search any datasets in the global WO network using this application.

5   Conclusion and Future Plan

We describe a framework called WDFed to enable efficient sharing and consuming of Big Data on the Web. The framework keeps metadata of datasets using the DCAT vocabulary, which are mapped to a self-descriptive REST API to enable automatic dataset discovery and access. Further DCAT are mapped to Schema.org markups to incorporate searching ability of most search engines. Instances built following this framework can be easily federated by combining their DCAT documents. Most functionalities of this framework can be implemented as native database operations. We demonstrate the advantages of this framework via several real-world use cases.

The framework relies on adoption of Web standards. As those standards mature, more complex tasks can be automatically performed by applications. By the time this paper is written the proposed framework still uses a few nonestandard link relations, and the next step of our work is either to find standard alternatives, or to promote the current ones into standard relations. Besides, it is relatively easy to build vocabularies that provide detailed information of other structured databases, and such vocabulary can be very helpful to enable more complex autonomous behaviours.

References

1. Alexander, K., Cyganiak, R., Hausenblas, M., Zhao, J.: Describing linked datasets on the design and usage of VoID, the "Vocabulary of Interlinked Datasets". In: Proceedings of the Linked Data on the Web Workshop (LDOW), At the International World Wide Web Conference (WWW) (2009)
2. Amundsen, M.: RFC 6573 - The Item and Collection Link Relations (2012)
3. Archer, P., Smith, K., Perego, A.: Protocol for Web Description Resources (POWDER): Description Resources (2009)
4. Berners-Lee, T.: Linked data-design issues (2006)
5. Clinton, D.: OpenSearch 1.1 draft 5 (2014)

---

[6] http://socpub.cloudapp.net/w3o/wo_list.html.

6. Difranzo, D., Erickson, J.S., Gloria, M., Luciano, J.S., McGuinness, D.L., Hendler, J.: The web observatory extension: facilitating web science collaboration through semantic markup, pp. 475–480, April 2014
7. Dzmanashvili, I.: RFC 6861 - The "create-form" and "edit-form" Link Relations (2013)
8. Fielding, R.: Architectural styles and the design of network-based software architectures. Ph.D. thesis, University of California, Irvine (2000)
9. Görlitz, O., Thimm, M., Staab, S.: SPLODGE: systematic generation of SPARQL benchmark queries for linked open data. In: Cudré-Mauroux, P., et al. (eds.) ISWC 2012. LNCS, vol. 7649, pp. 116–132. Springer, Heidelberg (2012). doi:10.1007/978-3-642-35176-1_8
10. Hardt, D.: RFC 6749 - The OAuth 2.0 Authorization Framework (2012)
11. Harris, S., Seaborne, A.: SPARQL 1.1 Query Language (2013)
12. Maali, F., Erickson, J.: Data Catalog Vocabulary (DCAT) (2014)
13. Mitchell, I., Wilson, M.: Linked data: connecting and exploiting big data. Technical report, Fujitsu (2012)
14. Nottingham, M.: RFC 5988 - Web Linking (2010)
15. Nottingham, M., Sayre, R.: RFC 4287 - The Atom Syndication Format (2005)
16. Snell, J.: RFC 4946 - Atom License Extension (2007)
17. Tiropanis, T., Hall, W., Hendler, J., de Larrinaga, C.: The web observatory: a middle layer for broad data. Big Data 2(3), 129–133 (2014)
18. Tiropanis, T., Hall, W., Shadbolt, N., De Roure, D., Contractor, N., Hendler, J.: The web science observatory. IEEE Intell. Syst. 28(2), 100–104 (2013)

Metadata Management in an Interdisciplinary, Project-Specific Data Repository: A Case Study from Earth Sciences

Constanze Curdt[(✉)]

University of Cologne, Cologne, Germany
c.curdt@uni-koeln.de

Abstract. This paper presents an approach to manage metadata of (research) data from the interdisciplinary, long-term, DFG-funded, collaborative research project 'Patterns in Soil-Vegetation-Atmosphere Systems: Monitoring, Modelling, and Data Assimilation'. In this framework, a data repository, the so-called TR32DB project database, was established in 2008 with the aim to manage the resulting data of the involved scientists. The data documentation with accurate, extensive metadata has been a key task. Consequently, a standardized, interoperable, multi-level metadata schema has been designed and implemented to ensure a proper documentation and publication of all project data (e.g. data, publication, reports), as well as to facilitate data search, exchange and re-use. A user-friendly web-interface was designed for a simple metadata input and search.

Keywords: Metadata · Metadata schema · Research data management · Data repository · Interdisciplinary project

1 Introduction

Research that is conducted in an interdisciplinary context (e.g. across institutions or research groups) requires the extensive sharing and exchange of data. This is only achievable, if the involved data is provided in a structured and well-documented manner [1]. Consequently, the documentation of data, i.e. description with metadata, is a key component in data infrastructures such as research data management systems or repositories [2, 3], in particular for long-term studies [4]. Metadata enables scientists to search for data, browse for specific values, as well as cite, (re-)use and make them understandable [5, 6]. Thus, it is essential to provide metadata in a good quality to facilitate the interaction with data (collections) [7]. In this context, it is essential to apply extensive standards and schemes for the documentation, as well as for discovery and management of the scientific data. Moreover, these standards have to be modified or adjusted according to the specific requirements of considered data [8]. Hence, most research data repositories or services apply or are in compliance with one or several existing standards that are common for their discipline or that meet the requirements of their data and data providers, as presented in a study by [9] or such as [10–14]. Likewise some data repositories apply multi-level approaches for data documentation [15].

© Springer International Publishing AG 2016
E. Garoufallou et al. (Eds.): MTSR 2016, CCIS 672, pp. 357–368, 2016.
DOI: 10.1007/978-3-319-49157-8_31

Additionally, according to [16] data and metadata management systems should provide metadata input, management and versioning. Furthermore, such systems should be conform to requirements of a specific discipline or community and their needs.

This paper presents the metadata management of (research) data within the framework of the interdisciplinary, long-term research project 'Patterns in Soil-Vegetation-Atmosphere Systems' funded by the DFG (2007-2018). Initially, an overview will be given about the research project and the project data. The established data repository will be introduced. Afterwards, the metadata management will be described in detail including the self-designed metadata schema and the implementation in the user-friendly web-interface. Finally, the last section provides some discussions and conclusions.

2   Project Background and Data

The presented study is conducted for the Collaborative Research Centre/Transregio (CRC/TR) 32 'Patterns in Soil-Vegetation-Atmosphere Systems: Monitoring, Modelling, and Data Assimilation' (www.tr32.de) funded by the German Research Foundation (DFG) since 2007. The CRC/TR32 is an interdisciplinary, long-term research project in the area of earth sciences with several research groups from the German Universities of Aachen, Bonn and Cologne, and from the Research Centre Jülich. The involved scientists focus their work on exchange processes between the soil, vegetation and atmospheric (SVA) boundary layer. Their aim is to yield improved numerical SVA models to predict water, energy and CO_2 transfer by calculating patterns at various spatial and temporal scales [17]. To achieve this goal, scientists from the field of soil- and plant sciences, hydrology, geography, geophysics, meteorology, remote sensing and mathematics are involved. The scientists conduct their research within the river Rur catchment, mainly situated in western Germany.

Since 2007, the involved scientists (e.g. Postdocs, PhD students, Master and Bachelor students) have created a large amount of heterogeneous data. These have been collected in various field measurement campaigns in the Rur catchment (e.g. airborne campaigns, hydrological or meteorological monitoring), as well as by laboratory analysis or modelling approaches. As an outcome of these studies, all involved scientists generate publications, conference contributions (e.g. posters, presentations) and PhD reports. Overall, the created data is provided in various file formats, file sizes (kb to GB per single file), as well as in different spatial and temporal scales.

3   The TR32DB Data Repository

Since the project start in 2007, several research data management (RDM) services have been established and implemented to support the involved CRC/TR32 scientists during the entire research life cycle [18]. These services comprise e.g. practical training, support and guidance in RDM for project scientists. Moreover, to handle the data of the scientists, the CRC/TR32 project database (TR32DB, www.tr32db.de) was established in 2008. This data repository was developed according to requirements of the multi-disciplinary project participants (e.g. heterogeneous data, variety of data formats, different file sizes up to several GB per single file) and the DFG recommendations (e.g. cooperation with a local library or computing centre, re-use of existing infrastructures and tools). With these requirements in mind, the TR32DB was established within the local infrastructure of the Regional Computing Centre of the University of Cologne. The TR32DB has been in online operation since 2008. In detail, the TR32DB system architecture [18] is implemented in a three-tier (client server) architecture comprising various components. These include for instance the file-based data storage and backup provided and operated by the distributed, network based Andrew File System. The TR32DB data storage is organized according to the project structure (e.g. different project clusters and sections) and supports the storage of six data type categories (geodata, data, publications, pictures, presentations, pictures). The data types emerged after analyzing the variety of data, predicted by the project participants at the beginning of the project funding in a survey. Moreover, the scientist requested the storage of the identified data types in the same, centralized system. The corresponding descriptive metadata of each data file are currently stored in a MySQL database. Further administrative details of the TR32DB are stored in the database such as user information and permissions.

The access to the TR32DB is provided via a user-friendly graphical web-interface. This interface supports common features of repositories such as data search, selection of datasets, views of metadata details and data download (if access is permitted). Moreover, the web-interface provides several web mapping components, such as a map based data search. The general features are accessible for every visitor of the web site, just functions such as data upload, metadata provision and application of DOIs are only permitted for TR32DB users. TR32DB features related to metadata management will be described in detail in Sect. 4.2. Since 2008, the TR32DB is under continuous development due to changing requirements and needs of the project participants, as well as due to technical modifications. The latest updates of the TR32DB public features have been announced in the news category of the TR32DB homepage. These include e.g. changes on the data search and implementation of TR32DB data statistics. Until the project end in 2018 the system will be prepared to run in a container solution in the infrastructure of the Regional Computing Centre to ensure availability of the services and data for the future.

4   Metadata Management Within the TR32DB Data Repository

The main focus during the design and establishment of the TR32DB was on metadata with the aim to describe the project data in detail, as well as make them easily findable, exchangeable and re-usable by other scientists. Hence, a metadata schema (Sect. 4.1) was developed for the TR32DB to accurately describe all data of the repository. Furthermore, corresponding features were implemented in the TR32DB web-interface (Sect. 4.2) supporting the metadata handling. Both will be described in the following sections.

4.1   The TR32DB Metadata Schema

With the aim to describe all project data, grouped in six data type categories (*geodata*, *data*, *publications*, *presentations*, *pictures*, *reports*), with accurate, standardized, interoperable metadata, the TR32DB Metadata Schema [19] was designed and implemented in the TR32DB data repository. As no standard or schema was available, which supported the description of all chosen data categories, a new schema had to be established. The design of the schema was conducted hand-in-hand with the TR32DB user interface implementation.

Initially before designing the metadata schema, the project framework conditions (e.g. project background, predicted data) and the current state of art of metadata schemas and standards were studied. Additionally, existing metadata schemas of project-related RDM system in the earth sciences were inspected (e.g. [20–22]). As an outcome of this study, the decision was made to use a 'mix and match' approach [23] by combining various existing schemas and standards with the aim to establish a multi-level metadata schema that enables the description of all TR32DB data types. Therefore, three levels of detail where defined which should be covered by the schema. (i) The general properties should enable to describe the data with basic information with the aim to make the data findable and citable. (ii) The data type specific properties should facilitate to describe the data with specific properties of a certain data type category. Finally, (iii) the project specific properties should allow to describe the data with details focusing on the CRC/TR32 project background and demands of the scientist. A schematically overview of the TR32DB Metadata Schema is presented in Fig. 1, showing the coverage of the general (red/left), the data type specific (blue/bottom & right), and the project specific properties (green/right).

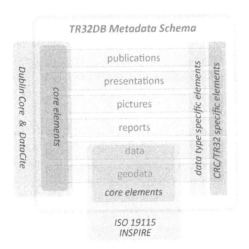

Fig. 1. Simplified overview of the TR32DB Metadata Schema showing the involved data types and the coverage of the general, data type and project specific metadata properties [18]

As a basis of the schema, metadata elements of Dublin Core [24] were applied since it is a widely accepted and simple standard that enables the description of all kind of data. Additionally, DataCite Metadata Schema Version 2.2 [25] was added to meet the demands of the CRC/TR32 participants in order to provide DOIs for project data. With regard to the data type specific properties, further elements of metadata standards and schemes such as ISO19115 Metadata Standard [26], INSPIRE [27], as well as elements of the Bibliographic Ontology [28] and the Event Ontology [29] were complemented and mapped. Furthermore, the TR32DB Metadata Schema was extended with further own project specific properties related to the background of the CRC/TR32 and demanded by the scientists (e.g. specific keywords with regard to SVA, used measurement instruments and modelling methods).

The TR32DB Metadata Schema for the documentation of research data in the TR32DB [19] is well-described in a detailed documentation. The schema specifies a defined number of metadata properties, including a core set of mandatory properties (e.g. creator, title), optional properties (e.g. identifier, relation) and automatically generated properties (e.g. metadata creator and date). In addition, available and TR32DB-specific controlled vocabulary lists are supported and listed in the appendix of the documentation. Controlled vocabulary lists are partly required by some metadata standards. Likewise such lists facilitate the metadata provision and prevent spelling mistakes. Furthermore, a mapping to the applied metadata standards is provided for interoperability.

In detail, the schema is organized in a structured list of metadata properties, arranged in two layers: (i) the general and (ii) the specific layer, presented in Fig. 2.

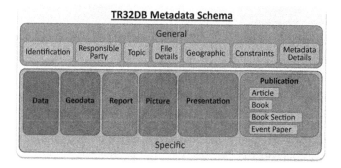

Fig. 2. Structure of the TR32DB Metadata Schema set-up in a general and a specific layer [19]

The general layer has the purpose to describe all data with the basic details such as a title, description, creator or subject. It is required and uniform for all data types. All involved metadata properties of the general layer are classified into seven categories (e.g. *identification, responsible party, topic*). The second layer is the specific layer. This layer enables the documentation of all data type categories with their specific metadata properties. Consequently, the layer is sub-divided in the six supported data type categories: *data, geodata, report, pictures, presentation,* and *publication*. The category *data* includes specific properties such as temporal extent (start/end date), lineage, used measurement instrument/modelling method, and the corresponding parameters. The category *geodata* involves specific properties such as temporal extent (start/end date), the applied reference system or spatial resolution. The category *report* contains properties such as report date, report type (e.g. fellow report, PhD report), report city and instructions. The category *picture* included properties such as recorded date (start/end date), location, method and details about the associated recording event (event type, name, location, web site). The category *presentation* contains properties such as presentation date, event (e.g. event type, name, location, we site), presentation type and presenter. Finally, the *publication* category makes an exception. This category is subdivided into different publication types, because various details are required for an accurate proper citation of a publication. Consequently, the sub-category *article* involves the properties article type (e.g. journal, magazine), publication source, publisher, volume, issue, pages and page range. In contrast, the sub-category *event paper* specifies information about the event (e.g. event name, location and period), where the paper was presented. Additionally, details about the proceedings title and editor can be specified.

A detailed description of each property and a corresponding sub-property of a specific category is described in detail in tabular form in the TR32DB Metadata Schema documentation [19]. Each property and sub-property is described with certain attributes. This include the identifier number, the property name, a short definition, the occurrence (0-n, 0-1, 1-n, 1), the obligation (Mandatory, Optional, Automatic) and further notes (e.g. allowed values (free text, date, controlled vocabulary), examples). Figure 3 presents an exemplary extract of the schema documentation of the property *title* and the sub-property *titleType*, part of the general layer and the category *identification*.

2.2.1   TR32DB 'GENERAL' Metadata Properties

ID	TR32DB property name	Definition	Occ	OB	Notes (allowed values, examples, other constraints)
Identification					
1	Title	A name given to the dataset. Typically a title will be a name by which the dataset is formally known.	1-n	M	Free Text *Example:* Enhanced Land Use Classification of 2008 for the Rur catchment
1.1	titleType	The type of Title.	1	M	*Controlled List:* **TitleType** See appendix for values and definitions.

Fig. 3. Extract of the TR32DB Metadata Schema documentation presenting the attributes of the property *title* and the sub-property *titleType* [19]

A.2.41   TitleType

ID	Value name TR32DB	Value name (dcterms)	Value name (DataCite)	Value name (ISO)	Definition
1.	mainTitle				Main title of the dataset.
2.	alternativeTitle	alternative	alternativeTitle	alternateTitle	Alternative title of the dataset.
3.	subtitle		subtitle		Subtitle of the dataset.
4.	translatedTitle		translatedTitle		Translated title of the dataset.

Fig. 4. Extract of the TR32DB Metadata Schema documentation presenting the controlled vocabulary list *TitleType* including the mapping to the applied schemes [19]

The associated controlled list of the sub-property *titleType* (Fig. 4) is presented in the appendix of the schema documentation. The listing includes the used TR32DB value and a definition, as well as a mapping to values of other applied schemes and standards (e.g. DataCite, ISO).

Besides of required and partly mandatory controlled lists provided from involved schemes and standards, also project related controlled lists were established. For instance controlled lists are available like an institution list, a TR32 keyword list, a creator list, an instrument list or a measurement location list. They are presented in the appendix of the schema documentation or at the TR32DB website (e.g. www.tr32db.de/listing/instrument.php, www.tr32db.de/listing/keyword.php).

4.2   The TR32DB Web-Interface

The TR32DB Web-Interface (www.tr32db.de) is the access platform for the users and visitors. Several services are provided with regard to metadata management. These include, for example, metadata provision and editing, data search via metadata and representation of detailed metadata of a selected dataset.

For the provision of metadata of a specific dataset, a user-friendly input wizard (Fig. 5) was designed and implemented. This wizard guides the user through the input. It is arranged according to the TR32DB Metadata Schema and divided into eight tabs. Initially, a template feature is provided that enables the re-use of existing metadata of

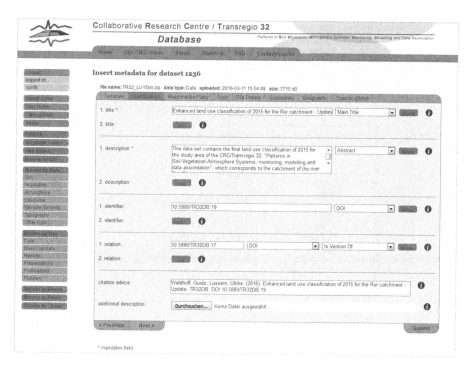

Fig. 5. Input of metadata for a dataset of the data type category *data* by using the input wizard

another dataset. This feature facilitates the metadata input for similar datasets such as time series of a specific measurement instrument. The following six tabs (*Identification, Responsible Party, Topic, File Details, Constraints, Geographic*) include the input of the general metadata properties and sub-properties. This includes the general description with a title, creator, abstract, as well as linking of datasets via relations. Depending on the selected data type, the eighth tab will be changed. This tab enables the input of the specific metadata properties according to a specific data type. Thus, the users follow the same input workflow for each data type.

In general, mandatory properties respective input fields are marked with an asterisk. Four types of input fields are provided such as text boxes, text areas, drop-down menus and calendar features. The latter prevent spelling mistakes by the users. Drop-down lists are always mandatory, if the corresponding input field or area is already filled. The submission of incomplete or invalid content will result in an error message. All required or incomplete input fields will be highlighted. Further details about specific input fields

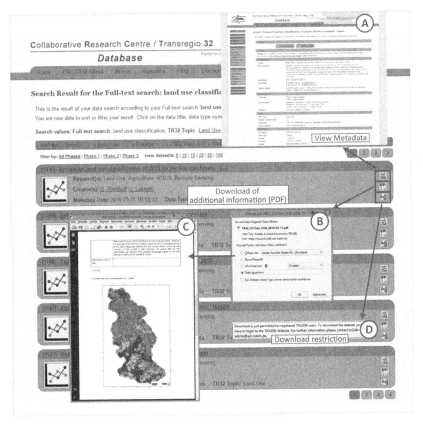

Fig. 6. Result of an advanced data search combining the values 'land use classification' (full-text search), 'land use' (TR32 topic) and 'Remote Sensing' (Keyword). Some features are provided for each dataset: (A) metadata viewing, (B) download of additional PDF file and (C) viewing and (D) presentation of download information/restriction

or drop-down lists are provided in the info-fields. They provide advanced information and examples for the specific properties. After a successful submission of all requested metadata properties and sub-properties, the dataset will be on-the-fly available in the TR32DB. The creator of the metadata is able to modify and re-submit the metadata as often as needed.

The TR32DB web-interface provides several features to search for data by means of their metadata. A map based search enables a spatial search for data with geographic information. This functionality was recently enhanced with a new feature that enables the visualization of a geographic coverage of a selected dataset on the map. A predefined data search ('browse by' function) according to selected values such as topics, data types or project sections is available. Moreover, an advanced search is provided that combines various search queries such as drop-down lists (e.g. data type, project section, creator) and a free text form. This function was also recently updated with new features such as a new search request for data of a certain period of time. The input of the search query was extended with auto-complete functions for a simplified selection of the allowed values (e.g. for keyword, creator, instrument, parameter). Additionally, a multiple selection of project sections was implemented. As a result of each data search, a list of datasets will be provided (Fig. 6). The selected search request values are displayed on top of the dataset list, which was just lately complemented. Each data type is displayed with a specific logo. Several dataset features are available. This includes the viewing of detailed metadata after selecting the title, logo or metadata view button of a specific dataset. The metadata overview page displays all available metadata of the selected dataset arranged according to the structure of the metadata schema. The geographic coverage of a dataset is now displayed on a map window, if geographic information is available. Furthermore, an optional additional descriptive PDF can be downloaded as well as the dataset itself, depending on the given permissions (e.g. download only for project section members, for all TR32DB users, free). Corresponding download information will be provided in case the download is not permitted.

5 Discussion and Conclusion

In this paper an approach was presented to manage research data and their corresponding descriptive metadata of an interdisciplinary research project with focus on soil, vegetation and atmospheric data. For this purpose, the TR32DB data repository was established. To cover all demands and data created by the project scientists, a project specific metadata schema was established and implemented in the TR32DB. This schema enables the accurate, interoperable documentation of all supported data types (e.g. geodata, data, publications, reports).

It is also an experience of other repositories, which support cross-disciplinary data that available metadata standards do not meet the requirements [10–12, 14]. Sometimes there is no one fits all metadata standard or schema available. As a consequence own metadata schemas or applications were established for some repositories such as, for example, for the Dryad repository [10] or the Datorium repository [11]. Both use the Dublin Core as basis for their schema, as it was also applied in the TR32DB Metadata

Schema. Moreover, the establishment of a multi-level approach for data repositories is quite common. This kind of solution is, for instance, used in the institutional repository of the University of Southampton. Its metadata model distinguishes between the three levels: core metadata, discipline metadata, and project metadata [15]. Similar to other existing metadata schemas and standards the TR32DB Metadata Schema provides and supports common features such as mandatory, optional and automatically generated properties and sub-properties. This ensures that TR32DB users are not overburdened in the metadata provision process. Usually TR32DB users tend to provide only mandatory properties. Controlled vocabulary lists are an important component of schemas and standards as, for example, also applied in the DataCite Metadata Schema [25] or in ISO 19115 [26]. These predefined values prevent spelling mistakes and are much more comfortable for users. Finally, a well-documented metadata schema, such as the TR32DB schema, supports its usage and the understanding of the single properties and sub-properties. With regard to further developments of the TR32DB Metadata Schema it is planned to update the used standards with recent versions as well as to integrate and map further existing metadata schemes such as DIF (Directory Interchange Format) or EML (Ecological Metadata Language). Moreover, the TR32DB Metadata Schema will be mapped to an XML structure to enable metadata import and export to the TR32DB, as well as storage of the metadata in XML format. Additionally, it is planned to expand the exchange of the TR32DB metadata. Currently, only metadata of TR32DB data with a DOI are available and accessible in other systems such as the DataCite Metadata Store. For the future it is planned to provide the TR32DB metadata also in other systems such as the European data infrastructure EUDAT or KomFor (Competence Center for research data in the earth and environmental science).

The TR32DB web-interface provides common features of data repositories with regard to metadata input, search and download of datasets. As requested by [16], web-interfaces should be designed user-friendly and with a lot of help functionalities. In addition, it is important to establish a user-friendly and well-designed metadata-input to streamline the metadata creation process. This approach should facilitate the use of drop down lists or auto completion [30]. With this in mind, a user-friendly input wizard was arranged for the TR32DB, which guides the user through the input process and provides several help options. Moreover, several search functionalities were establishes according to demands of the project participants. The user feedback for the usage of the web-interface is positive since the scientists were involved at an early stage, also in the design. On a regular basis practical training workshops or personal training have been conducted where the users usually provide/d suggestions for changes on the system.

Overall, the implementation of the TR32DB and its metadata management was successful. As of June 2016, around 600 GB of data are in the TR32DB storage related to around 1400 metadata records. By means of the metadata of the single datasets also some statistics were recently set up for the TR32DB (available online since February 2016 at http://www.tr32db.de/site/Statistics.php). These visualize, for example, the current distribution of the TR32DB data according to the project section, the data type categories, TR32 topics or measurement/modelling regions. Moreover, the top ten downloads of single datasets, distinguishes by the 'real' data file and the descriptive metadata PDF, are available at the statistic website. Additional statistics and charts of the TR32DB data are currently only available for system administrators or on user request. These include, upload and download statistics of the project data. These statistics show that internal project data (download only for TR32DB users permitted) have been downloaded and re-used by other project members. Likewise open accessible data have been downloaded from interested parties.

Acknowledgements. I would like to thank all colleagues involved in the design and implementation of the TR32DB in recent years. In addition, I gratefully acknowledge financial support by the CRC/TR32 'Patterns in Soil-Vegetation-Atmosphere Systems: Monitoring, Modelling, and Data Assimilation' funded by the German Research Foundation (DFG). The author would like to thank the anonymous reviewers for their suggestions and valuable comments on the manuscript.

References

1. Corti, L., Van den Eynden, V., Bissell, A., Woollard, M.: Managing and Sharing Research Data: A Guide to Good Practice. SAGE Publications Ltd, Los Angeles (2014)
2. Ma, J.: Managing metadata for digital projects. Libr. Collect. Acquisitions Techn. Serv. **30**, 3–17 (2006)
3. Greenberg, J., Swauger, S., Feinstein, E.: Metadata capital in a data repository. In: International Conference on Dublin Core and Metadata Applications, DC-2013, 2–6 September 2013, Lisbon, Portugal (2013)
4. Karasti, H., Baker, K.S., Millerand, F.: Infrastructure time: long-term matters in collaborative development. Comput. Support. Coop. Work **19**, 377–415 (2010)
5. Berman, F.: Got data? a guide to data preservation in the information age. Commun. ACM **51**, 50–56 (2008)
6. Mayernik, M.S.: Metadata realities for cyberinfrastructure: data authors as metadata creators. iConference 2010, Illinois (2010)
7. Miller, S.J.: Metadata for digital collections: a how-to-do-it manual. Neal-Schuman Publishers Inc., New York (2011)
8. Qin, J., Ball, A., Greenberg, J.: Functional and architectural requirements for metadata: supporting discovery and management of scientific data. In: The Kuching Proceedings of DCMI International Conference on Dublin Core and Metadata Applications DC-2012, pp. 62–71 (2012)
9. Farnel, S., Shiri, A.: Metadata for research data: current practices and trends. In: DCMI International Conference on Dublin Core and Metadata Applications DC-2014–The Austin Proceedings, pp. 74–82 (2014)
10. Krause, E.M., Clary, E., Ogletree, A., Greenberg, J.: Evolution of an application profile: advancing metadata best practices through the dryad data repository. In: DCMI International Conference on Dublin Core and Metadata Applications DC-2015–The São Paulo Proceedings, pp. 63–75 (2015)
11. Wira-Alam, A., Dimitrov, D., Zenk-Möltgen, W.: Extending basic dublin core elements for an open research data archive. In: DCMI International Conference on Dublin Core and Metadata Applications DC-2012–The Kuching Proceedings, pp. 56–61 (2012)
12. Klump, J., Ulbricht, D., Conze, R.: Curating the web's deep past – migration strategies for the German Continental Deep Drilling Program web content. GeoResJ **6**, 98–105 (2015)

13. Gerstner, E.-M., Bachmann, Y., Hahn, K., Lykke, A.M., Schmidt, M.: The west african data and metadata repository: a long-term data archive for ecological datasets from west africa. Flora et Vegetatio Sudano-Sambesica **18**, 3–10 (2015)

14. Cordero-Llana, L., Ramage, K., Law, K.S., Keckhut, P.: LABEX L-IPSL arctic metadata portal. Data Sci. J. **15**, 1–11 (2016). doi:10.5334/dsj-2016-002

15. Takeda, K., Brown, M., Coles, S., Carr, L., Earl, G., Frey, J., Hancock, P., White, W., Nichols, F., Whitton, M., Gibbs, H., Fowler, C., Wake, P., Patterson, S.: Data management for all: the institutional data management blueprint project. In: 6th International Digital Curation Conference, Chicago, USA (2010)

16. Jensen, U., Katsanidou, A., Zenk-Möltgen, W.: Metadaten und standards. In: Büttner, S., Hobohm, H.-C., Müller, L. (eds.) Handbuch Forschungsdatenmanagement, pp. 83–100. Bock u. Herchen, Bad Honnef (2011)

17. Simmer, C., Thiele-Eich, I., Masbou, M., Amelung, W., Bogena, H., Crewell, S., Diekkrüger, B., Ewert, F., Franssen, H.-J.H., Huisman, J.A., Kemna, A., Klitzsch, N., Kollet, S., Langensiepen, M., Löhnert, U., Rahman, A.S.M.M., Rascher, U., Schneider, K., Schween, J., Shao, Y., Shrestha, P., Stiebler, M., Sulis, M., Vanderborght, J., Vereecken, H., Kruk, J.V.D., Waldhoff, G., Zerenner, T.: Monitoring and modeling the terrestrial system from pores to catchments: the transregional collaborative research center on patterns in the soil–vegetation–atmosphere system. Bull. Am. Meteorol. Soc. **96**, 1765–1787 (2015)

18. Curdt, C., Hoffmeister, D.: Research data management services for a multidisciplinary, collaborative research project: design and implementation of the TR32DB project database. Program **49**, 494–512 (2015)

19. Curdt, C.: TR32DB metadata schema for the description of research data in the TR32DB. Transregional Collaborative Research Centre 32, Project Section Z1/INF, Institute of Geography, University of Cologne (2014). doi:10.5880/TR32DB.10

20. Backes, M., Dörschlag, D., Plümer, L.: Landwirtschaftliche Geodaten - Nachhaltige Datenhaltung und -nutzung durch ISO Standards, eZAI 2005, pp. 18–23 (2005)

21. Göttlicher, D., Bendix, J.: Eine modulare Multi-User Datenbank für eine ökologische Forschergruppe mit heterogenem Datenbestand, eZAI 2004, pp. 95–103 (2004)

22. Shumilov, S., Rogmann, A., Laubach, J.: GLOWA Volta GeoPortal: an interactive geodata repository and communication system. In: Ehlers, M., Behncke, K., Gerstengarbe, F.-W., Hillen, F., Koppers, L., Stroink, L., Wächter, J. (eds.) Digital Earth Summit on Geoinformatics 2008: Tools for Global Change Research, pp. 363–368. Wichmann, Heidelberg (2008)

23. Duval, E., Hodgins, W., Sutton, S., Weibel, S.L.: Metadata principles and practicalities. D-Lib Mag. **8**, 16 (2002)

24. Dublin Core Metadata Initiative. http://dublincore.org/documents/dces/

25. DataCite Metadata Working Group: DataCite Metadata Schema for the Publication and Citation of Research Data, Version 2.2. DataCite (2011)

26. International Organization for Standardization (ISO). http://www.iso.org/iso/catalogue_detail.htm?csnumber=26020

27. Commission, E.: Commission regulation (EC) No 1205/2008 of 3 December 2008 implementing directive 2007/2/EC of the European Parliament and of the Council as regards metadata. Official J. Eur. Union L **326**, 1–19 (2008)

28. The Bibliographic Ontology. http://bibliontology.com/specification.html

29. The Event Ontology. http://www.motools.sourceforge.net/event/event.html

30. Foulonneau, M., Riley, J.: Metadata for Digital Resources: Implementation. Systems Design and Interoperability. Chandos Publishing, Oxford (2008)

End-to-End Research Data Management Workflows

A Case Study with Dendro and EUDAT

Fábio Silva, Ricardo Carvalho Amorim, João Aguiar Castro,
João Rocha da Silva, and Cristina Ribeiro[(✉)]

INESC TEC—Faculdade de Engenharia da Universidade do Porto, Porto, Portugal
ffjs1993@gmail.com, ricardo.amorim3@gmail.com,
joaoaguiarcastro@gmail.com, joaorosilva@gmail.com, mcr@fe.up.pt

Abstract. Depositing and sharing research data is at the core of open science practices. However, institutions in the long tail of science are struggling to properly manage large amounts of data. Support for research data management is still fragile, and most existing solutions adopt generic metadata schemas for data description. These might be unable to capture the production contexts of many datasets, making them harder to interpret. EUDAT is a large ongoing EU-funded project that aims to provide a platform to help researchers manage their datasets and share them when they are ready to be published. Data-Publication@U.Porto is an EUDAT Data Pilot proposing the integration between Dendro, a prototype research data management platform, and the EUDAT B2Share module. The goal is to offer researchers a streamlined workflow: they organize and describe their data in Dendro as soon as they are available, and decide when to deposit in a data repository. Dendro integrates with the API of B2Share, automatically filling the standard metadata descriptors and complementing the data package with additional files for domain-specific descriptors. Our integration offers researchers a simple but complete workflow, from data preparation and description to data deposit.

1 Introduction

An unprecedented growth in data production is compelling institutions to implement infrastructures to make these resources available in the long run [6], while funding institutions require projects to make data available as specified in Data Management Plans, which are becoming mandatory. The challenges range from enabling researchers to deposit and describe their data early in the research projects, to ensuring the long term preservation of project results upon their completion.

Although the expertise in managing publication records can be seen as a starting point when designing applications for data management, recent studies reveal that adapting the existing tools to the new requirements often yields

© Springer International Publishing AG 2016
E. Garoufallou et al. (Eds.): MTSR 2016, CCIS 672, pp. 369–375, 2016.
DOI: 10.1007/978-3-319-49157-8_32

limited capabilities that may render the infrastructure unfit for research data management (RDM) [2].

To promote interoperability, RDM platforms are often compliant with metadata exchange protocols and offer interfaces to enable the integration with other platforms. This is the case with Dendro, a platform to assist researchers in the organisation and description of their datasets. Dendro can export the prepared datasets to any institutional repository, ideally one that can leverage domain-specific metadata to improve data visibility and increase the potential for reuse of the datasets [7].

EUDAT[1] is an European initiative that aims to create a centralized solution for data management in several research settings, ranging from the publication of datasets in the long tail to storing and delivering large datasets in specialised high-performance environments. EUDAT offers modules for data management and communication, along with a comprehensive API to simplify integration with established infrastructures.

This work is focused on the long tail of science [5], where small research groups from diverse domains need straightforward processes for data deposit and long-term preservation. We describe the integration of Dendro with the EUDAT e-infrastructure to compose an RDM workflow that can be easily integrated into the regular research processes.

2   Data Management for Reuse

Along with the increasing open-access demands, research institutions can benefit from timely disclosure of their outputs. As with research papers, published research data can be cited and provide credit to their authors. Moreover, by enabling other researchers to reuse data, institutions contribute to research transparency and increase their own visibility. Data reuse implies that the researcher can fully grasp both the origin and the context of production for the dataset [3].

A workflow that covers the entire data lifecycle is therefore required, to couple data and metadata from the start and provide a clear record of the data production process. The initial stages are often characterized by datasets being created and updated, making flexible staging platforms ideal to manage such resources. Complementary tools such as electronic laboratory notebooks have also shown promising results in motivating researchers to actively describe their data [1].

On the final stages of the workflow—deposit and dissemination—there is concern with the existence of sufficient metadata, so that the dataset can be located, interpreted and reused. The main issues are often related to capturing the context of production of the datasets, which often means dealing with multiple metadata schemas, while ensuring the compatibility with domain-level metadata, a problem that is often undervalued in emerging platforms [2].

---

[1] https://www.eudat.eu/.

In some research areas infrastructures are already in place, with well established data sharing guidelines, often specified in Data Management Plans. Repositories for these scenarios are commonly tailored to existing local needs, and often rely on datasets that follow a consistent structure across diverse projects. This is not common in institutions that deal with several projects at the same time and have to cope with heterogeneous datasets as well.

Widely used platforms for institutional repositories—such as DSpace and ePrints—have been adapted to handle research data. This solution satisfies the data access requirements, as the existing dissemination protocols, namely the OAI-PMH, are natively supported. The main issue with data publication is description. In large disciplinary repositories, the task can be committed to a curator with expertise in the domain. This approach is not viable for repositories in the long tail, dealing with many domains. Here, the bulk of the description task has to be assigned to researchers and data creators. To add this to their regular activities, new tools and workflows are required.

Data repository platforms such as Figshare[2] and Zenodo[3] have come to offer simpler, yet extensive interfaces to allow data deposits to be completed by the researchers [2]. Involving researchers in the management of their data is a wise step, taking advantage of their knowledge on the domain to generate accurate description for the data.

3   Dendro and EUDAT

Providing researchers with data management tools for the whole research process is expected to improve the quality of their data, to generate more and more specific metadata, and to make more datasets reach the publication stage. In the proposed workflow, Dendro contributes to the data organisation and description components, while EUDAT is used as the publication platform.

3.1   Dendro

Researchers need data management tools early in the research workflow, namely to capture domain-level metadata. In some cases, data description is already a part of the research routine, sometimes including laboratory notebooks as means of personal organization. The laboratory notebooks hold valuable metadata records which are expensive to produce; they serve as inspiration for more efficient tools to capture metadata in increasingly digital workflows and processes.

Projects also involve teams of several researchers; digital platforms can provide researchers with collaborative environments where they can represent their domain-specific metadata into structured and standards-compliant records. Dendro[4] focuses on creating comprehensive descriptions with domain-specific terms and providing collaborative features. It supports researchers on their data

---

[2] https://figshare.com/.

[3] http://zenodo.org/.

[4] http://dendro.fe.up.pt/demo.

management routine [7], and can be seen as an extension of their workspace. When the project comes to an end—or anytime the researchers choose—Dendro exports the final package, containing both data and metadata, to almost every data repository.

Dendro has a data model based on ontologies that can be regarded as conceptual representations of domains and may group descriptors from several metadata schemas. The ontologies are built in collaboration with the research teams, assessing their description needs and capturing the domain terminology that will enable their peers to interpret the datasets [4].

3.2   EUDAT

EUDAT proposes an integrated environment that addresses several requirements of researchers with respect to data processing, description and deposit. It is an array of platforms, including modules for data processing and refinement, data preservation, collaboration and dissemination. The services are offered in compliance with European guidelines on open access and research data disclosure. These capabilities make EUDAT a strong candidate for institutions that need to provide a Data Management Plan when applying for European research grants, but also for those that are looking for a platform for daily use by researchers. The existing modules are:

- **B2Drop**—stores and synchronizes data, providing collaborative tools using a Dropbox metaphor.
- **B2Share**—facilitates data deposit by researchers or institutions in some of the major domain repositories, e.g. CLARIN for linguistics or GBIF for biodiversity; some fields are required for deposit, and will be used as metadata; depending on the target repository, some more specific fields can be added; a unique identifier is assigned to the dataset.
- **B2Safe**—replicates research data; its features include policy rules, management of identifiers and integrity checking.
- **B2Stage**—offers computational resources to help researchers refine their data; it handles the exchange of data between EUDAT's storage resources and High-Performance Computing workspaces.
- **B2Find**—supports data discovery; using the OAI-PMH protocol, it gathers metadata from external repositories and B2Share, and exposes the results to users through a search interface.
- **B2Access**—handles federated authentication across all modules.

EUDAT is provided as a service, an approach that can reduce the impact of deploying a data management platform when compared to institution-supported solutions. The growing support community and a broad network of partners all over the EU contribute to the visibility of EUDAT in the data management landscape. The additional modules for large-scale storage and computing might also help institutions without sufficient funds to access such capabilities.

3.3   Integrating Dendro with EUDAT

Dendro allows researchers to download selected project folders and deposit the generated package in B2Share, automatically filling in the required metadata fields. The researcher immediately obtains a URL for the deposited dataset, which also includes a unique identifier that can be cited. The gathered metadata is pre-processed to filter descriptors that are recognized by the platform—such as `title`, or `description`. Dublin Core descriptors are exported through the existing API. The complete metadata record with all descriptors is exported as an RDF file that can later be ingested by other platforms to facilitate the interpretation and use of the dataset. At this stage, Dendro can export the results to the EUDAT platform in two ways:

- **Via B2Share**—through Dendro's interface, the researcher exports the project. The API of B2Share is used. The researcher chooses to deposit data in a personal account (by providing a personal authentication token) or in the default one.
- **Via B2Find**—Dendro exposes project metadata via an OAI-PMH server, with varying levels of access to data and metadata. This is done automatically by a script that gathers metadata from projects and XML files and exposes them to OAI-PMH harvesters.

The first approach is appropriate when the project data and metadata can be disclosed. This is often a decision of the project manager and ensures that datasets remain closed and are only exported to the EUDAT platform when ready. The second case covers scenarios where researchers cannot directly disclose their data, usually during the research project or when embargo periods are in place. Only the metadata is exposed and any external access requires authorization of the researchers. In this case, the dataset remains on the Dendro platform, and three levels of access control were implemented to address these constraints:

- **Private**—neither project data nor metadata are to be shared, addressing scenarios where the dataset contains sensitive or private data;
- **Public**—metadata is exposed via OAI-PMH protocol. The project's URL redirects users to a page that allows them to see the project structure and download it. This can be useful for projects in the public domain or containing institutional information that requires datasets to be visible to the community;
- **Metadata only**—metadata is exposed via OAI-PMH. However, the project's URL redirects to a page where the user can request access to the project. This level is used, for instance, when it is interesting for the researchers to reveal the project status, associated contacts and other metadata, but disclosure is postponed.

The Dendro interface was adapted to accommodate the implemented features. In Fig. 1, number **1** illustrates the creation of a project, where the researcher can choose a privacy level, which can be updated at any time. Number

Fig. 1. 1. Defining project visibility. **2.** Dataset in view-only mode.

2 has a view of a project configured as public, as it is systematically updated by the project team. Along with the data, external parties can access the associated metadata record.

4   Conclusions

This work proposes a workflow for research data, from creation to description and publication. Dendro was used for data organisation and description, and EUDAT for data publication. The integration of Dendro and EUDAT raised concerns with respect to metadata. The EUDAT API provides a set of Dublin Core descriptors, leaving out other domain metadata recorded in Dendro. To overcome this limitation we associated the complementary descriptor values as a file in the dataset package. Although they do not contribute to search in EUDAT, the metadata record exported by Dendro is compliant with existing metadata schemas, and the used ontologies are also published. Metadata can therefore be used by more advanced systems or applications. The integration of Dendro with B2Share for deposit and B2Find for search gives researchers flexibility to disclose their data according to the permissions they have set[5].

The two integration paths manage to export data and metadata within the Dendro environment, using EUDAT, one of the current solutions for European public data repository. This experiment has responded to the pressing needs of researchers who need a simple process for data deposit and publication. Two lines of work are ongoing: the release of Dendro as open-source code and the test of a public data repository managed by the University of Porto. More experiments to test these solutions with researchers are required. Moreover, work in this line depends on the European infrastructures and on forthcoming national and European policies.

Acknowledgements. This work is financed by the ERDF—European Regional Development Fund through the Operational Programme for Competitiveness and Internationalisation - COMPETE 2020 Programme and by National Funds through the

---

[5] An example of a deposited dataset: b2share.eudat.eu/record/404.

Portuguese funding agency, FCT - Fundação para a Ciência e a Tecnologia within project POCI-01-0145-FEDER-016736.

References

1. Amorim, R.C., Castro, J.A., da Silva, J.R., Ribeiro, C.: LabTablet: semantic meta-data collection on a multi-domain laboratory notebook. In: Closs, S., Studer, R., Garoufallou, E., Sicilia, M.-A. (eds.) MTSR 2014. CCIS, vol. 478, pp. 193–205. Springer, Heidelberg (2014). doi:10.1007/978-3-319-13674-5_19
2. Amorim, R.C., Castro, J.A., da Silva, J.R., Ribeiro, C.: A comparison of research data management platforms: architecture, flexible metadata and interoperability. Univ. Access Inf. Soc., 1–12 (2016). doi:10.1007/s10209-016-0475-y
3. Assante, M., Candela, L., Castelli, D., Tani, A.: Are scientific data repositories coping with research data publishing? Data Sci. J. **15**, 6 (2016). doi:10.5334/dsj-2016-006
4. Castro, J.A., da Silva, J.R., Ribeiro, C.: Creating lightweight ontologies for dataset description: practical applications in a cross-domain research data management workflow. In: Proceedings of the ACM/IEEE Joint Conference on Digital Libraries (2014) doi:10.1109/JCDL.2014.6970185
5. Heidorn, P.B.: Shedding light on the dark data in the long tail of science. Libr. Trends **57**(2), 280–299 (2008). doi:10.1353/lib.0.0036
6. Rice, R., Haywood, J.: Research data management initiatives at University of Edinburgh. Int. J. Digit. Curation **6**(2), 232–244 (2011). doi:10.2218/ijdc.v6i2.199
7. da Silva, J.R., Ribeiro, C., Lopes, J.C.: The Dendro research data management platform: applying ontologies to long-term preservation in a collaborative environment. In: iPRES Conference Proceedings (2014)

Author Index

Printed in the United States
By Bookmasters